Pellegrino's Clinical Bioethics

Pellegrino's Clinical Bioethics

A Compendium

Edmund D. Pellegrino

Edited by G. Kevin Donovan, David G. Miller,
and Claudia Ruiz Sotomayor

The Catholic University of America Press
Washington, D.C.

The paper used in this publication meets the minimum requirements of
American National Standards for Information Science—
Permanence of Paper for Printed Library Materials, ANSI Z39.48–1992.

∞

Cataloging-in-Publication Data available from the Library of Congress
ISBN (paperback): 978-0-8132-3-7534
ISBN (ebook): 978-0-8132-3-7541

Contents

Editors' Preface

THIS VOLUME WAS CONCEIVED as a tool to give present and subsequent generations easy access to the thoughts and perspectives of a founding father of modern-day bioethics, Edmund D. Pellegrino, MD. Widely renowned as a physician, philosopher, and educator, Pellegrino's prolific output of books, papers, and presentations has influenced countless contemporary clinicians and bioethicists. Although there have been compilations of his writings in the past, none has focused exclusively on those papers that represent only clinical bioethics, an omission that we now attempt to correct. These papers should serve as a valuable resource for those being introduced to the field of bioethics, as well as those scholars who are familiar with both the field and the topics as considered by Pellegrino. Therefore, it was constructed to be a valuable tool for students in the clinical sciences, especially medicine and nursing, as well as those more advanced in their training who will benefit from such a collection of Pellegrino's perspectives in clinical bioethics. The thoughts and perspectives as presented here, by Pellegrino and other authors, were all appropriate to the clinical understanding at the time. Some clinical knowledge may change with time, but the approach to ethical evaluation of the facts should stay constant.

This volume was conceived by the faculty of the Pellegrino Center for Clinical Bioethics as a way of continuing to honor the legacy of Edmund Pellegrino, shortly after his death. Given the depth and breadth of clinical topics in bioethics addressed by Pellegrino, it seems incredible that anything of importance in clinical ethics could have been unexamined by him. Nonetheless, in preparing this volume, the editors identified important additional topics that needed to be included. They have, with the help of colleagues who knew him well, written new chapters that reflect his teaching and perspectives. All the faculty and staff, as well as several other friends and colleagues, who contributed to expanding those areas about which Pellegrino had been surprisingly quiet, were well acquainted with his views. A sincere effort was made in each case to faithfully represent what would be the Pellegrino perspective, and to incorporate it in the author's own work. We wish to offer our sincere thanks to all of them; without them this unique work could not have been produced. We also want to thank Dr. Pellegrino's last student scholar, Jeevan Puthiamadathil, MD, who first compiled all of the clinically relevant Pellegrino articles categorized by topic. Most importantly, our thanks must go to Marti Patchell, Dr. Pellegrino's longtime personal assistant through most of his professional life. Her contributions, not only to this volume, but to Pellegrino's entire corpus, would require an entire volume of their own to list and acknowledge. Without her, this work would not have been completed.

G. Kevin Donovan, MD
David G. Miller, PhD
Claudia Ruiz Sotomayor, MD, DBe
Editors

Acknowledgments

THE EDITORS WOULD LIKE TO THANK the Pellegrino family, particularly Alice A. Pellegrino, for the written permission to publish Dr. Pellegrino's work, insofar as they own the rights.

They commend Dr. Pellegrino's long-time assistant, Mrs. Marti Patchell, for her outstanding help in completing this volume, as well as the team at The Catholic University of America Press.

Thus follows a list of the original publications; all are reprinted with the permission of the Pellegrino family and/or the original publisher, as appropriate.

Chapter 2: Edmund D. Pellegrino, "The Personal Ethics of the Physician: Curing Medicine from Within," *The Mount Sinai Journal of Medicine* 58, no. 5 (1991): 452–54. © Mount Sinai School of Medicine

Chapter 4: Edmund D. Pellegrino and Daniel P. Sulmasy, "Medical Ethics," in *Oxford Textbook of Medicine*, 4th ed., vol. 1 (Oxford University Press, 2003), sec. 1–10. © Oxford University Press. Reproduced with permission of the Licensor through PLSclear.

Chapter 5: Edmund D. Pellegrino, "The Moral Foundations of the Patient-Physician Relationship: The Essence of Medical Ethics in Military Medical Ethics," in *Military Medical Ethics*, vol. 1 of *Textbooks of Military Medicine*, ed. Thomas E. Beam and Linette Sparacino (Washington, D.C.: Office of the Surgeon General, US Army, 2003).

Chapter 6: Reprinted with permission from Edmund D. Pellegrino and David C. Thomasma, "The Conflict Between Autonomy and Beneficence in Medical Ethics: Proposal for a Resolution," *The Journal of Contemporary Health Law and Policy* 3 (1987): 23–46.

Chapter 7: Edmund D. Pellegrino, "The Anatomy of Clinical-Ethical Judgments in Perinatology and Neonatology: A Substantive and Procedural Framework," *Seminars in Perinatology* 11, no. 3 (1987): 202–9.

Chapter 10: Edmund D. Pellegrino, "Beneficence, Scientific Autonomy, and Self-Interest: Ethical Dilemmas in Clinical Research," *Cambridge Quarterly of Healthcare Ethics* 1, no. 4 (1992): 361–69. Copyright © Cambridge University Press 1992.

Chapter 11: "Balancing Science, Ethics and Politics: Stem Cell Research, a Paradigm Case," *The Journal of Contemporary Health Law and Policy* 18 (2002): 591–611.

Chapter 12: William P. Cheshire, Jr. MD, Edmund D. Pellegrino, MD, Linda K.

Bevington, C. Ben Mitchell, PhD, Nancy L. Jones, PhD, Kevin T. FitzGerald, PhD, C. Everett Koop, MD, and John F. Kilner, PhD, "Stem Cell Research: Why Medicine Should Reject Human Cloning," *Mayo Clinic Proc* 78 (2003): 6–14. © 2003 Mayo Foundation for Medical Education and Research. Reprinted with permission from Elsevier.

Chapter 15: Edmund D. Pellegrino, "Withholding and Withdrawing Treatments: Ethics at the Bedside," *Clinical Neurosurgery* 35 (1989): 164–84. © Wolters Kluwer Health, Inc.

Chapter 16: Edmund D. Pellegrino, "Withholding Treatment, Surrogate Decisions and Rationing: Some Selected Aspects," ed. Richard F. Southby, PhD, and Harold L. Hirsh, MD, JD (The Harold and Jane Hirsh Symposium, The George Washington University Health Policy Series) *Health Care, Law, and Ethics* 3 (1987): 151–72. Courtesy of The George Washington University. © The George Washington University. All rights reserved.

Chapter 17: Reprinted with permission from Edmund D. Pellegrino, "Ethical Issues in Palliative Care" in *Handbook of Psychiatry in Palliative Medicine*, ed. Harvey M. Chochinov and William Breitbart (New York: Oxford University Press, 2000), 337–48. © 2000, Oxford University Press, permission conveyed through Copyright Clearance Center, Inc.

Chapter 18: Edmund D. Pellegrino, "Futility in Medical Decisions: The Word and the Concept," *Healthcare Ethics Committee Forum* 17, no. 4 (2005): 308–18. Reprinted with permission of Springer Nature BV, permission conveyed through Copyright Clearance Center, Inc.

Chapter 19: Edmund D. Pellegrino, "Decisions at the End of Life: The Use and Abuse of the Concept of Futility," in *The Dignity of the Dying Person*, ed. Juan De Dios Vial Correa and Elio Sgreccia (Vatican City: Libreria Editrice Vaticano, 2000), 219–41. Reprinted with the permission of the Pontifical Academy of Life and Libreria Editrice Vaticana.

Chapter 21: Edmund D. Pellegrino, "Doctors Must Not Kill," *The Journal of Clinical Ethics* 3, no. 2 (1992): 95–102.

Chapter 26: Edmund D. Pellegrino, "The Physician's Conscience, Conscience Clauses, and Religious Belief: A Catholic Perspective," Fordham Urban Law Journal 30, no. 1 (2002): 221–44.

Chapter 28: Edmund D. Pellegrino: "Is Truth Telling to the Patient a Cultural Artifact?" *Journal of the American Medical Association* 268, no. 13 (1992): 1734–35. Copyright © 1992 American Medical Association.

Chapter 32: Edmund D. Pellegrino, "Clinical Ethics Consultations: Some Reflections on the Report of the SHHV-SBC," *Journal of Clinical Ethics* 10, no. 1 (1999): 5–13.

Chapter 35: Edmund D. Pellegrino: "Prevention of Medical Error: Where Professional and Organizational Ethics Meet," in *Accountability: Patient Safety and Policy Reform*, ed. Virginia A. Sharpe (Washington, D.C.: Georgetown University Press, 2004), 83–98. Copyright 2004 by Georgetown University Press, reprinted with permission, www.press.georgetown.edu.

Chapter 36: Edmund D. Pellegrino, "Managed Care at the Bedside: How Do We Look in the Moral Mirror?" *Kennedy Institute of Ethics Journal* 7, no. 4 (1997): 321–30. © 1997 The Johns Hopkins University Press. Reprinted with permission of Johns Hopkins University Press.

Chapter 37: Edmund D. Pellegrino: "The Commodification of Medical and Health Care: The Moral Consequences of a Paradigm Shift from a Professional to a Market Ethic," *Journal of Medicine and Philosophy* 24, no. 3 (1999): 243–66. © Society for Health and Human Values, permission conveyed through Copyright Clearance Center, Inc.

Introduction

Myles N. Sheehan, SJ, MD

EDMUND PELLEGRINO WAS ONE OF THE TOWERING FIGURES of medical ethics. But he was more than that. A consummate gentleman, physician, teacher, faithful Catholic, inspired humanist, and wise philosopher, Pellegrino was at the center of the birth of modern bioethics in the twentieth century; and his legacy continues to shape how we consider and assess developments in medicine. Dr. Pellegrino held numerous prestigious positions in his long life; he directed the Kennedy Institute of Ethics at Georgetown as well as founded the Center for Clinical Bioethics at Georgetown University Medical Center (renamed the Pellegrino Center for Clinical Bioethics shortly before his death). Pellegrino served as president of the Catholic University of America. He chaired the President's Council on Bioethics during the George W. Bush administration. A prolific writer, Pellegrino is widely known for his volumes on the philosophy of medicine, virtues in medicine, the importance of beneficence for physicians, as well as multiple articles and public addresses. It is important to note that his good friend and collaborator, Dr. David Thomasma, was a major contributor to Dr. Pellegrino's work, especially their shared volumes on ethics, philosophy of medicine, and the virtues. Dr. Pellegrino died in 2013, a few days shy of his ninety-third birthday. In the years since his death, Pellegrino's influence on the field of medical ethics continues to be felt.

This compendium is meant as a resource for healthcare professionals, students in these fields, and those involved with bioethics, to have in a ready source a series of articles and reflections on clinical issues addressed by Dr. Edmund Pellegrino. It is being published along with a companion volume that addresses Dr. Pellegrino's theological perspectives in healthcare ethics. The goals of this volume are to encourage reflection on the topics addressed, provide convenient access to Dr. Pellegrino's writings, and, perhaps most importantly, provide individuals the background to bring forward Pellegrino's methods of inquiry and analysis, and his perspective on evolving and new issues of clinical healthcare ethics. This volume differs from the two excellent previously published compendia, *The Philosophy of Medicine Reborn: A Pellegrino Reader,* and *Physician and Philosopher. The Philosophical Foundation of Medicine: Essays by Dr. Edmund Pellegrino,* in its selection of clinical topics rather than primary exploration of Dr. Pellegrino's philosophical writings on health care.[1] It is my hope that clinicians and healthcare ethicists can use the current work to further the legacy of Dr. Pellegrino by encouraging not simple admiration of what Pellegrino wrote and taught, but use of his example for new reflection on emerging clinical issues as we move into the future. Hagiography has its purposes, but that is not one of the purposes of this compendium. Rather, to paraphrase Jaroslav Pelikan, citing authentic tradition as the living faith of the dead, I would encourage readers to use Pellegrino's living example of reflection to provide guidance for exploration of ethical issues that will remain relevant into the future.[2]

What Are the Central Elements of Pellegrino's Ethical Reflections on Clinical Care?

First, Pellegrino grounds his ethics within the patient-clinician encounter. He does not attempt to develop a universal normative ethic but looks to this encounter as the heart of medical ethics. This is where the moral nature of medicine is found, explicitly in the action of a doctor or nurse to provide the right healing action for the patient. That right healing action is defined by choosing the right biomedical treatment based on the clinician's understanding of the good as the patient would define it in the context of the clinical question at hand, the good of the patient as a person whose autonomy is to be respected and who is to be treated beneficently and with justice, and, finally and paramount, the person's definition of his or her own spiritual good.[3]

Second, what Pellegrino refers to as the "fact of illness" makes the relationship between clinician and patient unequal and creates obligations on the clinician not only to work to ascertain the patient's version of what is good, but also requires the clinician to move beyond self-interest.[4]

Third, the patient-clinician encounter is one where the clinician is obliged to act for the patient's good, and that good should not be compromised by business relationships, managed care restrictions, or other conflicts of interest that would interfere with the clinician's oath to care for the patient.[5]

Fourth, although the use of principles (such as respect for autonomy, beneficence, non-maleficence, and justice) is necessary to come to proper ethical decisions

1. Edmund D. Pellegrino, *The Philosophy of Medicine Reborn: A Pellegrino Reader,* ed. H. Tristram Engelhardt and Fabrice Jotterand (Notre Dame, Ind.: University of Notre Dame Press, 2008); Edmund D. Pellegrino, *Physician and Philosopher: The Philosophical Foundation of Medicine: Essays by Dr. Edmund Pellegrino,* ed. Roger J. Bulger and John P. McGovern (Charlottesville, Va.: Carden Jennings Publishing, 2001).

2. Jaroslav Pelikan, *The Vindication of Tradition: The 1983 Jefferson Lecture in the Humanities* (New Haven, Conn., 1987), 65.

3. Edmund D. Pellegrino, "Moral Choice, the Good of the Patient, and the Patient's Good," in *The Philosophy of Medicine Reborn,* 163–86.

4. Daniel P. Sulmasy, Foreword to Pellegrino, *Physician and Philosopher,* xiii–xiv.

5. Edmund D. Pellegrino, "Medical Ethics in an Era of Bioethics: Resetting the Medical Profession's Compass," *Theoretical Medicine and Bioethics* 33 (2012): 21–24.

within medicine, this is not enough. The good physician (or nurse, or other clinician) respects these principles but also desires to grow in skill, character, and wisdom in the profession. This means growth in the virtues central to caring for persons, a process that requires reflection as well as consideration of appropriate role models for emulation.[6]

Fifth, the moral enterprise of medicine occurs within a larger society. Clinicians seeking to perform the right healing action need to be attuned to the way that societal influences may be harmful to patients. Thus, clinicians should be concerned about and advocate for access to health care, and against the pernicious influence of systemic injustice and racism in medicine and other social institutions, and environmental threats, such as climate change, that not only negatively affect individual patients but also threaten human life on earth. Social justice and awareness is a critical element of Pellegrino's approach to ethics.[7]

Others will list additional elements of Pellegrino's thought, but this list suffices to add to my recommendation that readers of this compendium will not only find the material in its pages useful and worthwhile but also will encourage others to look to the essentials of Dr. Pellegrino's thought in exploring issues in their own practice as well as developing challenges as the future unfolds.

6. Edmund D. Pellegrino and David C. Thomasma, *The Virtues in Medical Practice* (New York: Oxford University Press, 1993).

7. Lawrence J. Prograis Jr. and Edmund D. Pellegrino, eds., *African American Bioethics: Culture Race and Identity* (Washington, D.C.: Georgetown University Press, 2004), xvi.

I

Virtues and the Internal Morality of Medicine

A. Moral Compass Points: Autonomy vs. Beneficence

1
Moral Compass Points

David G. Miller, PhD

EDMUND PELLEGRINO SAW THE PHYSICIAN-PATIENT ENCOUNTER or relationship as generative of a number of considerations that orient clinical medical ethics. In some of his works, such as the one included here, he referred to these considerations as compass points. These points are embedded in the status and roles of physician and patient and the context in which they meet. He believed that these essential and ineradicable characteristics, when acknowledged and properly understood, could orient the physician's character and behavior. These elements of the clinical encounter or relationship arise because a vulnerable patient is required to trust a physician, who has offered to use his or her knowledge and training to benefit the patient. Due to the power imbalance in the relationship, the physician must bear some moral responsibility for what occurs in the process of medical treatment. The physician is also embedded in another relationship, to the community of medical practitioners. That community, because it is based in specialized knowledge and practice, has a moral responsibility to ensure that practitioners meet that community's professional standards.

Those compass points can orient the physician in moral space and help him to keep from getting lost in the temptation to deflect moral responsibility by ascribing it to regulatory concerns, competition, commercialization, for-profit medicine, etc. They provide the template of a map, but they do not serve as a GPS (global positioning system). That is to say, they provide guideposts that keep the physician properly oriented when he finds himself in an actual clinical encounter. In Pellegrino's view, the good physician should always be orienting himself towards maintaining and respecting the physician-patient relationship. That relationship is of the utmost importance insofar as it arises prior to any potential weighing of goods, principles, or theories, and any exercise of shared decision making.

Much of Pellegrino's work consists in expounding on this germinal insight. In other works, some of which are collected in this compendium, Pellegrino explores these elements more closely and carefully. The vulnerability of patients drives Pellegrino's rejection of conceiving of patients as customers, consumers, insured lives, etc. Because they are entrusted with their patients' lives, physicians cannot and should not be conceived as service providers, case managers, etc. Because health is a prerequisite for people's ability to pursue whichever conception of the good life they may have, health care cannot be conceived as a mere commodity to be distributed solely on the basis of supply and demand. Because the physician is the common moral pathway through which decisions are made and actions taken, she cannot simply do whatever she wishes nor can she escape moral responsibility by doing whatever the patient asks. Because the practice of medicine is embedded in a community of practice that relies on non-proprietary but esoteric knowledge, physicians must police themselves as practitioners and as clinical researchers—with the latter role generating its own relationships and responsibilities (some of which may come into conflict with those of the physician-patient relationship).

The following reading introduces that moral orientation and its importance to the formation of the physician's character.

2

The Personal Ethics of the Physician: Curing Medicine from Within

Edmund D. Pellegrino, MD

COMMENCEMENT SPEAKERS HAVE TWO OBLIGATIONS—one pleasant, one much more dubious. The pleasant task is to congratulate the new graduates and wish them well. This I am happy and honored to do. I must also congratulate your families, faculty, and friends because in their way they too have contributed to your accomplishment. You should all be proud of your shared efforts in one of humankind's nobler accomplishments—the making of a new doctor.

Despite the gloomy portents all around us, I want to welcome the new graduates enthusiastically into what I still believe to be a high calling, and a most satisfying and useful way of life. Forty-five years in medicine have not blunted my enthusiasm for our profession or its practitioners. With all our faults, we are still among the few groups in our society committed to some degree of effacement of self-interest in the interest of our patients.

This assertion takes me directly to the second of my tasks, the less welcome one, the commencement address, that peculiar obsession of academia which dictates that the joy of getting your degree be touched with a little pain. I must administer the last penance your medical school requires of you. The only antidote is brevity, and the merciful amnesia that will quickly erase the speaker's name and topic even before you leave this ceremony. There is, you see, no danger of permanent brain damage.
In a few minutes, you will receive your degrees. This is the university's public declaration that you will possess special knowledge. You should be proud of that degree. But it is not sufficient to make you a doctor. You will really enter the profession when you take the Hippocratic Oath. This is your public declaration, your promise that you will use your knowledge for the sick and not primarily for your own interests. This is what sets you apart as a true professional.

Reprinted with permission from Edmund D. Pellegrino, "The Personal Ethics of the Physician: Curing Medicine from Within," *The Mount Sinai Journal of Medicine* 58, no. 5 (1991): 452–54. © Mount Sinai School of Medicine

Before you take this solemn oath, let me ask you to reflect on its significance for your life as a doctor. To take an oath is a solemn act. In the moments remaining before you do so, I ask you to reflect on what the oath symbolizes and why it has for 2,500 years called doctors to a higher set of obligations than we expect of most people in our society.

If you do not take the oath with sincerity today, the rest of your medical life will be a lie. If you do, you will have to give an account every day of your fidelity to its ethical imperatives. Every day of your professional life, your human inclination to power, pleasure, prestige, and profit will tempt you to place self-interest ahead of your primary ethical obligations. It is easy to lose your way when so many around you are making moral compromises.

Six Moral Compass Points

Let me suggest six unchanging realities about being a physician that will be your compass points—the reminders that will set you back on track when you begin to lose your way, when you are confused about what you are—healer, technician, gatekeeper, bureaucrat, corporate employee, and a dozen other roles being pressed on us. It is my conviction that much of the moral malaise, the loss of enthusiasm and self-worth I see in my colleagues is the direct result of a failure to take account of these compass points.

The Vulnerability of the Patient. The first compass point is the dependence, vulnerability, and exploitability of the patient who seeks your help. A sick person is anxious, in distress, and fearful. To be healed or helped, she is impelled to seek your help though she wishes she could avoid it. The patient is not free to pursue life's other goals until health is restored or symptoms alleviated. Will you forget the uniqueness and the vulnerability built into the human predicament of illness?

Inequality of Power. The second compass point is the inherent inequality of the relationship. You possess the knowledge the patient needs. The preponderance of power is in your hands. You can use it well or poorly, for good or evil, for service or self-interest. You have enormous power to modify, modulate, and manipulate the patient's decisions. Will you use that power to impose your values, your social or political philosophy?

The Covenant of Trust. The third point about medical relationships is the special fiduciary character of the healing relationship. In a state of vulnerability and inequality, patients are forced to trust physicians. Patients are ill equipped to evaluate our competence. They are forced to reveal their intimate selves—baring their bodies, their personal lives, their souls, and their failings to another person who is a stranger. They trust us with these invasions of their privacy, because without trust they cannot be healed or helped. Moreover, we invite their trust. We begin our relationship with patients with an invitation. We ask, "How can I help you?" Implicitly we are making a promise: "I have the knowledge you need. Trust me to have it, and to use it in your best interests." Will you be faithful to this trust, which is central to all codes of ethical medicine, in all times and cultures?

Our Nonproprietary Knowledge. Fourth, the physician's knowledge cannot be wholly proprietary. It is knowledge ordained to a practical end, to meeting certain fundamental human needs. Medical knowledge does not exist for its own sake. Moreover, society permits us to gain that knowledge through invasions of privacy that would otherwise be criminal. In order that physicians may be trained, medical students are permitted to dissect human bodies, attend and assist at autopsies and operations, and participate in the care of sick people. Society permits these privileges not primarily so we physicians can make a living, but because society needs an uninterrupted supply of skilled doctors and because there is no substitute for hands-on medical training. From the first day of medical school when you accept the privilege of a medical education, like it or not you entered a covenant society. Will you hold your medical knowledge in trust for those who need it? Or will you dispense it primarily for your own profit?

The Role of Moral Accomplice. The fifth unchanging reality of the medical relationship is that the physician is the final common pathway through which help and harm must pass. Nothing happens to our patient without our orders. No policy, no law, no regulation can be effective without our cooperation. We are the final guardians of the patient's interest and de facto moral accomplices in what happens to the patient. We write the orders no matter what the policymakers or the law may require. Will you do the Pontius Pilate act and lay the blame elsewhere?

Membership in a Moral Community. The sixth distinguishing characteristic of medical relationships is that the doctor is a member of a moral community; collectively we share the privileges and obligations of our special knowledge. We are granted a monopoly on medical knowledge and hold it collectively as well as individually in trust for those who need it. We are collectively responsible for its equitable accessibility, availability, and distribution. We are responsible to and for each other. Will you, as a physician, be a moral island sufficient unto yourself, unmindful of your responsibility as member of a moral community?

The Role of Personal Ethics

In the current rebirth of medical ethics we have overemphasized dilemmas, quandaries, and puzzles. We have given the impression that ethics consists only of the right decision in a complicated case. We have ignored the personal ethics of the physician and forgotten that ultimately everything depends on the physician's virtue and character. We do need healers who know how to analyze moral dilemmas. But it takes more than philosophy or ethics to dispose us to do what is right and good for a sick person.

Sometimes in your own professional lives each of you will enter into a Dark Wood you will have to face yourselves. You will have to answer for where you are, what you have, and who you have become. Will you be able to say you have been a faithful steward, or will you have betrayed your trust? I hope the answer does not come only when you too are ill and face death, dependence, vulnerability, exploitability, and the need for healing.

Our profession is in a quandary today. We point to malpractice, government regulation, competition, commercialization, advertising, our personal debts, pandemics, for-profit medicine, and blame them for our malaise, our loss of enthusiasm. Some even feel justified in striking back, refusing to care for certain patients, "making it" while we can. But our illness is an illness from within. And it is curable only from within; only by recapturing our moral integrity can we recapture our identity and the joy that comes from a life spent not in pursuit of self-interest, but in service to the sick.

That is why the most important prescription of all is the ancient one we write for ourselves—"Physician, heal thyself." The six compass points I have suggested are the ingredients of that prescription and that healing. Keep them in mind as you take the oath today; ponder them every day, and you will lead a life of satisfaction, service, and moral integrity.

B. Ethical Theory and the Ethics Work-up

3

Introduction: Virtues

Daniel P. Sulmasy, MD, PhD

EDMUND PELLEGRINO HELD THAT THE ETHIC OF MEDICAL CARE depends upon foundational ethical principles—just as much as the clinical practice of medicine depends upon basic science. He considered these ethical principles to be universal and timeless, and therefore independent of setting. Whether practicing military medicine or in private practice; whether in the United States or Sri Lanka; whether working in a correctional facility or an HMO, the foundational ethical principles, for Pellegrino, are invariant. In an era in which it is popularly thought that nothing is ethically wrong except telling someone else that what they are doing is ethically wrong, or that morality is in the eye of the beholder, or that ethics is always relative to cultural and political circumstances, Pellegrino's thesis must be considered bold. One might wonder how Hippocrates could still be relevant today, or how a physician in Brazil could be held to the same ethical standards as a physician in New York City.

For Pellegrino, the ground for this timeless and universal ethic lies in a few fundamental aspects of human nature—our propensity to become sick or injured and our common, almost instinctual, commitment to help our fellow human beings who are so afflicted. Medicine is best described, for him, as a relationship between the sick person and the one who sets out to help that sick person. This relationship constitutes the timeless, universal foundation for the ethics of medicine.

In his article on "The Moral Foundations of the Patient-Physician Relationship," Pellegrino argues that medicine is not well described as a science, or a mechanical application of science, or a business, or social service, but as a relationship between a person in need of help and a person who is helper and healer (see chap. 5). This relationship is based on the universal fact of illness, the pledge (profession) of the physician to be dedicated to helping ill persons, and the actualization of that relationship in the acts of diagnosis, prognostication, and therapy. From this foundation, certain virtues (such as compassion and fidelity) can be derived, setting forth the ethical goals for anyone who professes to help the sick.

This foundation also permits the derivation of certain ethical principles to guide medical practice, as described in the "Anatomy of Clinical-Ethical Judgments in Perinatology and Neonatology: A Substantive and Procedural Framework"[1] (see chap. 7). Chief among these is beneficence—a commitment to act for the good of the patient. Yet this commitment is qualified by commitments to act justly and to respect the self-determination of the patient. In "The Conflict between Autonomy and Beneficence," Pellegrino and his long-time collaborator David Thomasma propose that beneficence is the primary principle, and that respect for autonomy can be folded into a broader sense of the good of the patient that is medicine's chief aim (see chap. 6).

With these principles and virtues in place, grounded in the nature of the patient-physician relationship, Pellegrino is well-equipped to describe how to actualize these principles and virtues in practice. Ethics is practical. Medicine is practical. The chief concern is right action. In the case of medical ethics this means, "What is the right and good thing to do for this patient?"

When this question becomes difficult to answer or becomes contested, Pellegrino proposes that an ethically sound answer can be approached through an "ethics workup," proceeding from facts of the case to the identification of the precise ethical issue, to the elucidation of the principles at stake, to a decision justified by ethical principles.

In the article titled "Medical Ethics," Pellegrino and I proposed further ways of making these principles and virtues concrete by addressing important clinical questions such as whether the patient has the capacity to act as an autonomous moral agent, how to make decisions when the patient lacks decisional capacity, how to obtain a genuinely informed and valid consent, and how to approach the care of patients at the end of life.

Taken together, the articles in this section constitute a foundational approach to practical clinical ethics that can serve not only physicians, but also nurses, dentists, clinical social workers, and anyone else committed to the care of the sick.

1. Edmund D. Pellegrino, "The Anatomy of Clinical-Ethical Judgments in Perinatology and Neonatology: A Substantive and Procedural Framework," *Seminars in Perinatology* 11, no. 3 (1987): 202–9, eweb:74365, https://repository.library.georgetown.edu/handle/10822/710599.

4

Medical Ethics

Edmund D. Pellegrino, MD, and Daniel P. Sulmasy, MD, PhD

Introduction

Clinical ethics, like all ethics, is a practical discipline. Whatever theory it employs, its ultimate aim is a morally defensible decision that is in the patient's best interests. This was the central moral precept of the Hippocratic Oath: "I will follow that system of regimen which, according to my ability and judgment, I consider for the benefit of my patient, and I will refrain from whatever is deleterious and mischievous."

Today, this is known as the principle of beneficence, which derives its moral force from the special nature of the relationships between sick persons and health professionals. When patients seek help they are anxious, often in pain, fearful, dependent, and therefore vulnerable and exploitable. In that state, physicians ask then, 'How can I help you?' By that act, physicians invite the patient's confidence and trust that they are competent and will use that competence primarily in the patient's interests.

The relationship is therefore not a contract but a covenant of trust, to which physicians must be faithful even if it means some suppression of their own self-interest. The good of the patient is thus a moral compass with four directional guide marks for the physician: medical good (competence), the patient's good expressed in his own preferences (respect for autonomy), the patient's inherent good as a human being (respect for dignity), and the patient's ultimate good (respect for spirituality). To act for the patient's good requires integration of these four elements on behalf of this person who presents as my patient now.

This chapter concentrates on the questions that must be asked, and the conditions that must prevail at the bedside to assure that each of the four levels of the patient's good is attained. Two moral algorithms are provided: one to assess the moral validity of the decision maker, and the second, to provide a clinical framework or "work-up" by which to analyze the ethical issues.

Who Decides?

Since the good of the patient is far broader than the patient's biomedical good, clinical ethics requires the patient's participation in decision making. Figure 1 presents an algorithm for determining the morally valid decision maker. If the patient has sufficient decision-making capacity, the patient is the ultimate decision maker. If the patient's decision-making capacity is variable, the physician should be guided by the last decision made when the patient was capable. If the patient has never had decision-making capacity (e.g., an infant or patient with cognitive impairments), the physician must turn to a morally valid surrogate. If the patient has lost decision-making capacity in a reversible manner (e.g., through depression), and it can be restored through medical treatment in a timely manner, the decision should be postponed until the patient is treated and capacity restored. If the loss of capacity is irreversible or the decision too urgent, the physician turns to any anticipatory declarations by the patient such as a living will, or the designation of a surrogate decision maker through a legal document such as a durable power of attorney for health care. Lacking these expressions of the patient's prior wishes, the physician must engage a morally valid surrogate, that is, someone who can responsibly and knowledgeably represent the patient's wishes, and who has intact decision-making capacity and is free of significant conflicts of interest.

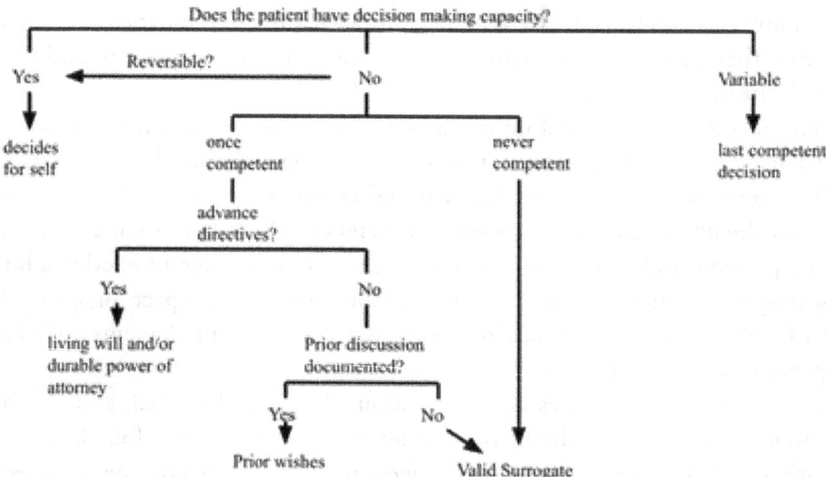

Decision-Making Capacity

Accurate determination of a patient's decision-making capacity is an essential clinical skill. In North American jurisprudence, the term "incompetence" refers to a judge's decision that an individual has lost all capacity to make decisions. The physician, however, is concerned with a narrower question: is this patient capable of making a decision about this clinical option in these particular clinical circumstances? The threshold of capacity will vary according to the gravity of the decision. For example, a patient suffering from suicidal depression might be allowed to refuse venipuncture, but not be allowed to sign out of the hospital against medical advice. Psychiatric consultation is often important, but the determination of decision-making capacity is generally the responsibility of the attending physician. The criteria for decision-making capacity are as follows:

(1) What is the patient's neurological status? If the patient is profoundly delirious, demented, obtunded, or aphasic, the patient will not have the capacity to participate in medical decision making.
(2) Does the patient have intact judgment? That is, is the patient free of impulsiveness, able correctly to assess the seriousness of situations, to plan, and to appreciate the connections between acts and consequences?
(3) Does the patient understand the nature of the procedure, its risks, benefits, and the consequences of deciding either to accept or to forgo the procedure?
(4) Can the patient explain the reasons for a decision in a way that is logical and also consistent with his or her life history and previously held values?
(5) Does the patient's decision remain relatively stable over time? Patients must certainly be free to change their minds, but a patient whose decision vacillates minute to minute or refuses to make any decision may not have intact decision-making capacity.

Informed Consent

Every physician must be able to obtain a morally adequate informed consent. This is not synonymous with obtaining a signature on a piece of paper. Informed consent is a process. It is one of the fundamental ways in which the physician shows respect for the good of patients as whole persons.

There are four basic elements in informed consent. The first element is that the patient must have decision-making capacity. The rudiments of how to assess decision-making capacity were described above.

Second, the decision by the patient must represent an autonomous authorization, that is, it must be free from coercion, or even subtle manipulation by the physician or by others. Information must be presented in a fair and balanced fashion. This does not imply absolute neutrality nor does it imply that the physician cannot make a recommendation or even try to persuade a patient if the physician thinks the patient is making a mistake. But a physician ought not, for instance, purposefully distort the facts or threaten to sever the physician-patient relationship if the patient does not follow the physician's advice.

Third, all relevant information must be disclosed to the patient. The content to be disclosed should generally include the indications and the nature of the procedure, its potential benefits and risks, and the alternatives, including not having any procedure.

The final element of informed consent, but certainly not the least, is comprehension. The patient must not merely have been told; the patient must understand. The common clinical practice of asking, "Do you have any questions?" is probably inadequate. If one is seriously interested in being sure that the patient has understood, it is better to ask the patient to explain back in his or her own words the information just disclosed.

Limits to Autonomy

While the good of patients as autonomous agents must be respected, patient autonomy is not absolute. Autonomy is limited, for instance, when there is a probable threat of serious injury to an identifiable third party or parties. An example would be a demand for confidentiality by a patient testing positive for HIV who refuses to tell a sexual partner. Autonomy is further limited by the intellectual integrity of medicine as a practice. For instance, a patient cannot demand a treatment that has been proven ineffective, such as laetrile for cancer. Autonomy can also be limited to protect public health, such as in mandatory vaccination in an epidemic of lethal infection. Finally, autonomy is limited when it violates the moral integrity of healthcare professionals as individual moral agents whose freedom of conscience must not be violated, for example, a physician opposed to euthanasia ought not to be forced to comply with a patient request even in settings where this is legal.

The Ethics Workup

Once the appropriate decision maker has been identified, and the conditions for her autonomy assured, the physician must turn to analysis of the ethical dilemma and its substantive resolution. Ethics committees and consultations sometimes help, but ultimately clinicians are accountable for what they do or agree to. Every clinician is obliged to master the "work-up" of the ethical problems just as surely as that of a clinical problem like coma, jaundice, or edema.

The analytical approach we use consists of the following seven steps.

1. Secure the Facts

Good ethics begins with reliable clinical and social data, with as accurate an assessment as possible of factors such as diagnosis, prognosis, effectiveness, benefits, burdens of treatments, brain function, patient preferences, and life situations. Each and all may be ethically relevant.

2. Define the Ethical Issue

The specifically ethical issue must be identified among the communication, interpersonal, and inter-professional problems which usually intermingle, especially when conflicts arise. The first step in resolving conflict is clarity in the statement of the issues.

3. Frame the Issue

By applying generally accepted ethical principles, one can better understand the important moral dimensions of the issue. These principles are: (i) preservation of the good of the patient as a whole person (the principle of beneficence), and (ii) respecting the goods and interests of others (the principle of justice). Beneficence for persons, as noted above, demands an examination of all four aspects of the patient's good—the biomedical good, the good of the patient's autonomous choices, the patient's good as a person, and the patient's own beliefs about the ultimate good. So, for example, a severely anemic Jehovah's Witness might refuse blood transfusion. Transfusion would serve the biomedical good of the patient, but it would violate the patient's autonomy and idea of the higher or ultimate spiritual good. Or a patient's family might demand continued treatment in the intensive care unit when such care was futile or unnecessary. This might impede other patients' access to intensive care and thus violate the principle of justice.

4. Situate the Issue

It is helpful to place the case in relation to one's personal experience and that of the profession. This method of moral analysis is called "casuistry" and asks whether the case at hand is analogous to any paradigmatic case for which a broad moral consensus has been reached. If so, one could reason by analogy to that case. For example, consider a novel case, such as whether one ought to allow a prisoner who has donated one kidney to his daughter to donate his remaining kidney to her after she has rejected the first kidney. It is useful to ask how this case compares with a more familiar case in which there is a broad moral consensus. For example, this case might be likened to the case of a man who jumps in front of a car to save his daughter's life. Such a man would be considered a hero. Casuistic analysis would ask how analogous these cases really are. What is the same about these two cases? What is different? What is the moral relevance of any similarities or differences?

5. Identify the Options

In almost every case, a variety of clinical options are possible. Ethics involves selection of the morally correct choice. It is therefore necessary that all the available clinical options be identified and considered from a clinical as well as a moral point of view. This is where the technically correct and the morally good should intersect for the patient's good.

6. Reason

It is important to weigh all the facts of the case critically and rigorously in light of one's ethical framework and clinical experience. One must interrelate the facts, the relevant principles, and any paradigm cases. Physicians should also play "devil's advocate" and examine possible objections to their own positions. They should also seek colleagues' input if time permits. An ethics committee or an ethics consultation service may be useful at this juncture. Once resolved, a retrospective critique of the reasoning employed in the case is helpful in preparing for the next time such a situation arises.

7. Decide

In clinical ethics, as in all other aspects of medicine, a decision must be made. Taking all of the aforesaid into account, a choice must be made even in the throes of uncertainty. There is no formula that guarantees the right choices. The answer will require a judicious combination of clinical judgment, practical wisdom, and common sense. In the final analysis, the decision rests with the physician's character and commitment to the good of the patient.

The Ethics of End-of-Life Care

The most common ethical issues faced by clinicians arise at the end of life. Here the good of the patient might include decisions to withhold and withdraw life-sustaining treatments, decisions not to resuscitate, and the use of potent opioid analgesics that may hasten death.s

The moral propriety of withholding or withdrawing life-sustaining treatment may be analyzed systematically by examining the proposed treatment for its effectiveness, benefit, and burdens.

Effective treatments are those that alter the natural history of an illness or alleviate an important symptom. Hippocrates counseled that physicians should "refuse to treat those who are overmastered by their diseases, recognizing that in such cases medicine is powerless." When, to a reasonable degree of medical certainty, it can be determined that a treatment will not be effective in securing the goals of treatment mutually determined by the medical team and the patient or the patient's surrogates, that treatment can be called clinically ineffective or "futile." In general, there is no moral obligation to provide futile treatment, although allowances must often be made for the psychological unpreparedness of the patient or family to accept the idea of futility.

If it is determined that a medical treatment is biomedically effective, the next question is whether it is beneficial. Beneficial treatments are those that bring some good to the patient beyond the biomedical good. Beneficial treatments serve not only the body, but the good as the patient chooses it, the good of the patient as a human person, or the patient's ultimate sense of the spiritual good. For instance, antibiotic treatment of pneumonia in a patient dying of malignancy might be effective, but it might not be beneficial if it merely postpones dying when the patient sees no benefit in it.

Both the effectiveness and benefits of a treatment must be weighed against their burdens—physical, financial, or emotional. When the burdens are disproportionate to effectiveness and benefits, treatment can be withheld or withdrawn. Planned reexamination of the three variables at previously agreed time intervals will avoid much of the confusion that surrounds Do Not Resuscitate (DNR) orders. Cardiopulmonary resuscitation is a treatment which is ineffective, burdensome, and without benefit in many terminally ill patients. When this is the case, a DNR order is morally licit.

It is incontrovertible that there is a moral mandate to treat pain. Nonetheless, some physicians might hesitate to do so adequately because there is a risk that this might unintentionally hasten the death of the patient. The centuries-old Rule of Double Effect may be invoked in such cases. According to this rule, a physician completely opposed to euthanasia can act with clear conscience in administering a drug like morphine to a dying patient if several conditions are met.

First, the physician must sincerely intend pain relief, not the death of the patient. Second, the dose must be consistent with a plan to relieve pain through the analgesic effects of morphine, not through causing respiratory arrest and death as the means of relieving pain. Finally, the need for pain relief must be great compared with the risk of respiratory arrest and death in that patient.

For example, if a patient is dying of metastatic breast cancer and is in severe pain, the potential benefit of intravenous morphine would seem overwhelming compared with the small risk that morphine might contribute to hastening an already imminent death. If these conditions are fulfilled, a physician should be able to control pain with a clear conscience, even knowing that death may unintentionally be hastened as a side-effect.

At present, there is significant controversy about whether physicians should be authorized to hasten the death of the patient intentionally through euthanasia or assisted suicide, actions not permitted by Western medicine since the Hippocratic ethic became dominant many centuries ago. Legal bans on these practices are being challenged through legislative initiatives and civil suits. Almost all professional organizations remain opposed.

Conclusion

This chapter has focused on the heart of clinical ethics, that is, acting for the good of the patient. This is the physician's central moral obligation, from which he cannot be relieved since he is bound in a covenant of trust to respond to the sick person who is in need of his medical knowledge.

We have illustrated this moral theme by analyzing the four levels of the good of the patient at the bedside in several ways: first, through defining the appropriate decision maker, the assessment of capacity, and the elements of informed consent; second, through the explication of an ethical work-up for specific cases; and third, by analyzing several important ethical decisions in caring for patients at the end of life.

We acknowledge the great importance of many emerging ethical issues such as those raised by genetics, preventive medicine, and information technology. We also recognize that many ethical issues now arise in the context of team care, cost containment, managed care, and in organizational settings in which the physician is simultaneously an employee, a manager, and perhaps even an investor. These issues are too important for superficial treatment. We would only point out that in these instances too the physician's primary responsibility is the good of the sick person. If physicians default on this commitment, the last moral safeguard of the sick will have been compromised to the peril of us all.

5

The Moral Foundations of the Patient-Physician Relationship: The Essence of Medical Ethics in Military Medical Ethics

Edmund Pellegrino, MD

Introduction

In medicine, whether in the civilian or military setting, medical ethics begins and ends in the patient-physician relationship. The conception we hold of that relationship shapes the decisions we make in every clinical situation. It sets the standard for right and wrong, good and bad professional conduct. It is the final arbiter of the moral status of every policy affecting the health of individuals or the public. Even public health, military, and penal medicine, which incorporate societal goals, must balance those goals against the realities of the relationship of a patient and a physician. How we see that relationship will determine the kind of society we are, have become, or want to be.

For these reasons, this chapter is devoted to the moral foundation of the conduct of the patient-physician relationship. Such a foundation, if it is to be adequate as the keystone of the edifice of medical ethics, must at a minimum answer certain key questions: Is there anything morally special about the patient-physician relationship, and if there is, what is it? What does the special nature of the relationship entail with respect to the duties physicians and patients owe each other, the virtues they should exhibit, or the rules, principles, and attitudes that should guide their interactions in the clinical encounter? These questions are implicit in the later chapters, which define the special nature of the patient-physician relationship in the clinical and the military context.

Reprinted with permission of the Pellegrino family from Edmund D. Pellegrino, "The Moral Foundations of the Patient-Physician Relationship: The Essence of Medical Ethics in Military Medical Ethics," in *Military Medical Ethics*, vol. 1 of *Textbooks of Military Medicine*, ed. Thomas E. Beam and Linette Sparacino (Washington, D.C.: Office of the Surgeon General, US Army, 2003). Dr. Pellegrino was section editor for Medical Ethics.

This chapter confines itself to the ethical aspects of the patient-physician relationship. Its focus therefore is on professional ethics—the ethics of the physician as a professional (and, by analogy, to other health professionals, e.g, nurses, dentists, clinical psychologists, social workers). The content of bedside ethical decisions—the ethics of particular clinical dilemmas—is discussed in later chapters. The religious and theological foundations of medical ethics are not included, even though, for many Americans, they are the ultimate source of all morality, in general or in the professional life. Finally, we must not forget that patients and physicians meet each other in an intricate matrix of psychosocial, cultural, and sociohistorical phenomena that can modify the expression of medical ethics.[1] These factors notwithstanding, there is a foundation for the duties of all health professions that is relatively constant across cultures, history, and national boundaries.

Is a Foundation for Medical Ethics Possible?

Historically, the Hippocratic Oath (Exhibit 5-1) and ethos were not universally accepted as the foundation for medical ethics by most ancient Greek physicians.[2] They originated with a small group of physicians who were eager to distance themselves from the majority of their contemporaries who were itinerant journeymen, businessmen, and craftsmen. In later antiquity, the Hippocratic ethic found favor with the three monotheistic religions and, through their influence, became widely disseminated.[3] At least from the late Middle Ages on, and well into the modern era, the moral precepts of the Hippocratic ethic were the standard for the ethical conduct of physicians.[4]

Our concern is not with the evolution of medical ethics as a historical or social epiphenomenon. It is the deeper moral phenomena upon which it has been based that are of importance. It is the existence of these phenomena, which the Hippocratic physicians and their successors grasped intuitively, that accounts for the durability of their ethic across so many centuries, countries, and cultures.

A quarter of a century ago the question of whether or not a foundation for medical ethics was possible would have seemed a naive question. At that time the Hippocratic

1. E. Shorter, "History of the Doctor-Patient Relationship," in *Companion Encyclopedia of the History of Medicine*, vol. 2, ed. W.F. Bynum and R. Porter (London: Routledge, 1993): 783–800.

2. L. Edelstein, "The Hippocratic Physician," in *Ancient Medicine: Selected Papers of Ludwig Edelstein*, trans. C.L. Temkin, ed. O. Temkin and C.L. Temkin (Baltimore, Md.: Johns Hopkins University Press, 1967), 87–110; P. Carrick, *Medical Ethics in Antiquity: Philosophical Perspectives on Abortion and Euthanasia* (Dordrecht & Boston: D. Reidel, 1985), 69–94; R. Baker, "History of Medical Ethics, in *Companion Encyclopedia of the History of Medicine*, vol. 2, ed. W.F. Bynum and R. Porter (London: Routledge, 1993), 852–87.

3. O. Temkin, *Hippocrates in a World of Pagans and Christians* (Baltimore, Md.: Johns Hopkins University Press, 1991); D.W. Amundsen, *Medicine, Society, and Faith in the Ancient and Medieval Worlds* (Baltimore, Md.: Johns Hopkins University Press, 1996); S.S. Kottek, J.O. Leibowitz, and B. Richler, "A Hebrew Paraphrase of the Hippocratic Oath (from a Fifteenth Century Manuscript)," *Med Hist.* 22, no. 4 (1978): 438–45.

4. R.J. Bulger, ed., *Hippocrates Revisited: A Search for Meaning* (New York: Medcom Press, 1973).

(Exhibit 5-1)
The Oath of Hippocrates

I swear by Apollo Physician and Asclepius and Hygieia and Panaceia and all the gods and goddesses, making them my witnesses, that I will fulfil according to my ability and judgment this oath and this covenant:

To hold him who has taught me this art as equal to my parents and to live my life in partnership with him, and if he is in need of money to give him a share of mine, and to regard his offspring as equal to my brothers in male lineage and to teach them this art—if they desire to learn it without fee and covenant; to give a share of precepts and oral instruction and all the other learning to my sons and to the sons of him who has instructed me and to pupils who have signed the covenant and have taken an oath according to the medical law, but to no one else.

I will apply dietetic measures for the benefit of the sick according to my ability and judgment; I will keep them from harm and injustice.

I will neither give a deadly drug to anybody if asked for it, nor will I make a suggestion to this effect. Similarly I will not give to a woman an abortive remedy. In purity and holiness I will guard my life and my art.

Image from Wikimedia, "Herm with the head of Hippocrates, 1 century A.D." Photo credit: Yair Haklai

I will not use the knife, not even on sufferers from stone, but will withdraw in favor of such men as are engaged in this work. Whatever houses I may visit, I will come for the benefit of the sick, remaining free of all intentional injustice, of all mischief and in particular of sexual relations with both female and male persons, be they free or slaves.

What I may see or hear in the course of the treatment or even outside of the treatment in regard to the life of men, which on no account one must spread abroad, I will keep to myself holding such things shameful to be spoken about.

If I fulfill this oath and do not violate it, may it be granted to me to enjoy life and art, being honored with fame among all men for all time to come; if I transgress it and swear falsely, may the opposite of all this be my lot.

ethics, exemplified by the oath, the so-called "deontological" books of the Hippocratic Corpus (dealing with the oath, precepts, the law, decorum, and the physician),[5] and its congeners in the Code of the American Medical Association (AMA)[6] and dozens of other codes of medical ethics and variations,[7] were taken for granted as the source and

5. Hippocrates, "Precepts," in *Hippocrates*, vol. 1, trans. and ed. W.H.S. Jones, Loeb Classical Library (Cambridge, Mass.: Harvard University Press, 1972): 299–301, 313–33; Hippocrates, in *Hippocrates*, vol. 2, trans. and ed. W.H.S. Jones, Loeb Classical Library (Cambridge, Mass.: Harvard University Press, 1972): "Law," 263–65; "Decorum," 279–301; "Physician (Chapter 1)," 311–13.

6. American Medical Association, Council on Ethical and Judicial Affairs, *Code of Medical Ethics, Current Opinions with Annotations*, 1996–1997 ed. (Chicago: American Medical Association, 1997).

7. R. Crawshaw, D.E. Rogers, E.D. Pellegrino, et al., "Patient-Physician Covenant," *JAMA* 273, no. 19 (1995): 1553.

foundation for the ethics of the patient-physician relationship. Today this foundation is no longer secure. An increasing number of ethicists, physicians, and even the public, believe not only that the Hippocratic ethic is out-of-date but that the whole idea of a stable foundation for ethics is no longer tenable.

Several challenges singly, and in combination, have brought about this present state of affairs. Four that seem most important are (1) the upheaval in social values in the 1960s; (2) the interest in medical ethics by professional philosophers; (3) the transformation of medical ethics into "bioethics"; and (4) the "postmodern" turn of philosophy in general, and moral philosophy in particular.

The first serious contemporary challenge to the Hippocratic foundation was sociopolitical. Beginning in the late 1960s in America, for various reasons (as discussed further in John Collins Harvey, "Clinical Ethics: The Art of Medicine," [chap. 3 of *Military Medical Ethics*, vol. 1—Ed.]) all traditional values and sources of moral authority were challenged—religion, the family, parents, teachers, all holders of authority, and all the professions. This was the era of participatory democracy, which saw the rise of the consumer movement, civil rights legislation, the Patient's Bill of Rights, and a rash of student protests against academic tradition and authority. In such a climate, physicians, medicine as a privileged profession, and medical ethics were especially vulnerable. They were seen as elitist, monopolistic of power, and self-aggrandizing.

The second challenge came from professional philosophers who for the first time in their history took serious interest in medical ethics. To be sure, the ancient philosophers often referred to medicine, but neither they nor their modern counterparts ever wrote serious treatises on medical ethics. Only in the last quarter of the twentieth century did philosophers examine the moral presuppositions of the traditional ethic. They did so using the conceptual tools of a variety of established moral systems. Each system introduced its own perspective on the relationships between physicians and patients. For example, the followers of the philosophy of Kant, placed their emphasis on patient autonomy; those who followed J.S. Mill chose utility maximization; and followers of W.D. Ross turned to prima facie principles (see David C. Thomasma, "Theories of Medical Ethics: The Philosophical Structure," [chap. 2 of *Military Medical Ethics*, vol. 1—Ed.]). As a result, the Hippocratic tradition of benevolence and beneficence was reinterpreted as authoritarian, insensitive to social ethics, and even unjust. Other major precepts, such as the prohibitions against abortion, breaches of confidentiality, and sexual intercourse with patients, were relaxed. Currently, the prohibitions against assisted suicide[8] and euthanasia are under attack. Pressures have steadily mounted for an oath and code more congruent with contemporary mores.[9]

8. J. Kevorkian, *Prescription-Medicide: The Goodness of Planned Death* (Buffalo, N.Y.: Prometheus Books, 1991).

9. S.H. Wanzer, D.D. Federman, S.J. Adelstein, et al., "The Physician's Responsibility toward Hopelessly Ill Patients: A Second Look," *N Engl J Med.* 320, no. 13 (1989): 844–49; E.D. Robin and R.F. McCauley, "Cultural Lag and the Hippocratic Oath," *Lancet* 345, no. 8962 (1995): 1422–24.

The third challenge arose out of the progressive intrusions into medical ethics by law, politics, economics, psychology, and culture. Beginning in the 1980s, a larger view of medical ethics emerged under the new rubric of "bioethics." Bioethics extended beyond the bedside to social and public policy, ecology, and the environment. In the 1990s, scholars in the social sciences entered this broader field. Those outside the field of philosophy challenged philosophical ethics as a rational discipline. They judged it too abstract and insufficient to encompass the full complexity of the moral life. Alternative theories and models of ethics such as casuistry, narrative, virtue, feminist, and caring ethics have been proposed.[10] To remedy these presumed deficiencies, ethics itself has often been reduced to issues of public policy and procedure rather than patient-physician relationships.[11]

The fourth challenge in the erosion of traditional medical ethics arose in the attack on philosophical ethics by the "postmodern" critique of philosophy itself. This critique centers on the claims of reason, itself, to arrive at moral truth. The postmodern critique challenges the traditional pretensions of philosophy to achieve moral truth through reason alone. Postmodernism declares philosophy to be "dead."[12] Secular bioethics is particularly vulnerable to this critique because it built its endeavor on the post-Enlightenment project of ethics free of metaphysics and religion and dependent only on an autonomous rationality. Postmodernism has become a "deteriorated version of the Enlightenment."[13] Post-modernism deprives contemporary bioethics of its rationalist underpinnings and denies it access to any foundation or overarching theory. In this view, any foundation for medical ethics such as the Hippocratic ethic, or the one this chapter shall describe, is ipso facto intellectually suspect.

This surely is not the place to attempt the complex task of refutation of the postmodernist thesis. But as Rosen argues, postmodernism reduces philosophy to ideology. This places the ideology of linguistic fashion in the place formerly occupied by philosophy.[14] We may try to eliminate foundations, but there is always a position of last resort beyond which we cannot retreat. Call it what we will, this position of last resort is in fact a "foundation." Thus anti-foundationalism is the postmodernists' position of last resort. What is of relevance to this chapter is that contemporary medical ethics faces an important choice with very practical consequences. (For a more detailed discussion of postmodernism and deconstructionism, please see John

10. E.D. Pellegrino, "Bioethics as an Interdisciplinary Enterprise: Where Does Ethics Fit in the Mosaic of Disciplines?" in *Philosophy of Medicine and Bioethics: A Twenty-Year Retrospective and Critical Appraisal*, ed. R.A. Carson and C.R. Burns (Dordrecht, Boston, & London: Kluwer, 1997), 1–23.

11. G.C. Meilaender, *Body, Soul, and Bioethics* (Indianapolis, Ind.: University of Notre Dame Press, 1995).

12. T. Docherty, ed., *Postmodernism: A Render* (New York: Columbia University Press, 1993); R. Rorty, *Philosophy and the Mirror of Nature* (Princeton, N.J.: Princeton University Press, 1979).

13. S. Rosen, *The Ancients and the Moderns: Rethinking Modernity* (New Haven, Conn.: Yale University Press, 1989), 20.

14. Rorty, *Philosophy and the Mirror of Nature*; Rosen, *The Ancients and the Moderns*, 176.

Collins Harvey, "Clinical Ethics: The Art of Medicine" [chap. 3 of *Military Medical Ethics*, vol. 1—Ed.]) If medical ethics chooses to go the postmodernist route, it must accept a variously interpreted and deconstructed ethic, one malleable by social and linguistic construction. Profession and patients will fragment further and further into smaller and smaller communities with different and contradictory moral values.[15] A uniform set of moral precepts binding all physicians will no longer be possible. Each therapeutic encounter will become a new negotiable event with its own rules, duties, and principles, or the whole of bioethics will be left to social consensus.[16] Any notion of a foundation for ethics based in respect for human life will be replaced by a technological determinism.[17]

This chapter takes a different pathway—the way of reconstruction of the ethical foundation for medical ethics, not its deconstruction. This does not imply a simplistic reaffirmation of the Hippocratic ethic. A true "reconstruction" means retaining what is valid in the old and enlarging it by new insights. This is not the same as changing ethics to accommodate social mores. The beginning, a quarter century ago, of formal philosophical reflection on the Hippocratic moral precepts, uncovered a genuine need for their justification beyond mere assertion. This has been salubrious because it changed medical ethics from a set of free moral assertions into a respectable ethical enterprise. This chapter undertakes a reconstruction of medical ethics out of the empirical phenomena of the clinical encounter and the experiences of illness and healing. These are the universal phenomena that underlie the relationships of patients and physicians across temporal and cultural barriers. These are the relationships perceived by the Hippocratic physicians but never formally or systematically argued.

Some Current Models of the Patient-Physician Relationship

Medicine is a multi-varied societal phenomenon in which physicians may play a variety of roles simultaneously. Each role elicits a particular kind of relationship with the patient and entails a particular kind of ethic. One of these roles, the role of healer, is primary; the others are subsidiary. Before turning to the reconstruction of this primary role, it is important to examine some of the alternative models and the ethics they entail. Pedro Laín Entralgo has written most perceptively about the history of the patient-physician relationship.[18] He summarizes the relationship in terms of the physician's motives under four general headings: (1) physician as

15. H.T. Engelhardt Jr., *The Foundations of Bioethics*, 2nd ed. (New York: Oxford University Press, 1995).

16. J.D. Moreno, *Deciding Together: Bioethics and Moral Consensus* (New York: Oxford University Press, 1995).

17. P. Singer, *Rethinking Life and Death: The Collapse of Our Traditional Ethics* (New York: St. Martin's Press, 1995).

18. P. Laín Entralgo, *Doctor and Patient*, trans. F. Partridge (New York: McGraw-Hill, 1969); P. Laín Entralgo, *La Relación Médico-Enfermo: Historia y Teoría* (Madrid: Revista del Occidente, 1964).

technical helper; (2) physician as seeker of knowledge; (3) physician as functionary of an institution; and (4) physician as a seeker of profit.[19] Elements of these motives are intermingled in each of the more specific models to be examined below.

The Physician as Clinical Scientist

One prominent model, often emphasized in medical schools, is the patient-physician relationship as an exercise in applied biology.[20] In this model, the relationship is a means for attaining knowledge and also for applying existing knowledge to solve a patient's diagnostic or therapeutic problem. The ethic governing this kind of relationship is the ethic of good science, the rules of which are objectivity, honesty in recording data, technical competence, and so forth. The patient is the object of study seen as a concrete instance of the universal laws of biology and pathology. This model does not deny the existence or importance of psychosocial and personal elements in the genesis or treatment of the illness. But these elements are not considered properly as in the domain of medicine or the physician. They belong to social workers, psychologists, and pastoral counselors. It is the physician's task to take note of these subjective elements but to refer them to others for treatment.

The Physician as Body Mechanic

A variant of this model of the physician as clinical scientist is to see the patient-physician relationship as a mechanical event equivalent to the owner of a defective automobile bringing it in for repair or replacement of a part.[21] In this model, the physician is the mechanic and the patient is the owner of a part to be fixed. Psychosocial and personal elements are really irrelevant. Because they are not mechanically fixable, they are not part of the physician's task. The ethic of this relationship is the ethic of technical competence, impersonality, and fulfillment of a service contract.

The Physician as Businessman

In the business model, health care is a commodity to be bought and sold on the open market for profit.[22] Its price, availability, distribution, and quality are dependent upon competition. The patient is a consumer who shops for care as he shops for other needed goods. The patient is a source of gain for the physician who

19. Laín Entralgo, *Doctor and Patient.*

20. D. Seldin, "The Medical Model: Biomedical Science as the Basis of Medicine," in *Beyond Tomorrow: Trends and Prospects in Medical Science* (New York: Rockefeller University Press, 1977).

21. "Physicians as Body Mechanics," in *Concepts of Health and Disease: Interdisciplinary Perspectives*, ed. A.L. Caplan, H.T. Engelhardt Jr., and J.J. McCartney (Reading, Mass.: Addison-Wesley, Advanced Book Program, World Science Division, 1981), 665–75.

22. M. Hall, "The Ethics of Health Care Rationing," *Public Aff Q.* 8 (1994): 33–50.

competes for patient "business." The ethic in this model is the ethic of business and the "ethic" of the marketplace. In the market, patients are players whose welfare depends upon what they can command in the way of resources and what they can negotiate in trade. Solicitude or concern for the "loser" is important only if it makes for better business. If someone makes a wrong choice, or lacks the wherewithal to enter the market in the first place, this is unfortunate but not the concern of the physician, or those for whom the physician works.

Two variants of the market relationship model are the entrepreneur and the managed care models.[23] These roles may be combined when physicians are simultaneously caregivers, "providers," and investors, owners, or risk-sharers in managed care organizations or healthcare facilities. Here providers compete for capitation contracts to provide care for large target populations, preferably those with few medical needs who will pay their premiums and will not need much in the way of care. The physician is an employee of an organization or one of its owners. The dominant ethic is the ethic of competitive business and corporations. This is a minimalist ethic always at risk of compromise if profit margins drop. The patient becomes a customer and client, a source of gain, or a unit of care—an "insured life" who can be "traded" in mergers or contract negotiations.

In these technical and market models, the physician regards the practice of medicine primarily as an occupation, a way to make a living rather than a means of service to others. Practicing medicine is a job like any other. There is no requirement to extend oneself beyond the job description. The ethics implicit in this kind of relationship is the ethic of the employee whose aim is to satisfy the patient so the patient will return and will recommend the physician to others. Only as much kindness and compassion as are needed for success need be offered. There is no commitment beyond strict working hours. Choice of physician is not important because physicians are interchangeable. The patient often is seen not as a personal responsibility of the physician, but of the organization.

The Physician as Social Servant

Another model increasingly being pressed upon today's profession is the physician as a social servant, as primarily an instrument of societal, or fiscal, good. Medical knowledge is thus directed to some purpose beyond, in addition to, or along with, meeting the needs of the individual patients. Examples of this genre are the physician acting as a rationer in a managed care system, physicians as employees of a penal institution, physicians in military service, and public health physicians. The physician's primary orientation is toward the good of the population in general, or a specific population in a social institution. The ethic implicit in this model is a population-based

23. E.J. Emanuel and N.N. Dubler, "Preserving the Physician-Patient Relationship in the Era of Managed Care," *JAMA* 273, no. 4 (1995): 323–29; H.T. Engelhardt Jr. and M.A. Rie, "Morality for the Medical-Industrial Complex: A Code of Ethics for the Mass Marketing of Health Care," *N Engl J Med.* 319, no. 16 (1988): 1086–89.

Table 5-1. Models of the Patient-Physician Relationship

Model	Physician	Patient	Ethic
Applied biology	Clinical scientist: uses knowledge to solve medical problems	Biological object harboring a disease	Good science, truth, objectivity, technical competence
Body repair	Body mechanic: fixes biological problems	Owner of defective body part	Technical competence, fulfillment of service contract
Commodity transaction	Businessman: competes for clients	Consumer of medicine and source of gain for the physician	Business, the laws of the marketplace
Investment opportunity	Businessman: views self as an entrepreneur	Unit of care, which can be sold or traded by contract	Competitive business, marketplace concerns, profit margins
Managed care industry	Businessman: functions as an employee	Client whose consumption of care must be controlled	Corporate goals, economic concerns
Social utility	Social servant: uses medicine as instrument of societal good	Client who is a societal subunit	Population-based needs of the many versus the individual
Professional	Helper, healer: uses medicine for the patient's benefit	Person to be helped	Covenant of trust

ethic. These are the roles of the physician as bureaucrat or functionary using medical knowledge for purposes other than the good of the individual patient.

The Physician as Helper and Healer

The most traditional model, and ethically the most demanding, is the model of the physician as helper and healer, a committed professional whose primary obligation is to the good of his patient.[24] In this model, the physician is committed to something other than self-interest, advancement of career or occupation, or even the good of society. This model is based ethically in the specificity of the role of physician as healer, helper, and curer. This ideal is not always actualized to be sure, but it has been at the heart of the Hippocratic ethic and its many variations. It is the model that this chapter proposes as the foundation for the ethics of the healing professions.

Summary of the Models

Table 5-1 lists the models of patient-physician relationship that we have discussed in this chapter. These are models that are now in vogue and often

24. M. Balint, *The Doctor, His Patient, and the Illness* (New York: International Universities Press, 1964).

competing for primacy. Each implies a different theory of medicine, a different interpersonal relationship between physician and patient, and a different ethic. In each of these models, except the healing model, the physician uses medical knowledge for what can be a good or bad purpose. These other purposes are not intrinsically evil, but neither are they distinctive of medicine because medicine is defined by its healing purpose. When purposes extrinsic to medicine itself conflict with the end of a healing relationship, ethical dilemmas arise in which priority must be given to patient welfare.[25] The nature of these conflicts (e.g., the physician as military officer, public health official, or forensic psychiatrist) will be treated in other chapters. This chapter focuses on the "end" of medicine as medicine, on what distinguishes it as a special kind of human activity with its own internal morality.

Healing and Helping: The "End" of Medicine

Medicine and physicians are a part of the social and historical fabric of the cultures within which they live and function. Medicine therefore is in part economics, business, societal purpose, and function. But it is not primarily any of these things. If a true foundation for medical ethics is to be found, it must be sought in what is unique to medicine; and this is the healing relationship between the patient and the physician. It is from this uniqueness that an ethic specific to medicine can be defined. The ethical implications of this uniqueness may be derived externally, by some form of social construction, or by applying some preexisting system of morals and ethics to the phenomena of medicine. Alternatively, the ethics of medicine may be derived internally, by a study of the phenomena of medicine itself. We will examine both approaches methodologically and substantively.

External Morality

An externally determined ethic is most often derived from a preexisting system of moral philosophy with origins outside medicine but applied to the activities peculiar to medicine. This is generally a "top-down" approach in which medical ethics draws upon principles, duties, or rules and action guidelines developed outside medicine to define morally appropriate conduct, or choices. Some examples are the derivation of duties of physicians and patients from the deontological ethic and categorical imperative of Immanuel Kant,[26] the principle of utility maximization of John Stuart Mill,[27] natural virtue ethics of Aristotle,[28] or the Christian virtues as

25. E.D. Pellegrino, "Societal Duty and Moral Complicity: The Physician's Dilemma of Divided Loyalty," *Int J Law Psychiatry* 16, nos. 3–4, (1993): 371–91.

26. I. Kant, *Groundwork of the Metaphysics of Morals*, trans. H.J. Paton (New York: Harper, 1964).

27. J.S. Mill, *Mill's Ethical Writings*, ed. J.B. Schneewind (New York: Collier, 1965).

28. Aristotle, *Nicomachean Ethics*, in *The Complete Works of Aristotle: The Revised Oxford Translation*, vol. 2, trans. W.D. Ross, rev. J.O. Urmson, ed. J. Barnes (Princeton, N.J.: Princeton University Press, 1984), 1229–1867.

exemplified in Thomas Aquinas.[29] Similarly, an external source of medical ethics might be drawn from a religious tradition, as in the theological ethics of Thomas Aquinas, the Catholic casuist tradition,[30] or an updated conception of natural law,[31] the Protestant tradition,[32] or the Jewish halakhic tradition.[33]

In recent years, systems of externally derived ethics have had their origins in sociocultural mores, in social constructivism, or coherence theories. Here the justification for judgments of right and wrong are determined by societal consensus, or coherence with other accepted beliefs and principles, or by "reflective equilibrium," a dialogue between general principles and intuitive judgments.[34] Existential,[35] narrative,[36] caring,[37] and feminist ethics[38] are further examples of external systems for the derivation of right and wrong. These and other ethical theories have been applied to medicine to justify what ought and ought not to be done in particular clinical situations.

The foregoing "external" sources of medical ethics are formulated in other chapters in this book and will not be given further consideration here. In any contemporary study of medical ethics, they deserve serious consideration. They express moral truths of various relevance to, but not necessarily determinative of, right and good conduct for physicians and other health workers. They relate to, but are not determined in the first instance by, the nature of medicine as a special kind of human activity. In one way or another, they leave a gap between ethical theory and the realities of the moral world of physician and patient. To close this gap, it is necessary to move more closely into the lived worlds of physician and patient—to a more internally determined ethic.[39]

29. T. Aquinas, *Summa Theologiae*, trans. and ed. W.D. Hughes (New York: McGraw-Hill & Blackfriars, 1969), 62, 137–49.

30. A.R. Jonsen and S.E. Toulmin, *The Abuse of Casuistry: A History of Moral Reasoning* (Berkeley: University of California Press, 1988).

31. J. Finnis, *Natural Law and Natural Rights* (Oxford: Clarendon Press, 1980).

32. J.F. Gustafson, *The Contribution of Theology to Medical Ethics* (Milwaukee, Minn.: Marquette, 1975).

33. I. Jakobovits, *Jewish Medical Ethics: A Comparative and Historical Study of the Jewish Religious Attitude to Medicine and Its Practice* (New York: Bloch Publishing, 1975); *Jewish Bioethics*, ed. F. Rosner and J.D. Bleich (New York: Hebrew Publishing Co, 1979).

34. J. Rawls, *A Theory of Justice* (Cambridge, Mass.: Belknap Press of Harvard University Press, 1971).

35. J.P. Sartre, *Cahiers pour une Morale* [Notebooks for an Ethics] (Paris: Gallimard, 1983).

36. M.C. Nussbaum, *The Fragility of Goodness: Luck and Ethics in Greek Tragedy and Philosophy* (New York: Cambridge University Press, 1986).

37. N. Noddings, *Caring: A Feminine Approach to Ethics and Moral Education* (Berkeley: University of California Press, 1984).

38. E. Frazier, J. Hornsby, and S. Lovibond, eds., *Ethics: A Feminist Reader* (Cambridge, Mass.: Blackwell, 1992).

39. E.D. Pellegrino, *The Lived World of Doctor and Patient* (New Haven, Conn.: Yale University Press, 2000).

The Ethics Internal to Medicine

There are several senses in which an ethic may be internal to medicine. One is the ethic expressed in ethical codes elaborated within the profession by physicians, for physicians. Examples would be the Hippocratic Oath,[40] the ethics of the Chinese physician,[41] the Indian Code,[42] the ethical "code" of Thomas Percival prepared for the physicians at the Manchester Infirmary,[43] the AMA Code of 1847, its many revisions since then, and their expansion in the Opinions of the Council on Ethical and Judicial Affairs of the AMA.[44] The ethical codes of the World Health Association, the British Medical Association, and a multitude of others would all be internal in this sense. These codes were prepared for, and by, members of the profession without significant input from those outside the profession.

These internal codes generally turn out to be statements of moral belief mixed with etiquette. They exhibit little in the way of formal "ethics," because their moral foundations are taken for granted and not derived or justified by analysis or argument. Their moral content is surprisingly similar to that expressed in the Hippocratic Oath and ethic. These codes all express certain moral truths that have shaped the ideals of professional behavior and the commitment of the community of physicians to patient welfare. They should not be discounted, as some suggest, simply because they were prepared *by* physicians and *for* physicians.[45] Their final test is not who composed them, but whether or not they contain arguable or demonstrable ethical truths.

Laws pertaining to medical practice are external to the internal morality of medicine. Laws are external because they promulgate statutes governing the obligations of physicians to patients, the source of which is legislative action outside of medicine. Nonetheless laws are responsive to the special nature of the medical relationship. The laws of torts, contracts, and fiduciaries, for example, recognize the special nature of the patient-physician relationship. In this latter sense, these laws accurately reflect the special nature of the therapeutic relationship and the vulnerability of patients who, as a result, are in need of legal protection. Laws governing medical practice, thus, get their moral force from their recognition of the realities of the patient-physician encounter.

40. Hippocrates, "Precepts," 313–33.

41. P.U. Unschuld, *Medical Ethics in Imperial China: A Study in Historical Anthropology* (Berkeley: University of California Press, 1979).

42. "Caraka Samhita," 3.8.13–14, as cited in D. Wujastyk, "Indian medicine," in *Companion Encyclopedia of the History of Medicine*, vol. 1, ed. W.F. Bynum and R. Porter (London: Routledge, 1993), 762.

43. T. Percival, *Medical Ethics, or A Code of Institutes and Precepts Adapted to the Professional Conduct of Physicians and Surgeons*, reprinted from the 1803 version (Birmingham, Ala.: Classics of Medicine Library, 1985).

44. American Medical Association, *Code of Medical Ethics*.

45. R.M. Veatch and C.M. Mason, "Hippocratic versus Judeo-Christian Medical Ethics: Principles in Conflict," *J Religious Ethics* 15, no. 1 (1987): 86–105.

Also situated somewhere between the internal and external boundaries of medicine are theories of ethics or models of the relationship with strong sociological and psychological foundations. One example would be Laín Entralgo's formulation of friendship (*philia*) as the foundation of the relationship.[46] Other examples would be the notion of caring, the patient's or physician's life story or narrative, or the experience of practice itself. Each concept has roots in the actualities of the patient-physician relationship; none is sufficient in itself to be the basis of a normative ethic for that relationship. Each is important but best incorporated in the view of a healing ethic.

Elements of an Internal Morality for Medicine

For purposes of this chapter, the term internal morality will be used more narrowly to signify a foundation for medical morality arising within the phenomena peculiar to medicine, those that define it as a special form of human activity, and by that fact generate specific moral responsibilities binding only on those who profess medicine or the other health professions. The three phenomena specific for the patient-physician relationship are (1) the fact of illness; (2) the act of profession; and (3) the act of medicine. Together they comprise the healing relationship, the end of which is the good of the patient.

The Fact of Illness

The most fundamental fact about medicine is that it exists because humans become ill. This and mortality are the two most universal characteristics of human existence. They transcend culture, history, and all other differences between and among humans. Illness is a subjective existential state in which the patient's sense of well-being or accommodation with existing disease is threatened or compromised by some new symptom or sign. The person who recognizes himself as ill enters a new stage of existence in which his humanity is diminished in several specific ways.

First, there is the loss of freedom to do what one wishes because of pain, disability, discomfort, and so forth. Being ill creates anxiety, fear of mortality, and disability. Illness may or may not correspond to objective pathology (i.e., disease). Illness threatens the image of one's physical and emotional integrity. The illness then becomes a center of concern and diverts energy and attention from other pursuits. It creates a disorganization and disequilibrium of the whole of the person's existence.

Ill persons may tolerate this state of disequilibrium for a long time but ultimately most decide they need help. That is when they become *patients*—persons bearing a burden of suffering. (The word "patient" is derived from the Latin *patior, patiens, pati*, meaning "to suffer, bear a burden."[47]) When people become patients,

46. Laín Entralgo, *La Relación Médico-Enfermo*, 149.

47. R.G.W. Glare, ed., *Oxford Latin Dictionary* (Cambridge: Oxford University Press, 1983), 1308–9.

they realize they need the knowledge and power of others to be healed. Patients then no longer treat themselves but are compelled to seek out a health professional in whom eventually they must place trust. They must enter a relationship of inequality because the health professional possesses the knowledge and skill the patient needs. Thus when well persons become ill, by that very fact, they become patients—vulnerable, suggestible, and exploitable. They experience a change in existential state that is not exactly parallel to any other state. Illness is a unique universal phenomenon of human existence and it is that uniqueness that generates its moral orientation.

The Act of Pro-fession and Promise

In this special state of vulnerability, the patient seeks out someone who professes to be a healer. The physician or other health professional asks what is wrong: "How can I help?" In this question, physicians invite the patient's trust that they possess the requisite knowledge, that they will use it to help and not to harm (i.e., to act in the patient's best interests, not their own, and not in the interests of others). When physicians voluntarily offer to help, they make an implicit promise. They offer themselves as healers, helpers, and caregivers. They generate expectations they promise implicitly to fulfill. They voluntarily bind themselves, by that very fact, to act beneficently. Their assigned societal role and their possession of special knowledge require them to help. Physicians thus automatically enter into a covenantal, trust relationship when they offer to care for a patient. A covenant is more than a casual promise. It is a mutual agreement with something sacred about it because it is made in the presence of need by one capable of meeting that need.

This promise to help is an act of profession ("profession" derives from the Latin *profiteor/profiteri* meaning "to acknowledge openly, to avow," and *professio*, "an open declaration of an intention"[48]), that is to say, it is a solemn promise that binds *this* physician to *this* patient in a way that makes the physician an accomplice if harm comes from the relationship. The patient may dissolve this bond unilaterally by discharging the physician. But if the patient is ill, the physician can end the relationship only after another physician has agreed to undertake the patient's care.

The Act and End of Medicine

It is the act of healing, helping, and curing (which is what the patient seeks, needs, and expects from the physician's promise of help and the physician's invitation to trust) that initiated the relationship. Help, healing, care, or cure are the immediate ends of medicine. To be authentic, this end must be defined in terms of the good of the patient, that which restores health, if that is possible, or provides comfort and care if restoration of health is not possible. The good of the patient is a complex concept, multi-layered and highly personalized. It consists of at least four

48. Ibid., 1475–76.

components: (1) the medical good, (2) the good as perceived by the patient, (3) the good of the patient as a human, and (4) the good of the patient's spiritual nature.[49]

The first (and lowest on the scale) component is the medical good—that which the competent application of medical knowledge can achieve—treatment, cure, comfort, or containment of the disease. This is the most objective level susceptible to scientific apprehension. It is the level at which diagnosis, prognosis, and therapy function.

The second component of the patient's good is the good as perceived by the patient—what constitutes a "quality" life, the trade-offs the patient may wish to make among the options for treatment, the amount of risk, pain, discomfort, and disability that will be accepted as a price of treatment. The patient's perception of good is subjective, individualized, and personalized. It may or may not correspond with the medical good as perceived by the physician. It can only be defined by the patient.

The third component of patient good is more general. It is the good of the patient as a human being, as a being with inherent dignity possessed of reason and will, free to choose, to plan one's own life with a minimum of coercion by others or by events. It is this inherent dignity that entitles all humans to respect for their own decisions, and it is from this good for humans that the principle of autonomy derives. Justice, likewise, is grounded in who and what we are, as possessors of a common humanity. Justice requires equal treatment of patients and retribution for harm done to them. Keeping promises, such as the physician's promise of healing and helping, is also a matter of justice—something owed to all persons.

The fourth component, and the highest good of the patient, is whatever pertains to that individual's spiritual nature—beliefs about the nature and destiny of human life, its meanings, purposes, and relationships to sources of morality beyond human determination. This good is also grounded in our humanity as beings capable of commitments to ideals and beliefs beyond the needs of our material bodies. This is the realm of religious belief, or nonbelief, the ultimate source of morality for most patients when confronted with their own finitude, suffering, or despair. It is also the ultimate source of mortality for many physicians.

The immediate telos, or end, of medicine is to advance the good of a particular patient on all of these four levels. This is what healing means, that is, to help the patient heal himself, to become whole again to the extent possible within the limitations imposed by the patho- or psychophysiological aberration that brought him to the physician in the first place. To achieve this end will require in the immediate term a *right* and *good* decision, one which is scientifically and technically correct, and one which conforms with the four levels of good as they present in this patient. Morally valid medical and clinical decisions therefore fuse the technical and the moral dimensions in the moment of clinical decision. It is through the immediate end of a right and good decision made with, and for, the patient that the

49. E.D. Pellegrino and D.C. Thomasma, *For the Patient's Good: The Restoration of Beneficence in Health Care* (New York: Oxford University Press, 1988).

broader end of healing, and ultimately, the even broader ends of health of the individual and of society are attained.

A right and good decision must also be carried out safely, efficiently, prudently, and with a minimum of pain and discomfort. These obligations arise out of the seriousness of the promise to protect and advance the well-being of the patient. The vulnerability of the patient, and the trust patients must ultimately place in the physician's skill, are the foundation for the obligation to be competent in performance as well as in knowledge. Competency in psychomotor skills is, therefore, a moral requirement. Thus the internal morality of medicine rests on the relationships of three phenomena that characterize the clinical encounter: the fact of illness, the promise to help, and the healing act of medicine. Schematically, they can be represented in this way:

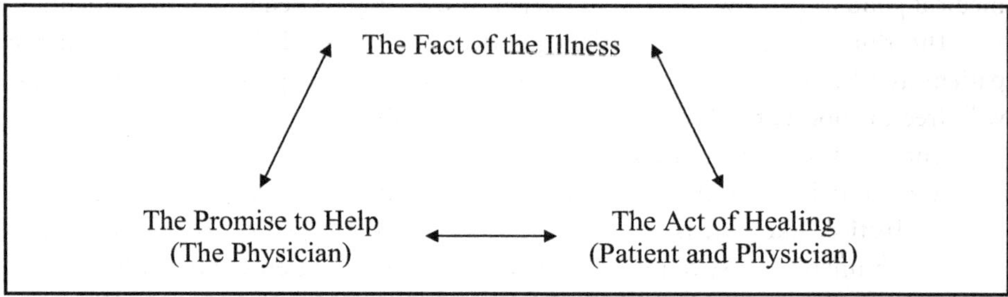

These are the ineradicable phenomena of the (human experiences of being ill, being healed, and professing to heal. They are universal human phenomena. No matter which culture, historical era, or national boundaries frame them, they are the same. They are the same phenomena experienced by ancient Greek patients and their physicians as well as today's patients and physicians. They will be the same in the next millennium and beyond because they are elements of the human condition. Medicine will become more highly technical than it is now, and more will be done by computers and automated means of diagnosis and treatment, but the need of sick persons for human interaction, intercession, and counsel will remain. Indeed, as machines take over the procedures of medicine, the need for the human touch and the ethical dimension of clinical decisions will be correspondingly greater.

The Clinical Encounter

Principles and Duties

Up to this point, I have distinguished *external* morality, as any system of ethics derived from outside of medicine (like the ethics of Aristotle, Kant, Hume, and Beauchamp and Childress), from *internal* morality, the derivation of duties, obligations, and principles from the phenomenology of medicine itself. However, the internal morality of medicine is not disconnected from the specific principles, rules, guidelines, and virtues that characterize external systems of morality applied to medicine. The difference in an internal morality is that they are derived from the empirical phenomena of the clinical encounter. They are, therefore, not dependent upon preexisting ethical theory or the resolution of disagreements between and among these theories.

The four principles of Beauchamp and Childress[50] (i.e., beneficence, nonmaleficence, autonomy, and justice) are one example of an external theory of medical morality that can be grounded more securely in the empirical realities of the clinical encounter. These principles have a firm foundation because they are necessary to achieve the end of medicine and fulfill the covenant of trust. Thus beneficence and nonmaleficence become duties because they are promised when the physician offers to help in the way specific to his profession. They are prima facie principles because no patient seeks professional help to be harmed; patients seek to be helped. No one expects to be used to advance the physician's or someone else's good. Subjects in clinical research may give free and informed consent to participation to advance the good of others but even the experimental subject's good must always be protected.

Autonomy, as argued above, is a good of the patient as a human being. It is one of the distinguishing characteristics of being human that we can make plans, make choices, and control significant parts of our lives. To ignore, override, or manipulate this decision-making capacity is to violate the good of the patient, to create harm and thus to defeat the end of medicine, which is healing and not harming. To violate autonomy is thus a maleficent act. Beneficence, nonmaleficence, and autonomy are not in opposition but reinforce each other. Autonomy is not absolute, however. The conditions that restrict are subjects for other chapters.

Justice, like autonomy, is a good of the patient as a human being. To violate justice is to violate an essential feature of human existence (i.e., what is owed to each human simply by virtue of being human). Like violations of beneficence and autonomy, violations of justice are maleficent and, therefore, frustrations of the end of medicine, which is the good of patients.

The Virtuous Physician

Like the prima facie principles of medical ethics, the virtues physicians should exhibit are linked to the ends of medicine and the phenomena that characterize the healing relationship. Virtues in general are defined as "the state of character which makes a man good and which makes him do his own work well."[51] When the concept of virtue is incorporated into medical ethics, it refers more specifically to those traits of character that make a physician or nurse (or other health worker) a good physician or nurse—one whose intention and action optimizes attainment of the ends of the healing relationship.[52]

50. T.L. Beauchamp and J.F. Childress, *Principles of Biomedical Ethics*, 4th ed. (New York: Oxford University Press, 1994).

51. Aristotle, *Nicomachean Ethics*, in *The Basic Works of Aristotle*, ed. R. McKeon (New York: Random House, 1941), 1106a:22–24.

52. E.D. Pellegrino, "The Virtuous Physician and the Ethics of Medicine," in *Virtue and Medicine: Explorations in the Character of Medicine*, ed. E.E. Shelp, Philosophy and Medicine 17 (Dordrecht, Holland: D. Reide Publishing Co., 1985), 237–55.

Essential to the notion of a virtue from its earliest definition in Plato and Aristotle is the idea of perfection (*areté*) in achieving a purpose. The virtuous physician or nurse is one who exhibits excellence in those character traits that enable one to come as close as possible to the healing purposes of the patient-physician (or patient-nurse) relationship. There are certain virtues or character traits that are particularly crucial; indeed, so crucial that they are entailed by the ends of medicine and without them those ends cannot be achieved. Some of these virtues are as follows[53]:

- *Fidelity to trust* is inescapable in the real world of a sick person seeking help. Without trust in the good intention of the physician and the physician's capability to perform competently, healing becomes difficult or impossible. The physician invites trust and must therefore be faithful to that trust, lest the offer to help be a lie and a deception.
- *Benevolence,* namely the predisposition habitually to wish to act for the patient's good, is the virtue that disposes the physician to do good, specifically to do good *for* the patient. This disposition is present even when it costs the physician something in time, frustration, loss of income, interference with personal plans, and so forth. Benevolence is a requisite virtue even in those difficult and frustrating cases of patient noncompliance, or abuse of health practices, or nonpayment of legitimate bills for service.
- Benevolence entails another virtue, namely *effacement of self-interest.* This is the disposition to serve the good of the patient even at some loss of personal self-interest. This virtue has limits. But where those limits are set is a highly personal matter. Heroic sacrifice is not required, but some degree of self-effacement is essential to attaining the ends of medicine. Without it a professional loses that which distinguishes his work from a mere occupation.
- *Compassion and caring* are equally relevant virtues. Compassion as a virtue is the habitual disposition to enter into the predicament of the sick person, to feel something of that predicament with him, and, as a result, to wish to help. Without entering the patient's predicament to some extent, it is not possible to heal in any full sense of that term. Care is a virtue closely related to the virtue of compassion. It may mean caring for the patient, that is, taking a personal interest in the patient's fate, or taking care in the way we carry out our professional duties, or taking care of the patient's needs and concerns. However interpreted, caring is an essential virtue integral to any morally satisfactory healing relationship. It is not, however, sufficient by itself to constitute a normative theory of ethics.
- Both care and compassion must be combined with *objectivity* if they are not to be harmful. Objectivity allows for an assessment of the actual physical state, diagnosis, and prognosis. It is united with compassion by putting all the

53. E.D. Pellegrino and D.C. Thomasma, *The Virtues in Medical Practice* (New York: Oxford University Press, 1993).

factual data into the lived world, life situation of this patient. Objectivity and compassion complement and balance each other.

- *Courage* is one of the four cardinal virtues from antiquity (the other three being temperance, justice, and prudence). It disposes physicians to take the personal risks necessary to care for the sick in times of emergency, disaster, or war; to expose oneself to contagion when necessary; and to take a moral stand when cooperation with what is morally wrong must be resisted.
- *Intellectual honesty* is a virtue insufficiently emphasized. Medicine and medical knowledge are powerful tools. They can be used for good and harm, or for control over others. Recognizing what one does not know, admitting it to oneself, to the patient, and to one's colleagues, is an essential safeguard for the vulnerable patient. Intellectual honesty is the antidote to the vice of intellectual hubris to which all professionals, and especially physicians, are so easily prone.
- In addition to intellectual honesty, a more general disposition to *humility* is required. This lies in a sober appreciation of the limitation of medicine as art and science, and of the physician, himself as an instrument of the patient's healing. It is an awesome thing to offer oneself to help or "heal" another. Merely to contemplate the demands on the health professional's knowledge, compassion, and understanding of the predicament of illness is to impart a sense of unworthiness on any responsible professional. Nonetheless, it is through fallible human beings that the knowledge and skill of medicine must be employed if the sick are to be helped. Physicians and other health professionals cannot permit themselves to be overwhelmed by their importance, nor by their sense of impotence and inadequacy. What humility requires is a calm and moderate assessment of the dangers of both indecision and presumption. A knowledge of the limitations of one's own person and of the art itself is gained only by careful, sustained, lifelong self-examination of the potential for good and harm in arrogating to oneself the title of "physician," "nurse," "psychologist," "social worker," and so forth.
- *Finally, and one of the most important of the clinical virtues, is prudence.* This is not the modern exercise of self-protective caution, which avoids risks to one's own welfare and does not venture to do good if it means a loss of self-interest. *Phronesis,* the Greek word for prudent judgment or practical wisdom, was, for Aristotle and Aquinas, the link between the intellectual (the capacity to know) and the moral virtues (the capacity to act well).[54] It encompassed the capacity of practical wisdom, knowing how to choose the appropriate means in a complicated situation so as best to serve the good ends of the healing

54. Aristotle, *Nicomachean Ethics,* 1144b30–1145a6; J. Pieper, *The Four Cardinal Virtues: Prudence, Justice, Fortitude, Temperance,* trans. R. Winston and C. Winston (New York: Harcourt Brace & World, 1965); J.M. Cooper, *Reason and Human Good in Aristotle* (Indianapolis, Ind.: Hackett Publishing Co., 1986), 63–64.

> relationship. Prudence is the power of discernment. In the clinical context, it is akin to clinical judgment—knowing how, when, and in what way, to act in the face of uncertainty, in a situation we have never encountered before, or one in which the virtues themselves appear in conflict.

Obviously, there are other character traits that can be entailed by the realities and ends of the patient-physician relationship but these just listed are indispensable. In their absence it would be difficult or impossible to assure a healing relationship that met minimum standards of ethical propriety.

Virtues do not by themselves constitute a whole moral philosophy for medicine.[55] They lack the specificity and concreteness of principles, rules, and axioms as action guidelines. Virtues also are subject to a multiplicity of definitions and orderings. They may conflict with each other because they are tied to the definition of the patient's good, and there may be differences about how to define that good for, and with, a particular patient. The tendency to subjectivism is accentuated by the circularity of the logic that ordinarily accompanies virtue theories, that is, virtuous persons do what is good; the good is what virtuous persons do. However, by grounding the virtues in the empirical realities of the patient-physician relationship we can avoid some of this circularity and most of the shortcomings of virtue theory in general.

The Virtuous Patient

The ethics of any human relationship implies reciprocal duties, principles, and virtues. In medical ethics, it is the duties of physicians that are emphasized. Given the balance of power in the physician's favor and the vulnerability of the patient, this is the morally proper ordering. Nonetheless, some mention of the patient's obligation and virtues is necessary if a full account of the internal morality of the healing relationship is to be provided.

If the end of medicine is to be attained, patients must participate in their own healing, and must facilitate the physician's pursuit of this end. This requires, at a minimum, that patients must be honest in the facts they provide in their histories of their illness. They must not withhold, misrepresent, or manipulate the facts for some ulterior motive. Patients should also cooperate in carrying out the treatment plan by following directions and reporting changes promptly. Without this minimal cooperation, the physician cannot fulfill his moral obligation to attain the healing ends of medicine.

People, in addition, have responsibilities to preserve health even before they become patients. Smoking, dietary and alcohol excesses, sedentary habits, failure to receive appropriate vaccinations, and similar behaviors thwart the "end" of medicine. Moderation (or temperance as it is sometimes called), another of the ancient cardinal

55. E.D. Pellegrino, "Toward a Virtue-Based Normative Ethics for the Health Professions," *Kennedy Inst Ethics J.* 5, no. 3 (1995): 253–77.

virtues, is a requisite virtue on the part of patients if health is to be maintained and the effects of disease are to be mitigated or prevented.

Failures on the patient's part are, however, not ipso facto a warrant for refusing to treat the patient who does become ill by failing to follow the physician's advice or because of poor health behavior. Physicians are not judges of the patient's virtue and are not empowered to punish patients by withholding their ministrations.

Patient autonomy is not absolute, however. The good of the physician as a human being entitles him to respect for his autonomy as well as the patient. Thus, if a patient requests a treatment that is futile, violates the canons of rational medicine or the religious beliefs of the physician, or poses a definable, grave, and probable harm to an identifiable third person, the physician is obliged to refuse. The physician, unless discharged by the patient, may withdraw from care of a sick person only when another physician whose values are more congruent with the patient's is willing to assume care. Until that time, the physician must care for the patient but must also do so in accord with his own conscience. The physician is a moral agent and as such must take responsibility for his actions.

When no emergency is present, physicians may refuse to care for a patient who threatens physical harm to others, consistently violates the physician's instructions, or endangers the life of the physician. Examples would be the violent drug addict, the sociopath, or the psychotic paranoid patient who threatens the physician or the physician's staff. Withdrawal can also be justified when the patient's repeated behavior makes achievement of the ends or purposes of medicine totally impossible. This decision must be taken with caution, without vindictiveness, and with a readiness to help again if the patient changes this behavior or presents in an emergency seeking assistance.

Another reciprocal duty of patients is to recognize their own finitude. There is a point in the natural history of any serious illness at which it becomes futile to continue, or to add, treatments. Hippocrates recognized this patient obligation when he said that patients should not expect medicine to cure them when they are overmastered by the disease.[56] This can now be stated as a principle: There is no obligation to treat when treatment is futile (in other words, ineffective, non-beneficial, or overly burdensome in relation to benefit or effect). This is an obligation too often ignored by patients and families who demand that *everything* be done even when death is inevitable, the patient is in a permanent vegetative state, and further treatment is without value, or when the burdens outweigh the benefits. Physicians and patients have mutual obligations to recognize when treatment is no longer effective or beneficial. Together they should then decide to desist from treatment.

56. Hippocrates, "On the Art," in *Hippocrates*, vol. 2, trans. and ed. W.H.S. Jones, Loeb Classical Library (Cambridge, Mass.: Harvard University Press, 1981), 193.

Medical Ethics and Social Responsibility

This chapter has focused on the individual patient-physician relationship. Other chapters will deal with institutional and social roles. However, it is important to indicate that this emphasis on the internal morality of medicine does not preclude, nor excuse, physicians from societal obligations. The physician is a steward of medical knowledge who has been allowed certain privileges by society in the course of caring for sick persons. These privileges include hearing confidential information, seeing and touching patient bodies (sometimes in very intimate ways), as well as performing surgical procedures and using controlled substances to alleviate suffering. Acceptance of the privileges of a medical education and possession of medical knowledge generates obligations to make them available for the betterment of society. Medical students enter into a similar covenant with society when they accept the privileges of a medical education. These privileges include the right to dissect human tissue, to participate in the care of patients as a student or resident, and to learn to carry out medical procedures. These privileges are sanctioned by society so that a continuous supply of medically trained personnel can be assured for society.

Ethical issues arise when the physician is forced to choose between the good of an individual patient and the needs of society. Specific conflicts of this kind, as they occur in the military service and in battle conditions, constitute a large part of this textbook and will not be covered here. Suffice it to say that except in the most extreme exigencies, the physician remains a physician always. To depart from the internal morality of medicine is to repudiate what it is to be a physician. Persons who enter any kind of relationship with a physician expect, and have a right to expect, fidelity to the fundamental ethic of the profession. Any compromise with this expectation for reasons of social or national exigency must be closely scrutinized if medicine and physicians are not to be used as means to political, social, or economic purposes not their own.

Managed care is becoming a paradigm case of this issue. Physicians in managed care are urged to become *gatekeepers*, to act as agents to conserve society's resources, and to take the needs of other patients into account in deciding who gets what care, and how much. Presumably physicians are expected to deny needed care so that those more needful may have access to that care (e.g., to pay for child health, to extend coverage to the uninsured), or to cut costs and yield profits for investors.

On the covenant model of medical ethics detailed in this chapter, physicians should not act as gatekeepers. If there must be rationing, then it should be *explicit* rationing, that is, rationing through decisions on benefits made societally but not by individual physicians in individual cases. All patients need to know the limitations society places in their care. With explicit rationing physicians can still serve the patient's interest with the confines of externally imposed limitations. But with explicit or with implicit rationing, the physician must reserve the right to refuse to obey a social policy if it is harmful to his patient.

Physicians should make a societal contribution or cost containment. First, they must practice the most rational medicine, providing only what is effective, beneficial, and not excessively burdensome. When two treatments are equally effective, the less expensive should be chosen. Another way to contribute to societal welfare is to provide expert testimony to policy makers so that benefit packages can be based on effectiveness and benefit, not cost. A third way is to act together as a profession for the welfare of the sick, especially for the underprivileged, the poor, the disabled, and the elderly who do not fare well in market and competition-driven managed care plans. Finally, as a citizen, the physician has a duty to be informed of public policy, and to foster the welfare of the sick through lobbying for appropriate public policy. There will be times in managed care organizations when the pressure on physicians to serve interests other than those of their patients will so damage the trust relationship that virtuous physicians have a duty to refuse.

Medical Ethics, Culture, and History

Some may object that in a culturally pluralistic world like ours, the idea of a stable foundation for medical ethics binding on all physicians across national and cultural boundaries is an anachronism. It is true that responses to illness and disease by patients, physicians, and societies may vary widely. People in different times and cultures have different attitudes and behaviors in the presence of pain, suffering, and death. They value human life itself and the lives of the aged, disabled, or unborn in different ways. Their interpretation of the meanings and origins of illness vary, as do their therapeutic endeavors.

The same is true in ethics. The emphasis on autonomy, truth telling, and confidentiality is closely bound to Anglo-American beliefs in individual freedom, privacy, and self-determination.[57] In other cultures, decisions may be made by families, tribal chieftains, or by community discussion. Infanticide, abortion, and the rights of women may vary with historical era, ethnicity, or religious belief.

In the minds of some, all of these differences militate against the possibility of a universal ethic of medicine. Medical ethics, they would say, is whatever we make it to be. It can be socially constructed differently in different societies and times. What is morally right for one may be wrong for another. Pluralism is a fact, and it is anachronistic to seek a common foundation even for medical ethics. This is the thrust of antifoundationalism, the trend of so-called postmodern philosophy and ethics that denies the possibility of any stable set of moral precepts.

In medicine, at least, antifoundationalism flies in the face of the human experience of illness, which is common across cultures and time. The phenomena do not change. They are common to humans whether they lived in ancient Greece, live in the United States today, or will live in a space station in the future. A broken leg, a crushing chest pain, spitting blood, or chills and fever induce anxiety, fear, distress,

57. S.M. Glick, "Unlimited Human Autonomy—A Cultural Bias?" *N Engl J Med.* 336, no. 13 (1997): 954–56; A. Surbone, "Truth Telling to the Patient," *JAMA* 268, no. 13 (1992): 1661–62.

vulnerability, and a need and call for help. The Hippocratic physician, today's internist, tomorrow's flight surgeon on a space station, or the shaman in a distant era or country, each confronts a human in need of help. The methods may differ, but the end of medicine is the same in each case: healing, helping, relieving pain and anxiety, and curing when possible. These phenomena of being ill, being healed, and healing, itself, transcend time and culture. They ground the ethics of the patient-physician relationship in universal human experiences even though cultural and historical settings may differ.

A Common Ethics for the Health Professions

This chapter has concentrated on the ethics of the profession of medicine. Only analogically has it touched the ethics of the other health professions such as nursing, dentistry, medical social work, clinical psychology, pharmacy, and allied health. Each of these clinical professions confronts human beings in the state of illness; each deals with the same fundamental phenomena of illness and healing. The same virtues, principles, and duties that bind the physician bind these other clinicians in their clinical encounters with sick persons seeking help.

The common foundation for the ethics of the health professions is the empirical reality of the human relationship between patients and health professionals—the *internal* morality described above. This common ground of empirical fact speaks for a common ethic of the clinical, healing relationship. To be sure, upon this common base there will be certain additional obligations specific to each profession, and expressed in their different codes of professional ethics. But a common thread runs through all these codes. The ethics of the healing relationship is, in the end, the general ethic of the health professions. Its foundation will be the same for physicians, nurses, dentists, social workers, psychologists, allied health professionals, and healthcare administrators. That foundation will be, as always, the varied phenomena of the human encounter between one human in distress seeking help from another who professes willingness to help, possesses the technical and moral skill to do so, and promises to use them for the good of the person seeking help.

Conclusion

In the chapters immediately following this one [in *Military Medical Ethics*, vol. 1—Ed.], the rich theoretical foundation for the patient-physician relationship will be explored, with a particular focus on the clinical setting. Then, using the tools of research methodology, we will explore the many overall influences on the patient-physician relationship.

This chapter opened by noting that medical ethics begins and ends in the patient-physician relationship, whether that is in a civilian or military setting. Thus the point was made that in many respects military physicians do not differ from their civilian medical counterparts. The military physician, as a physician, is distinguished from other military personnel by his engagement in a special kind of human relationship that, of its nature, demands a certain level of moral commitment. That commitment must be the determinant of the physician's conduct even in the extraordinary circumstances of national defense and war. The extent to which these exigencies may shape those moral commitments is explored in the many other chapters in this work on the subject of military medical ethics. What is inescapable is the fact that the physician cannot avoid complicity if harm comes to his or her patient. The good of the patient is, as always, the gold standard of moral propriety.

6

The Conflict Between Autonomy and Beneficence in Medical Ethics: Proposal for a Resolution

Edmund D. Pellegrino, MD, and David C. Thomasma, PhD

1.0 Introduction

Three radical changes have occurred in the ancient edifice of medical ethics in the last two decades. Each promise to transform the nature of the physician-patient relationship with repercussions in the domains of law, society, and ethics. Each merits the most careful scrutiny by the profession and the public, because how we resolve the moral dilemmas they produce will determine not only our relationships with the medical profession, but what kind of society we are, or wish to be.

The three changes we consider most crucial are these: (1) The shift in the locus of decision making from the physician to the patient—a shift philosophically from the primacy of beneficence to the primacy of autonomy in physician-patient relationships; (2) An unprecedented expansion of medical technological capability, thus expanding enormously the range and complexity of clinical and policy decisions in health care; (3) The entry of economic considerations as primary forces in individual and policy decisions regarding health and medical care, thereby creating a conflict between the canon of economics and the canons of traditional medical ethics.

In this essay we will confine ourselves only to the first of this triad of changes—the shift from beneficence to autonomy in medical ethics. We recognize the interdependence of the whole triad of changes and that to dissect them from each other is difficult and somewhat misleading. Yet the scope of this essay forbids a full examination of the interdependence of autonomy, technological possibility, and economics in the evolution of contemporary medical ethics. We are preparing a more thorough examination of these relationships in a forthcoming book (E.D. Pellegrino and D.C. Thomasma, *For the Patient's Good* [New York: Oxford University Press, 1988]).

Reprinted with permission from Edmund D. Pellegrino and David C. Thomasma, "The Conflict Between Autonomy and Beneficence in Medical Ethics: Proposal for a Resolution," *The Journal of Contemporary Health Law and Policy* 3 (1987): 23–46.

This essay falls into two parts: First, we will examine the limitations of both the autonomy and paternalist models of the physician-patient relationship and second, we will outline a model of beneficence that promotes the good of the patient yet de-absolutizes autonomy and avoids the pitfalls of traditional paternalism.

2.0 The Autonomy Model and Its Limitations

2.1 Socio-historical Critique of Autonomy

The autonomy "model" of clinical decision making is firmly grounded in the dignity of human persons and the claim they have on each other to privacy, self-direction, the establishment of their own values and life-plans based on information and reasoning, and the freedom to act on the results of their cogitations. The historical origin of the principle of autonomy, as it is interpreted by many ethicists today, is of recent date. It is found mainly in the philosophical treatises of the French and English Enlightenments and the emergence of the doctrines of individual and political rights to freedom that undergird modern democracy.

The notions of individual rights and autonomy have been gathering strength in American public life since the founding of our country. They lagged behind in the medical relationship, however. Only in the last several decades have they become powerful enough to challenge the traditional paternalism that dominated the relations of doctors and patients for twenty-five hundred years—at least since the time of Hippocrates.[1] For our purposes, we need not trace this history in detail. It suffices to take note of the major forces that have nurtured the exponential growth of autonomy in the last two decades.

First is the expansion of political democracy to every sphere of civic life fostering in each of us the desire to participate in the decisions that affect our lives as individuals. This "democratization" carries with it a certain distrust of all authority, expertise, privilege and prerogatives—of the kind traditionally wielded by physicians, lawyers, and other professionals.

To this we must add the general improvement in the education of the public and the dissemination by the media of information about both the advances of medicine as well as the ethical and legal dilemmas those advances produce. As a result, the public appreciates that it makes a vast difference in our lives what decisions doctors make in the use of medical knowledge, and that those decisions increasingly involve value choices.

Lastly, the increasingly divergent moral pluralism in our society impels us to seek to protect our personal values against usurpation by others. This is recognized as a genuine danger in medical decisions which can involve our deepest convictions about life and death, abortion, euthanasia, genetic manipulation, and the like.

1. L. Newton, "The Patient as Responsible Adult: Consequences of the Revised Perspective," in *Proceedings of the Fifty-Fifth Annual Meeting of the American Catholic Philosophical Association* (Washington, D.C.; American Catholic Philosophical Association, 1982), 240–49.

These factors have converged to undermine the traditional model of the benign, paternalistic physician, who assumes full responsibility and authority to determine the patient's best interests and to act so as to advance those interests—if need be, without the patient's participation.[2] This is the conception of "beneficence" still dominant in the minds of many physicians and patients; it still shapes the ethos and ethics of medicine. It is the conception, too, that is the focus of criticism by the proponents of autonomy who equate beneficence almost entirely with medical paternalism.

It is true that, here and there, in the history of medicine there were suggestions of patient participation in clinical decisions. Plato, for example, distinguished the physician of the free man from the physician of the slave by the fact that the former educated and consulted with this patient, while the latter did not. It is true too, as Katz points out in more recent times, that some like Samuel de Sorbière tentatively suggested that patients ought to be part of the doctor's decisions.[3] Moreover, a closer reading of Percival's *Ethics* serves to qualify the standard account of the eighteenth-century physician as merely a condescending gentleman—authoritarian.[4] The same is true of some aspects of Worthington Hooker's treatise on patients and physicians—an American work on medical morals that deserves to be better known.[5]

But exceptions like these aside, paternalism was the dominant, and indeed the accepted, model of the clinical relationship for most of medicine's history. Paternalism was not as ethically dubious in times past as it would be today. For one thing, the educational gap between physician and patient was wider than it is now. Further, participatory democracy even in Greek times was not a reality for most people. Finally, almost all medical treatments were non-specific, as much dependent upon faith in the physician as on any genuine therapeutic potency. Indeed, most therapeutic efforts were either useless or dangerous. Ordinary patients were not aware, however, of the therapeutic poverty of most prescriptions.

"Aesculapian Power" was a major ingredient of cure. It rested on faith in the quasi-hieratic power and authority of the physician as a person. Indeed, the physician himself was part and parcel of the cure as Lain Entralgo points out.[6] Only in very recent times have these conditions changed so radically that the most objectionable features of medical paternalism have begun to be felt. Medicine's

2. Neither in the oath, the other deontological books, nor the whole body of the Hippocratic corpus is there evidence of an obligation to obtain patient consent or participation. Indeed, most of the direct references were to the contrary; see E.D. Pellegrino, "Toward a Reconstruction of Medical Morality: The Primacy of the Act of Profession and the Fact of Illness," *The Journal of Medicine and Philosophy* 4, no. 1 (1979) (Univ. of Chicago Press): 32–56.

3. J. Katz, *The Silent World of Doctor and Patient* (New York: The Free Press. 1984).

4. E.D. Pellegrino, "Thomas Percival, The Ethics Beneath the Etiquette," foreword to Thomas Percival, M.D., *Medical Ethics, on a Code of Institutions and Precepts Adapted to the Professional Conduct of Physician and Surgeon* (Birmingham, Ala., Classics of Medicine Library, 1985).

5. Worthington Hooker, *Physician and Patient* (New York: Baker and Scribner, 1849).

6. P. Laín Entralgo, *La medicina hipocrática* (Madrid: Revista Occidente, 1978); P. Laín Entralgo, *Doctor and Patient* (London: World University Library, 1969).

capacities to alter individual and social life are unprecedented. An educated public grasps this fact, and no image grants the physician unrestricted discretion in the use of his powers.

Eliot Freidson has detailed the negative effects of the professionalization of contemporary medicine.[7] His study is very cogent, but it undervalues some of the more positive aspects of professionalization. Anything so powerful as modern medicine requires a formal professionalization process to assure that high levels of competence, conduct, and accountability be maintained. This aspect of professionalization must not be lost in the current antipathy to paternalism.

In fact, the dangers of paternalism lie less in professionalization than in the irresponsible uses of power and the attributions of superiority that arise from the social ascendancy of medicine. As John Stuart Mill remarked: "Whenever there is an ascendant class, a large portion of the morality emanates from its class interests and its feelings of superiority."[8] This is the basis for Illich's mordant criticism, too, but he weakens his case by the polemics of social revolution with which he heavily flavors his charges against medicine.[9]

Thus, the revulsion many competent adults feel about medical paternalism may more properly derive from anger about physicians' pretensions as a superior class, than from any inherent property or professionalism itself. Women have often complained that gynecologists "pooh-pooh" their complaints (e.g., the premenstrual syndrome debate) on the grounds of sexist power struggles. Patients are forced to wait for the doctor in the waiting rooms, suggesting that their time, their values, their lives are not as important as the doctor's. Cartoons abound in which the public fears that doctors are too busy managing their money to care about patients. These are examples more of the role of power and class than of any inherent properties of professionalization. They are often at the root of the declining status of the physician recorded in serial AMA surveys on the public image of the doctor.[10]

The patient autonomy movement is an understandable counter-reaction to such class domination. Nonetheless, autonomy should not be viewed as an absolute model for the doctor-patient relationship itself because it is insufficient to claim, as the move to patient autonomy often does, that medical paternalism is a direct outgrowth of professionalization. Nor is paternalism a prima facie medical or social evil as Berlant supposes.[11] Modern medicine incorporates moments of patient choice as well as moments of necessary, beneficial paternalism. The former occur when the diagnosis and options are clear and well documented. The latter occur when not enough is known about a disease and its prognosis, when no therapeutic modality

7. E. Freidson, *Doctoring Together: A Study of Professional Social Control* (New York: Elsevier, 1975).

8. J.S. Mill, *On Liberty*, ed. E. Rapaport (Indianapolis: Hackett Publishing Co., 1978), 6.

9. I. Illich, *Medical Nemesis* (New York: Bantam Books, 1977).

10. Larry S. Freshneck, *Physician and Public Attitudes on Health Care Issues* (Chicago: American Medical Association, 1984).

11. Jeffrey L. Berlant, *Profession and Monopoly* (Berkeley: University of California, 1985).

has a clear edge, or when an existing therapy has marginal or dubious benefit. In these cases, physicians may be forced to recommend, even urge a course of action based on an intuitive assessment of the data. The important thing here is that the physician makes clear the uncertainties with which he must contend. This preserves autonomy, and so it is not paternalistic but it is beneficent.

Some philosophers who defend the absoluteness of patient autonomy on moral grounds neglect the fact that "decision making" among humans is an interpersonal transaction. They thus downgrade the bilaterality of the patient-physician relationship. Doctor and patient are existentially bound to each other in a way that makes moral atomism and absolute decisional autonomy unrealistic and undesirable goals for both parties. Moreover, the philosophical atomism on which such a notion is based is of dubious viability in a complex, interrelated, technologically driven society like our own.[12]

The patient autonomy model does not give sufficient attention to the impact of disease on the patient's capacities for autonomy. We agree with Eric Cassell that medicine should restore patient autonomy,[13] but one cannot assume that autonomy is fully restorable or preservable in cases of serious illness. Patricia Bradley formulates a telling objection to the position of Robert Veatch, one of the most prominent ethicists arguing for the patient autonomy model. As Bradley says, "Veatch argues that the relationship between doctor and patient is an equal one, ignoring . . . the fact of illness which places the patient in a potentially vulnerable relationship with his physician. . . . Based as it is on a wrong assumption, this model must be rejected when applied to the traditional doctor-patient relationship."[14]

Even the briefest experience with illness shows that ill persons often can become so anxious, guilty, angry, fearful, or hostile that they make judgments they would not make in calmer times. Patients become preoccupied with their diseases and their bodies.[15] The patient may see his body as an object that failed him. Patients are forced to reassess their values and goals. These primary characteristics of illness alter personal wholeness to a profound degree. They change some of our assumptions about the operation of personal autonomy in the one who is ill.[16]

Healing as a moral component of the physician-patient relationship is not given sufficient weight in the autonomy model. The physician's task goes beyond the prevention of harm. It includes restoration or improvement of biological function.

12. D. Thomasma, "The Goals of Medicine and Society," in *Studies in Science and Culture*, ed. H. Brock (Newark, Del.: University of Delaware Press, 1985).

13. Eric J. Cassell, *The Healer's Art* (Cambridge, Mass.: M.I.T. Press, 1985).

14. P. Bradley, "A Response to the March 1979 Issue of the *Journal of Medicine and Philosophy*," *J. Med. Philosophy* 5 (1980): 213–14. See also E.D. Pellegrino and David C. Thomasma, *A Philosophical Basis of Medical Practice* (Oxford, England: Oxford University Press, 1981).

15. Eric J. Cassell, "Disease, an 'It': Concepts of Disease Revealed by Presentation of Symptoms," *Soc. Sci. in Med.* 10 (1976): 143–46.

16. E.D. Pellegrino, "Being Ill, Being Healed, Some Reflections on the Grounding of Medical Morality," *Bulletin of the N. Y. Academy of Medicine* 56, no. 1 (1981): 70–79.

If health is in any degree to value the body as a biological organism, then the physician has some obligation to work toward this good, which is intrinsic to medicine. This is the case even when the patient, in the presence of life-threatening illness, may deny or reject health. In this view both the physician and patient are obliged to work towards restoration of the good of the body which is health. We have structured elsewhere the hierarchical relationship of this good and the good of self-determination when they are in conflict in, and between, patient and doctor.[17]

Distinguished ethicists like James Childress recognize the realities of these conflicts between autonomy and paternalism. They prefer to err, if they must, on the side of autonomy on what we consider erroneous meta-ethical grounds, namely that rights take precedence over goods. The central point or differences in our own position is that there are better arguments in particular instances for the ascendancy of goods over rights.

It is, of course, increasingly true that patients with chronic, unremitting, disabling disorders, like degenerative disorders of the nervous system, incurable neoplasms, or intractable pain, may wish to assert their right to an autonomous decision to discontinue treatment or to die. But, most often, life and health are still primary values most patients will assert over their moral and legal rights of autonomous refusal of medical care. The average patient usually assumes too that his health is still the physician's primary value as a physician. There is a real danger of harm to the patient if doctor and patient misunderstand each other on this point.

Patient autonomy models often have their origins in the civil and human rights movement, rather than in an ontology of relations specific to medicine and healing. Few would disavow the positive gains effected by the political struggle for human rights. But it does not follow that the adversarial presumptions of the rights movement are transferable, without modification, to the debate about the locus of medical decision making. The image of doctors as adversaries, sinister in their conspiracy against patients is a colorful but overplayed metaphor. It stimulates scrutiny of the abuses of power that can, and all too often do, characterize medical decisions. Nonetheless, there are still physicians of character motivated by an ethics of compassion and commitment that transcends post-enlightenment, rights-based ethical theories.[18]

2.2 Limitations of Autonomy, Philosophical Critique

Our objections to autonomy as an absolute principle in medical decision making, or as a replacement to beneficence when they are seemingly in conflict need a little more formal analysis than the foregoing. We find autonomy wanting in three dimensions—the contextual, the existential, and conceptual.

17. E.D. Pellegrino, "Moral Choice, the Good of the Patient, and the Patient's Good," in *Ethics and Critical Case Medicine*, ed. J. Moskop and L. Kampelman (Dordrecht: D. Reidel, 1985).

18. D.C. Thomasma, "The Basis of Medicine and Religion: Respect for Persons," *Linacre Quarterly* 47 (1980): 142–50.

2.2.1 Context Limitation

The autonomy model may not apply in some contexts of medical treatment. For example, paternalism may be appropriate when treating the aged and senile referred from nursing homes for urinary tract infections.[19] In such cases, physicians may have an obligation to disregard the patient's wishes until they are convinced that the patient is competent. This is an example of "weak" paternalism, different in moral quality from "strong" paternalism in which the objections of a competent patient are overruled, or deception is practiced to manipulate a decision.

Physicians may, therefore, act over the objections of patients, to preserve life or prevent serious harm, when patients are senile, confused, depressed, or otherwise incapacitated in their capacities to make autonomous judgments. The same holds for emergency room treatment. Here the uncertain prognosis, the urgency for an unambiguous decision, and the probability that the patient would want to be treated were he fully competent, would militate against unrestrained adherence to the patient's expressed wishes. The same may apply to hasty requests not to be resuscitated when clinical outcomes are still in doubt. On the other hand, when the clinical context is clearer, less urgent, and the patient's competence certain, autonomy can and should be given primacy.

Weak paternalism may be appropriate when treating children where competence is difficult to judge, or genuinely in doubt. We may presume that, were they autonomous, children would choose to be treated—provided there is sufficient benefit to be gained from treatment. This is true too with therapeutic research in children. Parents may act for the good of the child in the absence of the child's consent. The context—the calculus of benefits and burdens of experimental procedure—will determine the moral licitness of the research. This would not be the case with non-therapeutic research where a child would be exposed to risk, without the chance of benefit. One cannot under any doctrine of paternalism—weak or strong—presume or impose altruism on the part of another.

The clinical context may change day by day, hour by hour, even in the same patient and with it, the moral defensibility of any act of paternalism.[20] Senile patients may wax and wane in competence. Acutely and desperately ill patients with severe trauma and burns may vacillate in their desire to live. While the decision to treat may be very difficult, the presumption to treat is often morally defensible if there is an appreciable probability of a successful outcome. What constitutes "success" under such circumstances may be debatable but overly hasty decisions not to treat out of deference to the principle of autonomy may be more damaging to the patient's ultimate best interests than some degree of paternalism.

19. D.C. Thomasma, "Professional and Ethical Obligations Toward the Aged," *Linacre Quarterly* 48 (1981): 73–80.

20. D.C. Thomasma, "The Context as a Moral Rule in Medical Ethics," *J. Bioethics* 5, no. 1 (1984): 63–75.

This "variability of context" is, therefore, an important moral limitation on autonomy. It demands careful assessment in each case, even while we remain sensitive to the moral obligation to respect patient autonomy. Engelhardt also takes note of this variability and concludes that it is the product of rational disagreements about the risk-benefit calculus.[21] On his view the variability rests on the relativity of subjective values of the interpreting parties and not in the objective difference in contexts as we have argued.

Irrespective of its origins, context variability raises questions about any model of patient autonomy. It also underscores the need for some ranking of goods in medicine if we are to choose one ethical model over others, or one moral choice over others. Any mode of clinical decision making, by virtue of the fact that it is clinical, must take into account the particularities and uniqueness of each human being's experience of illness.

2.2.2 Existential limitations: The Fact of Illness

The effects of illness and disease on personal autonomy limit self-determination to variable degrees. That is why so many physicians report that patients really want them to make the decisions.[22] On this view autonomy ought not, therefore, be taken as a starting point or absolute ordering principle in medicine. Rather it should be seen as part of the goal of treatment, one of the goods of the patient, to be promoted but not to the total exclusion of all other goods.

If we take the impact of illness and disease seriously, we must modify the autonomy model. That model has four features: self-direction, establishing a life plan, deliberating about applying a life plan (reasoning and information), and acting on the basis of such deliberations.[23]

Becoming "sick" can modify each of these features. To "be sick" is to be subject to the patho-physiological effects of illness, pain, fear, and to the special professional and institutional environment in which decisions occur. Self-direction is marred by the way disease may disrupt the unity of the self, ego, and the body. Life plans are threatened by the finitude of human life revealed in illness. Deliberation and application are impeded by the distractions of pain and fear, or by the process of institutionalization. The extent to which the operations of autonomy can be impeded by being, and becoming, a patient is impressive.

Of course the autonomy of most patients is only mildly incapacitated by disease. We must not, therefore, use autonomy limitation as an excuse for all sorts of paternalism. On the whole, patients' choices can, and should, be accepted. On the

21. H.T. Engelhardt Jr., "Philosophy of Medicine," draft of seminar paper, Kennedy Center for Bioethics, October 17, 1980, 27ff.

22. Arnold Soffer, "Searching Questions and Inappropriate Answers," editorial, *Arch. Internal Medicine* 142 (1982): 1117–18.

23. T. Beauchamp and J. Childress, *Principles of Biomedical Ethics* (New York: Oxford University Press, 1979), 56–57.

other hand, people who are incapacitated by disease or trauma should not be abandoned to their autonomy, merely given the "facts" and asked to make a decision. This is a form of moral abandonment. The proponents of autonomy as the prime moral obligation should give more attention to the available data about the psychosocial impact of disease on personal and moral status. A Medline search in 1985 produced *eighty-six articles* directly related to the impact of disease on autonomy and life adjustments. These data strongly suggest that a) autonomy is limited by illness and disease, and that b) any model of doctor-patient relationship must take this limitation into account. Neither of these points is sufficiently appreciated in the current ethical debates about autonomy.

One might argue that we are merely talking about varying degrees of competence, and that the problem is one of determining competence. On this view the autonomy model would remain intact since incompetent patients could be treated paternalistically without violating the principles of autonomy. Such an interpretation would, however, offend those for whom autonomy has become an absolute principle of medical ethics.

2.2.3 Conceptual Limitation as a Model

The autonomy model, as a model, is also limited. It has been constructed in dialectical opposition to the paternalistic model. But neither paternalism nor autonomy correctly describes the full range of ethical norms governing the doctor and patient. What occurs between doctor and patient has many formulations. It can be seen as restoring autonomy, safeguarding the person,[24] respecting persons,[25] healing or restoring a lost whole-ness,[26] putting the patient's needs first[27] making a right and good medical decision,[28] or acting in the best interests of the patient.[29] Each formulation ascribes a somewhat different moral tone to the physician's obligations with respect to the patient's autonomy.

What occurs between doctor and patient in non-therapeutic research is not encompassed in any of these terms. The goal here is not primarily the benefit of the experimental subject but the discovery of knowledge. Moreover, the subject's autonomy is not affected by illness. In non-therapeutic research then, respect for

24. E.D. Pellegrino, "Treating the Patient as Person: Philosophical Groundings-Commentary on Alasdair Macintyre," in *Changing Values in Medicine,* ed. Eric J, Cassell and Mark Siegler (Frederick, Md.: University Publications of America, 1985).

25. D.C. Thomasma, "The Basis of Medicine and Religion: Respect for Persons," *Linacre Quarterly* 47 (1980): 142–50, reprinted from *Hosp. Prog.* (September 1979).

26. E.D. Pellegrino, commencement address, University of Illinois at the Medical Center, Chicago, June 6, 1980.

27. T. McGovern, "The Patient Comes First," *Forum Med.* 3 (1980): 596–98.

28. E.D. Pellegrino, "The Healing Relationship: Architectonics of Clinical Medicine," in *The Clinical Encounter, The Moral Fabric of the Patient Physician Relationship,* ed. Earl Shelp, Philosophy and Medicine 4 (Dordrecht: Reidel, 1983), 153–72.

29. F.J. Ingelfinger, "Arrogance," *NEJM* 303 (1980): 1507–11.

autonomy is mandatory as a moral principle. The duty of beneficence towards the patient would be at the lowest level of sensitivity—that is, non-maleficence. In therapeutic research, however, the good of the patient is involved, and the questions about autonomy and paternalism would be the same as the ones we have already raised.

2.3 Social-Ethical Limitations of Autonomy

Beyond its limitations in one-to-one medical decisions, autonomy as a dominant ethical principle in medical ethics has serious limitations in social ethics, as well. These are: (1) the movement it fosters from substantive to procedural ethics; (2) the move from concern for the common to individual good; and (3) the erosion of the concept of democracy itself. Only brief mention into these limitations is possible here.

2.3.1 From Substance to Procedure in Ethics

One very strong impetus for the trend to autonomy is the obvious moral pluralism of our society and the possibility that physicians and others in authority may override personal belief systems. One way to guard against this kind of moral trespass is to accept the irreconcilability of moral conflict, and turn the focus on the process of decision making. Emphasis is then placed on respect for the autonomy of the parties to a clinical decision. Most of the recommendations of the multi-volume reports of the President's Commission for the Study of Biomedical and Behavioral Research[30] place their emphasis on procedure—informed consent, anticipatory declarations of several kinds, the proper use of proxy and surrogate decision makers and the establishment of ethics committees. Many of the cases that have come to the courts have turned on the question of who shall decide, under what conditions, and by what criteria. Formalization of the decision-making process is reflected also in the recommendations of the Department of Health and Human Services[31] pertaining to the care of terminally ill, handicapped, and physically impaired infants. Tristram Engelhardt has argued particularly vigorously that the function of ethics itself in a morally pluralistic society that wishes to remain peaceable must be analysis and clarification. A particular set of values, he contends, cannot be propounded for the whole community; freedom is to be protected at all costs. Ethics therefore must not concern itself with moral content or normative prescriptions.[32]

This neat dissection of analytical from normative ethics is illusory. Analytical ethics does in fact make a normative assertion, namely, that autonomy is the first principle, and it overrides all others. Engelhardt's rigorous extrapolation of the logic of autonomy to its conclusions is the strongest argument against a purely procedural

30. President's Commission for the Study of Ethical Problems in Medicine and Biomedical and Behavioral Research, multi-volume series (Washington, D.C.).

31. "Child Abuse and Neglect: Prevention and Treatment Program: The Final Rule," *Federal Register* 50, no. 72 (1985): 14878–901.

32. H.T. Engelhardt Jr., *The Foundations of Bioethics* (New York: Oxford University Press, 1986).

ethic. Pragmatically attractive as it may be in such a morally pluralistic and democratic society as ours, procedure cannot be self-justifying. To assert freedom as ultimate would mean that the search for the good life and the good society must be abandoned. We are forced to retreat into private morality for the most meaningful questions humans ask. But this retreat brings its own serious problems with it.

2.3.2 Moral Atomism

The retreat to private morality eventually leads to a kind of moral atomism in which each individual's moral beliefs and actions—unless they disturb the peaceable community—are unassailable. Moral debate is not only frustrating but futile, since each person is his own arbiter of the right and the good. The traditional notion of ethics as reasoned public discourse in search of the common good is discarded. The sense of community identity that derives from some consensus on things that *ought* to be done and what *ought never* to be done, is lost.

The overall result is a defection from what John Courtney Murray called the "affairs of the commonwealth."[33] These affairs end up in courts, decided by legal adversarial procedure, and judicial opinion, which must inevitably mix moral substance with procedure. Court opinions are already deciding substantive moral issues without the accompanying moral debate requisite to ethically sound judgment.

For example: Is the decision to withhold elective treatment from a Down's syndrome child the private decision of its parents, based on their evaluation of its quality of life or their personal and social burden of raising a retarded child? Moral privatism of this sort challenges government intrusion, and dilutes the state's "interests." Can the obligation of the government to protect the rights of the weak, the vulnerable, the comatose, be abrogated on the plea of moral privacy and autonomy? The Baby Doe case in Bloomington, Indiana, and the Baby Jane Doe case in Stony Brook, Long Island, are cases in point.[34]

It is impossible to escape the burden of doing ethics since most people still want to know whether what is procedurally acceptable is also right and good. Freedom to make one's own choices must be protected but the flight into metaethics will not eradicate the equal need to engage as a society in the pursuit of some common moral goals beyond autonomy.

2.3.3 Erosion of the Idea of Democracy

Perhaps the most serious consequence of the absolutization of autonomy is the limit it places on the idea of democracy itself. Democracy is reduced to a procedure for settling otherwise irreconcilable differences among citizens, but without commitment to any common set of values except freedom of private judgment. Certainly one measure of a democratic society is the degree of freedom it affords for

33. John Courtney Murray, *We Hold These Truths* (New York: Sheed and Ward, 1960.

34. Marcial Angell, "The Baby Doe Rules," *New Eng J. of Med.* 314, no. 10 (1986): 642–44. See also "Child Abuse and Neglect Prevention and Treatment Program: The Final Rule," *Federal Register.*

divergent and contrary opinion. But those freedoms must serve some common community purpose as well.

Is not one of the traditional aims of our democracy to advance the cause of community rather than its atomization? This is our inheritance from the classical and Judaeo-Christian traditions. How far may this inheritance be squandered without corroding the idea of democracy itself? How logical is the separation between public and private morals? Is the purpose of government only to restrain unbridled self-interest or to promote the common good? The incommensurability of some of today's conflicting moral claims makes consensus unlikely. But without moral consensus private values become more selfish, more crude, more intensely combative. Power becomes the maker of morals if the common moral perspective is lost and democracy itself is weakened. Is there some other choice than Rousseau or Khomeni?

The socio-political and socio-ethical consequences of the move to autonomy are yet to be comprehended fully. Medical ethics—as the arena of some of the sharpest ethical debates in contemporary society—is the paradigm that brings these questions to our immediate attention. Their significance transcends medical ethics. And how we resolve them will determine what kind of society we shall have.

3.0 Limitations of the Paternalism Model

Just as formidable objections to the patient autonomy model can be raised, so too can objections be raised to medical paternalism. The foremost objection is that a physician often cannot heal a person just by curing a disease, especially if the physician systematically ignores or disregards the patient's view. Cassell's argument that restoring function, or curing, should be a secondary aim of medicine and that medicine's primary aim is to restore autonomy has much to recommend it.[35] It is a little extreme, however, and in its way absolutizes autonomy as Childress has pointed out.[36]

Healing does involve restoring autonomy. For this reason, Culver and Gert are correct to insist that strong paternalism always demands justification because it violates a moral rule. But Culver and Gert do not deal directly with the moral content of medicine itself. Strong paternalism is objectionable not only because it violates moral rules, but because it violates the architectonic aim of medicine, which is to heal the one who is ill. To violate a person's autonomy is not to heal, but to wound his humanity.

Strong paternalism is objectionable because it violates the humanity of the patient. The obligation is owed to rational beings to be free to decide about the conduct of their own lives. Indeed, such decisions are peculiarly human. To infringe on such a fundamental right clearly demands special justification. Medical paternalism fails because it overrides an essential element in deontological ethics, at the core of medicine, that is, respect for persons. To violate the patient's autonomy

35. Eric Cassell, *The Healer's Art* (Cambridge, Mass.: M.I.T. Press, 1985).

36. James C. Childress, *Who Should Decide: Paternalism in Health Care* (New York: Oxford University Press, 1982).

is to deprive him of one essential component of his own good, and thus to violate medicine's promise to act for the good of the patient.

Many physicians hold that the patient's rights to autonomy should not get in the way of their medical needs, that is, medical "indications" should dominate clinical decisions. But as we shall argue the case, the hierarchy of patient goods may not always place medical needs in the highest place.[37] Lack of respect for such a hierarchy of values is a major cause of patient complaints about physician paternalism.

Like the autonomy model, medical paternalism can fail to distinguish contexts and their role in medical and ethical decision making. As a consequence, medical paternalism tends to universalize a stance valid in one context but not necessarily in another. Generalization of one experience, like "saving" one patient through paternalism, into a universal moral posture, is not valid.

Perhaps the biggest failure of medical paternalism is its assumption that medical values or the medical good is the highest good, and that it has an absolute quality which overrides other values. Or, even less justifiably, a particular physician's preferences for one treatment among several may become an absolute. Some surgeons prefer radical mastectomy while others prefer limited resection and radiation for cancer of the breast. Some cardiologists prefer medical over surgical management in certain types of angina pectoris. Alternative procedures may lead to similar outcomes but with different risks and quality of life. Selection of one procedure over others depends as much on the patient's and the physician's values as on the scientific data. The patient, for reasons of great importance to her, may reject even the scientifically preferred therapy for one of lesser effectiveness.

Medical paternalism asserts that the physician unequivocally knows better than the patient what is "good" for him. It also subsumes all the patient's good under only one good—medical good—a point we shall develop in more detail later. Other dimensions of the good of the patient must also be considered. One of these is surely the preservation of the fundamental human good of making one's own decisions about the kind of life one wants to lead, or the risks one wants to take.

In what ways, then, is the paternalism model inadequate? We suggest three criticisms, parallel to the three we levelled at the autonomy model.

3.1 Context Limitation

Paternalism may apply in certain limited context, such as making decisions over the objection of a minor about what might be best for that child, but it cannot function as a universal or general principle of medicine, since it assumes something fundamentally flawed: not only that professionals know what is best for all patients, but also that they may override the patient's wishes in the pursuit of what is medically indicated. Paternalism has enjoyed such a long season because physicians can readily

37. E.D. Pellegrino, "Moral Choice: The Good of the Patient and the Patient's Good," in *Ethics and Critical Care Medicine,* ed. John C. Moskop and Loretta Kopelman (Dordrecht: Reidel, 1985), 117–38.

offer examples in which their expertise saved some patient from a truly disastrous decision. An internist of our acquaintance, for example, cites a "memorable" time when he convinced a rabbi to have a colostomy following colonic resection for cancer. The convincing took three days of vigorous debate and discussion. In telling this tale, the physician appealed to the metaphysical assumption that it is better to live than to die, to live with an impairment than to die without one. In the main, persons do accept this assumption. But not always. And not everywhere. To what extent he may, or may not, have unjustifiably exhausted the rabbi's will to autonomy is impossible to say. Given the rabbi's Talmudic training the dialectic must have been intense and the physician's "victory" not an easy one.

3.2 Existential Limitation

Not only is the context a limiting factor, but the concept of paternalism itself is limited. It is not just a matter of the fallacy of expertise. The expert does not always know more about what is "best" for the non-expert. Difficulties of prognosis and the harms/benefit calculus are obvious limitations. There is also no way to define clearly what is absolutely best for the patient in medical terms alone. That definition is always related to the values the patient professes, those the institution and society assume, and the culture holds to be important. Lacking any unequivocal definition of "benefit" the physician cannot presume to define the whole of the patient's good, without essential input from the patient.

Some patients reject the benefits of medical interventions simply for their own reasons. Mr. Barting, in a hospital intensive care unit in California, wished to be removed from the respirator even though he understood this would lead to his death. The Supreme Court of California upheld his legal right to his request. Some hospitals or other institutions may limit the medical benefits to be given to patients. For example, Catholic hospitals rule out abortions. Nursing homes may have policies which "interpret" benefit to patients to include limitations on cardiopulmonary resuscitation.

Society defines what benefits shall be given its citizens in general, and even which medical benefits may be offered. In England, there are limitations on the use of dialysis after a certain age. Respirators are removed after ten days if certain patients do not respond by that time. Finally, social values are in a constant state of definition and re-definition. At the moment our social values tend to accept the medical intervention model of beneficence. Other cultures, or subcultures within our own, may accept or reject this model.

All of these considerations put limits on paternalism. They mitigate the absoluteness of benefit from medical benefits that paternalism requires.

3.3 Conceptual Limitation as a Model

Paternalism as a model of patient-physician relationship is itself flawed. At the root of this limitation is the fact that authentic healing cannot take place in a paternalistic model, since paternalism overrides patient choices. Personal choice is essential to the processes of reintegration essential to healing. Undeniably, "cures" can often take place in a paternalistic relationship—for example, treating pneumococcal pneumonia in an elderly patient with a stroke. This is "effective" and *medically* indicated, but not necessarily beneficial treatment, if the patient is dying of metastatic malignancy.

The paternalistic model also fosters a certain detachment deleterious to patient "care" and it separates cures from care. The physician tends to apply the medically indicated course of action as if that patient were, indeed, the corpus upon which one practiced the medical craft. A "cure" might ensue. But the patient's most cherished value, his life plan, the kind of life he might wish to have, his relationships to others might be so violated as to vitiate the medical good. Wounding outweighs healing under such circumstances.

4.0 Beneficence in Trust: A Model for Medical Decisions

If both autonomy and paternalism have deficiencies, what model of medical decisions and clinical ethics would avoid the limitations of each—and also optimize their utility in advancing the good of the patient? We believe that a reinterpretation of the principle of beneficence can achieve these ends; and the remainder of this essay details the way we think it can do so in a model which centers on beneficence in trust.

4.1 The Principle of Beneficence: Degrees of Sensitivity to Good of Others

Since the model we wish to propound is grounded in the principle of beneficence, we must first outline our understanding of this essential first principle of medicine and healing. Beneficence, like the good of the patient which it presumably serves, is not a univocal or simple notion. There are levels of beneficence, as Frankena points out.[38] Just how and at what level one interprets beneficence will determine the moral obligations one feels one owes patients, and also the degree of altruism one feels obliged to practice.

The most minimal level is the level of non-maleficence, that is, the duty not to do direct harm to another. This is the level contained in the Hippocratic prescription in the *Epidemics*—"at least do no harm." This level of beneficence is expected in any civilized society. It is enjoined even in the most extreme libertarian moral philosophies.

A further step in beneficence is the duty to prevent harm to others, that is, to remove or limit the possibilities of harm. Here we move from passive non-maleficence to a more active intervention on behalf of others. At once, even at this minimal level, the possibility of paternalism begins to appear. For example, we might readily agree that there is a duty to remove an obstacle on a railroad track that would result in derailment and harm to other people. But this action might conceivably deprive some passenger who at that moment wishes to die and would prefer to do so

38. William K. Frankena, *Ethics* (Englewood Cliffs, N.J.: Prentice-Hall, 1963).

in a railway accident, of the privilege of doing so. Most people would discount such an unlikely prospect and agree that there was a positive moral obligation to remove the object in the interest of the majority of passengers who certainly do not want to die this way.

But what about protecting people against accidents by enforcing laws to "buckle up," or wear helmets, or goggles at work. This view of beneficence requires a more direct limitation of autonomy. While libertarians might resist this degree of paternalism, society seems to be giving certain measures like this a sanction on the thesis of harm to others caused by the economic costs and disability imposed on society when the injured go untreated—even if they would elect to ignore safety requirements.

One, however, can go further—and out of the principle of preventing harm, make the growing of tobacco illegal, or the preparation of distilled liquors. Here the intervention, while indirect so far as the one injured goes, is nonetheless significant for smokers and tobacco growers. Freedom and autonomy conflict in this case with beneficence in a way our society does not at this point countenance. Conceivably it might sanction such measures at some time in the future. To do so is to raise the interpretation of beneficence to a higher degree than simple non-maleficence.

Even further along the scale is to interpret the duty of beneficence as binding, even at some risk, discomfort, or pain to the benefactor. Law does not require anyone to risk life, limb or even inconvenience, even to save the life of another. But in medical encounters traditionally, some degree of effacement of personal self-interest has always been understood as a duty. One need only think of the expectation that physicians will treat patients with contagious diseases (such as HIV or Ebola) and the scorn heaped on physicians who desert their posts in times of disaster or epidemic. Camus's Rieux is an exemplary physician precisely because he did not desert the people of Oran even though he might have claimed the privilege since his wife was ill elsewhere in a sanatorium.

This level of beneficence is implicit in that "higher degree of self-effacement"—which Harvey Cushing termed the "common devotion" of the medical profession. It is admittedly a degree of beneficence above that expected by law, or the mores of other activities like business, relationships with neighbors, or professional colleagues. It is a level, we would submit, that is essential to medicine as a moral enterprise. Without some degree of self-effacement medicine ceases to be a profession in any traditional sense of that term and becomes only a trade or craft.

On the other hand, the degree of self-effacement expected is not of the heroic or sacrificial kind. It does not require the dedication of a Mother Teresa, Albert Schweitzer, or St. Francis. In them, we move beyond duty—at least in secular terms—to supererogation, to obligations one may feel out of religious or other altruistic motives. We enter here the realm of "agapeistic" ethics—one grounded in love and charity for others.[39]

39. Gene Outka, *Agape: An Ethical Analysis* (New Haven and London: Yale University Press 1972).

We would argue that beneficence in medical transactions should include some degree of effacement of the physician's self-interest in the interests of his patient. Just how much effacement is required cannot be defined in any absolute way. Many physicians today think this degree of beneficence is questionable and even objectionable. Physicians are asserting their "rights" to recreation, family life, social activity, time off, freedom to choose to treat only those who pay, the right to strike, and to work for investor-owned institutions. These may not be overtly unethical practices, but they are often at the moral margin where self-effacement would dictate some limit on the physician's personal interests or privileges.

For the purposes of the model we wish to advance, we will argue that the fact of illness, what it does to the sick person, and the kind of special relationship it entails with the physician dictate a degree of beneficence that goes beyond passive non-maleficence. It includes, in our view, some obligation to act in the patient's interests even at some cost to the comfort, power, prestige, or fiscal benefit of the physician.

With this interpretation of beneficence let us turn to the model which we feel most closely exemplifies the duties physicians owe patients.

4.2 Major Features of the Beneficence Model

Given the shortcomings we have pointed out with both the patient autonomy and medical paternalism models, is there an alternative that does not reduce to one or the other? We suggest there is, though we appreciate that the complexity of the physician-patient relationship can never be adequately described in a single model. One purpose in sketching the beneficence model is to circumvent the substantial problems with the models we have already mentioned. We do not claim that the physician-patient relationship is fully defined by this model either.

There are six major features of the beneficence model:

(1) The Aim of Medicine Is Beneficent

Medicine as a human activity is of necessity a form of beneficence. It is a response to the need and plea of a sick person for help, without which the patient might die, or suffer unnecessary pain, or disability. The obligation to help the sick is a general one involving humans, even those who are not professed healers. It is grounded in the claim that comes from the vulnerability, and suffering of a fellow human. One is impelled, even by the lesser degrees of beneficence, not to harm, and, even, to ease suffering.

When one is a professed healer one possesses knowledge and skill society has permitted one to acquire precisely because it can benefit others. One also promises to help and to act on behalf of the good of the patient when one offers oneself to another as a healer. Further, without the special knowledge the healer has acquired others would suffer so that in a sense all the sick have some claim on all healers.[40]

40. D.C. Thomasma and E.D. Pellegrino, "Philosophy of Medicine as the Source for Medical Ethics," *Metamedicine* 2 (1981): 5–11. Now *Theoretical Medicine.*

Beneficence is a prime requirement for medicine, and it has three specific obligations. First, the patient's problems and needs are the physician's primary concern, taking precedence, except in the rarest circumstances, over all other concerns. Second, harm must be avoided because the physician cannot fulfill the promise of helping if he intentionally harms the patient for any reason. Third, both autonomy and paternalism are superseded by the obligation to act beneficently, that is to say, the choice of whether one acts to foster autonomy, or acts paternalistically should be based on that which most benefits the patient and not the intellectual convictions or emotional impulses of the physician.

(2) Primacy of the Existential Condition of the Patient

The second feature of the beneficence model is the primacy of the existential condition of the patient rather than of traditional professional codes. A good example can be found in Mark Siegler's list of criteria for deciding the limits of autonomy to be accepted by a physician treating a seriously ill patient.[41] These criteria include the patient's ability to make rational choices about care; the nature and past values of the patient; age of the patient; the nature of the illness; the values of the physician who must make a choice in the care of the patient and the clinical setting, especially the diffusion of care. The first four items deal with the personal condition of the patient, and the last two deal with the healthcare professional and environment. Presumably, Siegler does not mean that age should be considered an independent variable in making decisions, but rather as his subsequent writings would suggest, age is a valuable marker of the condition of the patient.

(3) No Automatic Ranking of Values

Both the patient's autonomy and medical paternalism models emphasize single values which are always to be preferred. For example, the patient's right to autonomy is always to be preferred over other values in the patient's autonomy model. In the paternalism model, each patient must be treated as if he or she did not know what is best. By contrast, in the beneficence model, no such "automatic" ranking of values takes place. The elements of the beneficence model are not ranked in any pre-set hierarchy. Each patient must be handled individually not only for the medical but also for the moral implications. No ethical stance, other than acting for the patient's best interests, is applied beforehand. This model requires that patients and physicians become able to identify, rank, discuss, and negotiate values,[42] and define the particular good of a particular patient. This is not to say, however, that general ethical axioms applied to more than one patient are invalid.

41. M. Siegler, "Critical Illness: The Limits of Autonomy," *Hastings Center Report* 7 (1977): 12–15.

42. D.C. Thomasma, "Training in Medical Ethics: An Ethical Workup," *Forum on Medicine* 1 (1978): 33–36.

(4) Consensus

The fourth feature of the model is consensus. Because there is to be no imposition of values, or decisions made in the best interests of patients without their participation, a consensus with the patient and with other members of the healthcare team is needed. Admittedly, a consensus model takes time and energy, but it also wards off many agonizing hours of later conflict in the course of a serious illness. In fact, one of the seductions of the autonomy and paternalism models is their comparative ease of decision making: either the physician makes all the decisions, or the patient does so. Both models abandon the trials and rewards of a mutual dialogue and exchange between doctor and patient.[43] Both also can assault the moral agency of the patient or the physician.

A consensus reached at the beginning of a patient's care cannot be assumed to continue unchanged as new developments occur. The consensus must be monitored for its continued validity. This requires a continuing dialogue between the patient and his medical attendants.

(5) Prudential Moral Object

The fifth feature is a prudential moral object, that is, an attempt must be made to resolve difficult ethical quandaries by preserving as many values of both the patient and physician as possible. Ackerman has argued that this should be the goal of bioethics.[44] Whether or not one agrees entirely, it is a goal of a consensus-driven, patient-oriented approach in which prudential judgments are made on a patient-by-patient basis.

(6) Axioms

Explicit axioms comprise the sixth and last major feature of the beneficence model. Just as the physician examines each patient in light of generalized theories or categories of disease and health, his prudential judgment about each patient must adhere to a series of more general ethical axioms, or moral rules. These are necessary to avoid the moral pitfalls of the autonomy and paternalism models, and of situational studies.

Axioms of the Beneficence Model:

1) *Both doctor and patient must be free to make informed decisions to act fully as moral agents.* The values of both doctor and patient must be respected since each is a person deserving of respect as such. Value consensus results only if each can, without coercion or deception, express his or her own values in discourse and action. Neither can impose his or her values on the other;

43. J. Katz, *The Silent World of Doctor and Patient* (New York: The Free Press, 1984).

44. T. Ackerman, "What Bioethics Should Be," *J Med Phil* 5 (1980): 26–275.

neither can "use" the other for selfish ends; each must be free to withdraw from the relationship if value conflicts are not resolvable.[45]

2) *Physicians have the greater responsibility in the relationship because of the inherent inequality in information and power between themselves and those who are ill.* Physicians are obliged therefore to provide the information patients need to make genuinely informed decisions, and to use their power with due regard for the vulnerability, and exploitability of the sick. These obligations are rooted in the special nature of the healing relationship. The self-imposed moral aims of the profession and the expectations of society derive their force from this fact as well.
3) *Physicians must be persons of personal moral integrity.* The physician must have the capacity to make prudential judgments that factor in the particulars of each case, the general features of the disease and general moral principles. Ultimately the good of the patient depends as much on the physician's character as his capacity to make these judgments, as that is, on the extent to which he can be trusted to keep the good of the patient as his primary aim.[46] In a morally pluralistic society, there is a tendency to downplay moral character in the education of the physician. However, there are qualities of moral judgment that should apply to all physicians, and for this they will need to be educated. As Aristotle noted, "it is impossible, or not easy, to do noble acts without the proper equipment."[47] Skill in making ethical judgments must be taught in medical schools.[48] Yet, skills without moral integrity will not suffice in those moments when no one is there to watch, and the good of the patient hangs on the moral integrity of the physician.
4) *Physicians must respect and comprehend moral ambiguity yet not abandon the search for what is right and good in each decision.* By training and disposition, physicians are inclined to diagnostic closure and problem resolution. They are dismayed when there is no single "right" answer to a moral dilemma. Yet in the beneficence model this may often be the case since the good is defined by principles and individuals in their life contexts without standardized formulae. Physicians must avoid the pitfalls so aptly described by Alasdair Macintyre: "It is a central feature of contemporary moral debates that they are unsettleable and interminable . . . because no argument can be carried through to victorious conclusion, argument characteristically gives way to the mere and increasingly

45. E.D. Pellegrino, "The Anatomy of Clinical Judgments, Some Notes on Right Reason and Right Actions," in *Clinical Judgment: A Critical Appraisal,* ed. H.T. Engelhardt Jr., S.F. Spicker, and B. Towers (Dordrecht: Reidel, 1979), 169–94.

46. E.D. Pellegrino, "The Virtuous Physician and the Ethics of Medicine," in *Virtue and Medicine: Explanations in the Character of Medicine,* ed. Earl Shelp, Philosophy and Medicine 17 (Dordrecht: Reidel, 1985).

47. Aristotle, *Nicomachean Ethics,* book 1, chap. 8, 32–33.

48. D.E. Thomasma, "Report Number 2," in *Capstone Conference Workshops: Reflections on the State of the Art* (Washington, D.C.: Institute on Human Values in Medicine, 1982), 66–79.

shrill battle of assertion with counter assertion."[49] No matter how frustrating moral "debates" may be, the physician must still make moral decisions with, and for, his patients. It is incumbent upon him, therefore, to learn how to deal with the reality of moral ambiguity. He has not the scholar's luxury of "on the one hand" and "on the other hand," etc. He just acts, and to act is to choose among alternatives—moral as well as technical.

5.0 Conclusion

The values of patient welfare and patient autonomy—which translate into the corresponding moral duties of beneficence and respect for persons—may come into tension with each other. In our view, however, these duties cannot remain in conflict if medicine is to achieve its goal of healing.

But healing, as we define it, is a form of assistance in making the patient whole again by working through his body. If the values of patient welfare and patient autonomy remain in conflict, then authentic healing cannot take place. A physician, therefore, must become a moderate autonomist, and a moderate welfarist, at once. This can be accomplished in a beneficence model like the one we suggest.

Another way of arriving at this position is to consider the principle of respect for persons. This principle leads to two moral duties. The first is to respect the self-determination or autonomy of others. The second, often-neglected duty is to help restore that autonomy or help establish it when it is absent. Looked at this way, beneficence is seen to be a direct consequence of a fundamental moral principle and the guiding duty of medicine. If this is true, then the autonomy model is necessarily incomplete.

Beneficence is the principle that prompts physicians to cite their moral commitments and personal support for patients beyond just respecting their rights. It is beneficence, not authoritarianism, as he incorrectly supposed, that prompted Ingelfinger to argue that doctors must recommend a course of action, not just lay out alternatives and abandon patients.[50] It is beneficence too, not just respect for autonomy that properly protects patients' rights. It is the primary duty of beneficence, and not paternalism, that has historically been the guiding norm of medicine.

To be sure, beneficence can be and has been subverted into paternalism. But if our task is "proposing revised values" as Callahan asserts,[51] then it is important to focus on the virtue of benevolence (or the principle of beneficence) rather than the rule of autonomy. This virtue is consistent with the ethical tradition of persons united in community. This is a tradition more in keeping with the ethical roots of medicine than one that stresses autonomous individualism.

49. A. Macintyre, "Why Is the Search for the Foundations of Ethics So Frustrating?" *Hastings Center Report* 9 (1979): 16–22.

50. F.J. Ingelfinger, "Arrogance," *NEJM* 303 (1980): 1507–11.

51. D. Callahan, "Shattuck Lecture: Contemporary Biomedical Ethics," *NEJM* 302 (1980): 1128–33.

7

The Anatomy of Clinical-Ethical Judgments in Perinatology and Neonatology: A Substantive and Procedural Framework

Edmund D. Pellegrino, MD

CLINICAL ETHICS IS DISTINGUISHED BY THE FACT that it must focus on practical decisions. In the "moment of clinical truth," a decision must be made, often under pressures of time, and when the facts are incomplete or uncertain. Retreat into the back-and-forth debates of the classroom is neither possible nor ethically responsible. Despite the difficulties, every clinician must make decisions that are morally defensible. He or she must knowledgeably and skillfully confront the central moral question: "What is the right and good thing to do for this patient?"

Answering that question today is more difficult than ever. The issues are more complex than ever, the range of choices wider. There are more parties to the decision: the patient, the family, the institution, the law, public policy, and economics. All must be decided without the benefit of consensus on the most sensitive and fundamental moral questions.

For the perinatologist and the neonatologist, clinical-ethical decisions are further complicated by the vulnerability of their patients, who include the pregnant woman, the fetus, and the newborn infant. The skills of ethical analysis are as essential as the knowledge of pathology and physiology.

The objective of this report is to examine the anatomy of clinical-ethical judgments, and to offer a substantive and a procedural schema—a framework of issues and questions that should help in answering the practical question: "What shall I do in this case?"

Reprinted with permission of the Pellegrino family from Edmund D. Pellegrino, "The Anatomy of Clinical-Ethical Judgments in Perinatology and Neonatology: A Substantive and Procedural Framework," *Seminars in Perinatology* 11, no. 3 (1987): 202–9.

Three Ethical Obligations

The framework that will be described here is essential to fulfilling three ethical obligations that bind every clinician, no matter what his specialty may be. First is the obligation to understand and know the structure of one's own values, that is, to understand the foundation on which one builds moral choices in response to a concrete clinical dilemma. Second is the obligation to have some working knowledge of the formal discipline of ethics. In my own teaching, I often encounter clinicians who do not appreciate that ethics is a discipline, that like the basic and clinical sciences, it can be learned, and that it can provide an orderly, systematic, and rational approach to clinical ethical decisions. The third is the obligation to carry out the process of ethical decision making itself, and the decision reached thereby in a morally defensible way. A right decision derived or implemented unethically is as faulty as a morally wrong decision.

These three obligations are not spelled out in the Hippocratic Oath, the ethical guidelines of the American Medical Association, or any of the several codes that have emerged since the Nuremberg trials. They may be implied in some of these codes, but they must be made explicit. This is especially true for the neo- or perinatologist, who faces ethical dilemmas that could scarcely be imagined by the authors of our traditional ethical codes, even a decade ago, to say nothing of 2,500 years ago.

To fulfill his or her ethical obligations, the clinician needs two analytical structures: one substantive and one procedural. Both converge on the central question of what is the right thing to do in a particular case. The substantive structure consists of four conceptual questions that shape all our moral choices and the procedural structure consists of five steps that enable us to make the actual concrete decision.

The Substantive Structure

Four questions, or levels of consideration, comprise the skeleton on which all moral choices ultimately hang. Each question and its sub-questions shape our moral philosophy, our understanding of morality, what it requires, how we ought to behave in our professional lives, and why we do so. Each probes the levels of belief or concern on which we base our choices, but do not make explicit. Each is connected with the other in actual practice.

The four substantive issues are these: (1) the philosophy of the physician-patient relationship to which we subscribe; (2) the interpretation we place on the principles of ethics common to clinical ethical decisions; (3) the theory of ethics we favor; and (4) the ultimate source of our morality. Implicit assumptions in all these issues are buried in every clinical decision. They reveal themselves in any careful analysis of clinical-ethical judgment.

Ethical Models of the Physician-Patient Relationship

The first question concerns the nature of the physician-patient relationship. It is central to all clinical-ethical decisions. What we think is right and wrong in professional ethics depends on what we think is the proper moral role for the physician in meeting the patient's needs. This is true whether the patient is an adult or an infant. The infant is, of course, vastly more dependent on the physician than a competent patient. Decisions are made with surrogates who act on behalf of the infant. But the patient is, in the final analysis, the infant.

A series of analogical models are used in defining the physician-patient relationship, that is to say, we usually take a relationship from outside medicine and apply it to the medical relationship. These models overlap somewhat, but the question is which model is the primary one, the one we turn to in order to resolve conflicts in our obligation.

Some of the most frequently used models follow; each has fundamentally different ethical implications. The first is the physician-patient relationship as applied biology.[1] This is a dominant view for many academicians. Here the physician's major responsibility is technical and scientific competence. Competence is his primary obligation. The nonscientific elements in the relationship are not denied, but they are not intrinsically part of medicine. They can, and should, be left to others—social workers, nurses, psychologists—as the patient's needs may dictate. The prime good is the medical good of the patient and what is medically indicated is what should always be done.

The second model is the relationship as a contract for services, usually analogous to a legal contract.[2] Here, the physician's obligations are defined by negotiation and are not fixed. The physician is expected to perform as agreed, and failure to do so constitutes a breach of contract. With this model, ethics is usually given a legalistic caste and the relationship between physician and patient is based less on trust than on mutual agreement of specific performance criteria.

The third model is the covenant.[3] This suggests a quasi-religious relationship, a binding promise like that between the Lord and his people in the Old Testament. Here the physician's obligation to help and heal transcends a contract or scientific relationship. Physician and patient are bound to each other in a trust relationship essential to healing. In this view, the obligations of a biological contract and most other models are morally sparse.

The fourth model is the commodity transaction.[4] Here medical care is likened to the purchase of another commodity in the marketplace. The physician is a businessman, or entrepreneur, seeking his own interests, which presumably will eventually redound to the benefit of all, the patient included. The ethical obligations are those of the businessman, tempered somewhat by the nature of the kind of activity involved in healing, but nonetheless essentially the same as the merchant's.

1. D. Seldin, "The Boundaries of Medicine," presidential address, *Transactions of the Association of American Physicians* 94 (1981): 75–86.

2. R.M. Veatch, *A Theory of Medical Ethics* (New York: Basic Books, 1981).

3. W.F. May, *The Physician's Covenant* (Philadelphia: Westminster, 1983).

4. R. Sade, "Medical Care as a Right: A Refutation," *N Engl J Med* 285 (1971): 1288–92.

Medical knowledge is simply another form of proprietary knowledge to be used for the physician's profit.

The fifth model is the social functionary,[5] in which the physician is seen as essential to maintenance of a properly functioning and productive society. He serves the patient because it is good for society to do so. Within this model, he may be a bureaucrat, or a proletarian, a worker in a corporate endeavor. In either case, he serves not solely as the patient's advocate, but also as the advocate for an institution. Conflicts of obligations, as gatekeeper or guardian of the public health and as patient advocate, are built into this model.

The sixth model is the one I favor. In contrast with the others, it is not analogical. It is derived from the nature of medicine as a special kind of human activity. It depends on the internal morality of medicine itself. This morality—the obligations imposed on the physician—arises in the vulnerability of the sick person, the nature of the physician's promise to help in the face of that vulnerability, and the nature of healing as the immediate aim of the relationship. I have developed this model in detail elsewhere.[6] The model focuses on the nature of illness, the inequality in knowledge and power between physician and patient, and the obligations that arise out of that fact and the existential state of the patient. My view emphasizes beneficence in trust, fidelity to promises, and effacement of the physician's self-interest. These obligations are built into the very nature of healing.

Although medicine and the physician-patient relationship are some of all of these things, the important question is: Which model do you choose when these several facets of the relationship conflict, as they often do, in concrete decisions? The answer to this question is the most pressing and fundamental problem in professional ethics today.

A question which, in a sense, precedes these is: Who is the patient? To whom is the physician primarily responsible? In the case of the competent adult, it is clearly the patient himself, not the family, society, or the institution. Each has some interest that deserves consideration, but the values and choices of the competent adult predominate.

In the case of infants, the answer is more difficult. Some would argue that the "patient" is the family and that their values and needs should predominate. Others, because of the economic impact of decisions with neonates, would place social concerns first. Still, others insist that the interests of the infant are primary and that his or her benefit must dominate when there are conflicts of interest. The complexity and reality of these conflicts are evident in the well-publicized controversies surrounding the cases of Baby Fae, Baby Doe, Baby Jane Doe, and the role of government in how such decisions are made.

5. T. Parsons, *The Social System* (London: Free Press, 1954).

6. E.D. Pellegrino, "Toward a Reconstruction of Medical Morals: The Primacy of the Act of Profession and the Fact of Illness," *J Med Philosophy* 4 (1979): 32–56.

Interpretations of Ethical Principles

Whatever model we choose must be fitted to the three major ethical principles that figure in most clinical-ethical decisions. These are beneficence, autonomy, and justice. Secondary principles that derive from these are confidentiality, truth-telling, and promise-keeping. While most physicians will accept these principles, the way they are interpreted varies considerably. As a result, there may be wide variations in our moral judgment about what is the right and good thing to do in a particular case.

The Principle of Beneficence

Beneficence, acting in the best interest of the patient, is historically the first principle of medicine. It is central to every code of medical ethics since the Hippocratic Oath. But beneficence may be interpreted in different ways and with different degrees of stringency. Thus, some limit beneficence to nonmaleficence, the negative ethical obligation that forbids harming the patient. However, others accept a positive ethical obligation to do good. But what level of doing good do we mean?—only if it means no inconvenience to ourselves, at some small or great inconvenience, or even danger? Or is sacrifice of an even higher degree required? Defining the level of beneficence to which we feel bound is one of the most debated questions today.

In addition to the obligation of beneficence, there are questions about what in fact is in the patient's best interests. In the case of competent adults, we have some idea of the patient's value system and wishes to guide us. But in the case of the infant, we cannot know what he or she would wish. Some argue that the safest course under these circumstances is medical indication; if there is medical benefit to be obtained then a treatment should be used. Others would interject the projected "quality of life" of the infant with severe congenital defects. Others would consider the "burden of life" in their decision. If that burden is too great, they hold that beneficence dictates non-treatment.

The definition one uses of beneficence will often depend on answers to the deeper levels of questioning, the ethical theory one espouses or the ultimate sources of one's morality. It is essential to clarify what we mean by beneficence to make rational decisions in particular cases.

The Principle of Autonomy

The principle of autonomy, respect for the patient's moral right to decide for himself and the physician's obligation to respect and enhance that decision, has come to challenge beneficence as the primary principle in medical ethics. This is perhaps the most fundamental change in physician-patient relationships in the history of medical ethics. Previously, beneficence was exclusively interpreted in terms of what the doctor thought best for the patient. Indeed, the patient's opinion was rarely sought. The whole thrust of Hippocratic ethics is toward the physician as a benign authoritarian or father figure.

A number of questions must be asked about the interpretation of autonomy. Is it unrealistic to expect someone who is ill, in pain, and anxious to be able to make objective choices? Or to understand enough about the medical decision to judge wisely? Can the parents of a really sick child make the difficult decision about what is best for that child? That is the view of the strong paternalists, those who think the physician always knows what is in the patient's best interests and is obligated, therefore, to do what is medically indicated.[7]

Others on the other hand hold a "weaker" view, namely that there are times when the physician should override the patient's autonomy, that is, when the patient is very ill, when the choice is between certain death and an effective treatment, or in the case of infants or incompetent adults, when the patient's surrogates have opted against beneficial medical treatment.

Should autonomy become an absolute principle to be abrogated only when it ends in direct harm to innocent parties as Engelhardt holds?[8] Or should you balance beneficence and autonomy as McCullough and Beauchamp suggest?[9] Or should a different concept of the relationship be examined, such as Thomasma and I suggest in a forthcoming book that focuses on the idea of beneficence in trust?[10]

The Principle of Justice

Justice is another of those words we all use to bolster our arguments. Yet, as Plato showed in the first book of his *Republic*, we do not always understand what we mean when we use this powerful and complex word.[11] Is it enough to say that justice is to render to others what is owed them? Do we render good for good and harm for harm? Do we temper justice with charity as the Christian Gospels teach? What kind of justice does the model of the physician-patient relationship we choose dictate? Can we justly reject or refuse to treat patients who are the victims of their own self-abuse, that is, some AIDS victims, the smokers, drinkers, drug addicts, sociopaths, and criminals?

What rule do we use to translate our concept of justice into a concrete decision? ls it equity of access to health care, need, merit, social worth, ability to pay, first come first served, citizens of your own country over others? Do we use the rule of similar treatment for similar people in similar situations?

These are all fundamental questions in peri- and neonatology, where we must deal with infants who cannot defend their own rights against the rights of others. Does justice permit letting an uncomplicated Down's Syndrome infant die because

7. J.F. Childress, *Who Shall Decide?* (New York: Oxford, 1982).

8. T. Engelhardt, *The Foundations of Bioethics* (New York: Oxford, 1986).

9. T.L. Beauchamp and L.B. McCullough, *Medical Ethics: The Moral Responsibilities of Physicians* (Englewood Cliffs, N.J.: Prentice-Hall, 1984).

10. E.D. Pellegrino and T. Thomasma, *For the Patient's Good: The Restoration of Beneficence in Health Care* (New York: Oxford, 1988).

11. Plato, *The Republic*, trans. Allan Bloom (New York: Basic Books, 1968).

his or her parents feel they cannot cope emotionally or economically with a retarded child? Is it just for the parents of a badly handicapped infant with a terminal disease to demand treatment to the bitter end when they are using up society's resources in money, facilities, or personnel?

Do the sick have any claim in justice on the rest of society? Do those more fortunate have an obligation to the less fortunate? Which of the two current theories of justice is the more pertinent to the care of neonates, Nozick's[12] or Rawls's?[13] Manifestly answering the question "What is justice?" is as difficult and as inescapable as Pilate's question "What is truth?"

Yet, if we are to make rational clinical-ethical judgments, we must arrive at some position on the way we interpret not only justice, but beneficence and autonomy, and how to resolve conflicts between them. But what we think of these three principles will have deeper groundings in the last two levels of the structure of ethical decisions: ethical theory and the ultimate source of morality we espouse.

Grounding of Ethical Principles

Clinical-ethical judgment and decision making cannot be rational processes if we do not understand the grounding of the principles we use, for it is from that grounding that we draw their interpretation.

There are a variety of ethical theories, most of which cannot be discussed here.[14] They all depend on some theory of the good. Only the two major theories, consequentialism and deontology, can be examined in this brief report.

Consequentialism

Consequentialist theories of ethics hold that the right and good action is determined by its consequences, by the amount of good versus harm, pain versus pleasure, cost versus benefit. Is the center of the good your own egotistical self-interest or the good of everybody? Are you a utilitarian who seeks to maximize utility or good for the greatest number? Beneficence and justice are often justified on consequentialist foundations.

Deontology

The deontological theories of ethics hold that the goodness of an act is inherent in the act itself. In this view, some acts are always right or wrong irrespective of the consequences they produce. Should the truth always be told, promises always kept, confidentiality always protected? Some answer yes, some no. The principle of autonomy is often based in a deontological foundation. Justice and beneficence can also be interpreted deontologically.

12. R. Nozick, *Anarchy, State, and Utopia* (New York: Basic Books, 1974).
13. J. Rawls, *A Theory of Justice* (Cambridge, Mass.: Harvard, 1971).
14. W. Frankena, *Ethics* (Englewood Cliffs, N.J.: Prentice Hall, 1973).

Deontological viewpoints are apt to be nonnegotiable, creating conflict when the participants in a clinical-ethical decision do not all share this theory of ethics. Even if they do, they might differ on whether the act in question is morally licit or not. Consequentialism has its problems too, since the "calculus" of harms and benefits is often difficult to carry out and frequently debatable in its conclusion.

These simple definitions of consequentialism and deontology do not exhaust the differences or similarities between them. Also, there is considerable overlap in actual practice. Usually, a blending of both theories occurs in concrete decisions. For the neonatologist and perinatologist, the question is where to place his own primary emphasis.

Ultimate Sources of Morality

This is the most fundamental level of ethical concern and often the most firmly held. It represents the set of moral beliefs most closely identified with the person, the rules by which he orders his own life, the beliefs he gives up last, if ever.

When all is said and done, what do you believe validates your moral values? The existentialists hold that morality is self-generated. Each person makes his own values as part of his own life project. Others hold morality to be whatever society, culture, or custom dictates. The right and good, therefore, change with geography, history, and society. Another view is that what is right is what we can make the strongest and most coherent argument for. Still others use ultimate criteria like moral sentiment, and biological or economic utility. Finally, a large number of people find the ultimate source of morality in religion, in beliefs revealed or commanded by a personal God, the cosmos, or a worldly spirit.

These sources can only be enumerated without evaluating them, which requires a full course in the philosophy of religion. What is important for the clinician is to recognize the source of his own and his patient's morality. In the case of neonatology and perinatology, these will be the values of parents, society, or institutions. These must be understood if conflicts are to be resolved in morally defensible ways.

In a morally pluralistic society, one in which ultimate sources of morality may differ sharply, it may be preferable to turn to the internal morality of medicine itself as suggested in the previous discussion of the sixth model of patient care. This is a common ground for all health professionals, and patients as well. This does not mean that the deeper sources of morality are to be ignored, but only that we must seek some ground for possible consensus in the practical world of clinical-ethical decision making.

The Purpose of Ethics

One question that underlies the whole idea of a schema for ethical decisions is what purpose ethics itself fulfills. The modern view is largely that the purpose is conceptual clarification and analysis, understanding the terms and logic of ethical judgments and the assumptions on which they are based. This contrasts with the view that ethics has normative content that can tell us what we ought to do and how we should live. The latter is the classical understanding of ethics as found from its beginnings in Plato and Aristotle and in the moral teachings of most of the world's religious and philosophical systems.

Many clinicians, with their scientific and pragmatic training, judge ethics to be useless and frustrating. It seems that ethics cannot arrive at proven conclusions that will convince everyone. There seem to be no "right" answers. Therefore, one view is as good as another. This is the road to moral relativism and ultimately moral chaos. It destroys the possibility of rational clinical-ethical judgments. It also defeats itself, since the view that ethics is useless is itself also no more than an opinion.

Most of the discussion in this report has tried to show that the resolution of ethical conflict and dilemma is possible, that an orderly and systematic approach can contribute to the quality of the decision. There may well be more than one well-argued case for resolution of some ethical issues, but this does not mean that all arguments are equally coherent. Some arguments are better than others. Some are patently indefensible.

There is no real alternative to ethics so long as we are rational creatures and wish to respect each other's moral values without submitting to them without examination. A democratic society is based finally on a continuing dialogue among its citizens on the moral values that make for a good society. Medicine and medical ethics are two of the most significant arenas for this dialogue today. A pluralism of well-founded and rigorously argued judgments is preferable to moral privatism in which each person retreats into the redoubt of his own beliefs and refuses to examine them critically.

All humans make moral judgments. Ethics provides the skills whereby those judgments can be understood, assimilated, and transmitted to others. It is indispensable to a liberal education and a free society.

The Ethical Work-Up: A Procedural Schema

Thus far, I have emphasized the substantive framework for clinical-ethical judgment. This is essential to fulfill two of the three ethical obligations of the clinician outlined at the beginning, namely, understanding one's own value system on four major levels, and possessing some knowledge of ethics as a formal discipline. The third obligation, making the actual clinical decision and implementing it in a morally defensible way, requires a procedural schema. This schema is a set of steps to be used in making the decision itself. This is the ethical work-up. It consists of five steps.

Establish the Facts

The first step in an ethical analysis, as in a clinical analysis, is to establish the technical facts as securely as possible. Confusion in the ethical decision is as much the result of uncertainty about the facts as about the ethical issues. Indeed, the issues are in large part defined by the particular features of the case in hand.

Clinicians are inclined to make authoritative statements, some that do not bear up under careful scrutiny. Clinical judgments about diagnosis and especially prognosis and treatment are probability statements. The clinician should be clear about the degree of certitude and uncertainty in the facts upon which he bases his ethical judgments.

The patient, or the surrogate decision maker in the case of infants, should be helped to understand the essential facts and the uncertainties as well. The clinician must communicate the diagnosis, the prognosis treated and untreated, and the potential benefits and side effects of treatment.

He must be certain as well about whether the patient is "brain dead," in a permanent vegetative state, or is terminally ill, and what he means by these terms. These judgments and judgments about the patient's competence figure importantly in most ethical decisions.

Determine What Is in the Patient's Best Interests[15]

While it is easy to assert that beneficence is crucial to medical ethics, it is much more difficult to say precisely what will be in the patient's best interests, since this is a value-laden judgment. Such things as the balance of harm and benefit, effectiveness, and relationship to the patient's concept of quality of life must be figured into the decision. The patient's best interests are not fully assumed in what is medically indicated. The patient's perception of what is good in terms of his own life plan, the values he wishes to protect, and his spiritual goals are even more important than medical benefit. Indeed, medical benefit ought to serve those other values.

The patient is a moral agent, and in the case of infants, the patients have moral agents who are responsible for what they decide. Their judgment of the infant's best interests may not always coincide with the physician's. Therefore, it is incumbent that the physician be as careful as possible not to interject his own values in determining what is "best" for the infant. Indeed, when we do not know the values of the patient, as in the case of an infant, best interest should be as close to medical indication as possible. Quality of life determinations of best interest are particularly dangerous.

Define the Ethical Issues and Principles

This step depends on the various levels of substantive ethics discussed earlier. It is essential to recognize where there are conflicts, how they might be resolved, and what principles, theories of ethics, and sources of morality and philosophy of physician-patient relationships are being used. This is the heart of the ethical analysis.

Dealing with ethical principles such as respect for autonomy and beneficence, it is often necessary to decide which carries greater moral weight in a particular case. The arguments we use reveal what we think are our ethical obligations to the patient. Indeed, we must appreciate that usually the ethical judgment involves not absolute

15. [When teaching on this subject, Pellegrino would put "Define Ethical Issues and Principles" before the next section, "Determine What Is in the Patient's Best Interests."—Ed.]

right and wrong, but a conflict between competing good things. This is the "tragic choice" that makes ethical analysis so painful at times.

State Your Decision in Concrete Terms

The essential feature of clinical ethics that distinguishes it from a classroom exercise is that, at some point, the discussion must come to an end, and a position must be taken. A moral choice cannot be avoided. This is painful, but inescapable. Ethics does not require that one be perfect. We must make the best decisions we can, given the complexity and uncertainties of the case. Clinical ethics like clinical diagnosis ends in action. It is a moral responsibility to choose that action as intelligently as we can. Our decision must be clearly stated. Ambiguity is acceptable in discussion, but dangerous when action is finally taken.

Justify the Decision

The decision is truly ethical only if it is justified in rational terms. Ethics is a formal discipline, a systematic, orderly, critical examination of the rightness and wrongness of a particular act with a view to determining our ethical obligations. It examines the meaning of the terms we use, the logic of our argument, and the assumptions from which we start.

This means we must also examine the arguments for and against our decision. In this way we can check our own logic and reexamine our own values and beliefs against those of others who may have good arguments to the contrary. The purpose of an ethical analysis is not to prove that we are right, but to arrive at the best defensible judgment in the interests of the patient.

As any clinician recognizes, these processes are fully analogous to the way we analyze clinical problems, conduct a differential diagnosis and arrive at a treatment regimen. The capacities to make clinical-ethical judgments and other clinical judgments are really the same, and the two are inseparable in clinical practice.

Carrying Out the Ethical Decision

Assuming that we have, through the two schemata described above, arrived at an ethically defensible clinical-ethical decision, we must conduct the process of decision making itself in a morally defensible way. We must deal ethically with all the participants in the decision. This usually centers on the question "Who decides?"

When the patient is competent, he or she decides. The limitations on a competent patient's autonomy are two: asking the physician to do something that violates the physician's moral beliefs, or something that produces direct and serious harm to a third party. Under these circumstances, the physician cannot presume to override the patient's wishes. But he must withdraw from the case, respectfully giving his reasons and avoiding recrimination.

In neonatal and perinatal medicine, these situations are not uncommon. Catholic physicians, for example, cannot be asked to perform abortions or to withhold treatments of benefit to the child simply because the parents think the child's quality of life or the economic or psychological burdens of raising the child are too much to bear. Another example would be the refusal of transfusions by Jehovah's Witnesses when such a treatment would be life-saving for an infant. In many clinical situations, the fundamental values of parents, or pregnant women, can conflict with those of the physician. These conflicts can be handled in a way that protects the infant and, if possible, respects the values of the parents.

In dealing with incompetent patients, there is a moral obligation to transfer the moral rights of the patient to a valid surrogate, someone who acts for, and presumably protects the interests of, the incompetent person. With once-competent adults or older children, we seek a "substituted" judgment, a decision, based on knowledge of the patient's values, and as close as possible to what the patient would have chosen were he or she able to make the choice.

In perinatology and neonatology, we can have no idea what the fetus's or infant's values might be or what he or she would consider a quality life. The principle of respect for autonomy simply does not apply. We must depend on the family or guardian for a surrogate decision. But we cannot assume that they will automatically share the same view as the physician about the best interests of the infant.

The situation may become a difficult one, because biologic relationship does not automatically confer moral competence or morally correct motivations. One of the unsettled questions of great importance in peri- and neonatal medicine is whether parents and physicians should be entrusted with final decision-making authority, or whether concerned third parties or government may intervene if they think an injustice is being done.

The physician, who should be the advocate of the patient, has a serious responsibility to make some judgment about the moral validity of a surrogate decision maker. He has a bond, or covenant, with the patient that compels him to act on the patient's behalf. He must be the patient's advocate when he perceives that these surrogates may not be acting in the patient's best interest because of some conflict of interest or a mistaken notion about what is best. The difficult fact is that, to some extent, the physician must make a judgment about the moral competence of the parents—an extremely vexatious matter as the cases of Baby Doe, Baby Jane Doe, and Baby Fae amply illustrate.

This is not to depreciate the interest and the importance of the parents; rather, it is to question whether their authority is absolute or can be questioned. That is the question behind the recent debates about the so-called Baby Doe regulations of the DHHS. Much more discussion on this issue is needed.

It is crucial to remember always that the fetus and the infant are peculiarly vulnerable. We cannot tell what their future values will be. We do not know what each will consider a life worth living. One can argue back and forth about the Baby Doe case and whether or not those parents acted in the best interest of the infant. Whatever one's substantive view may be, one ought to try to discern the motives behind the decision. Sometimes there is a serious conflict of interest. Also there can be a conflict among the surrogates, for example, when father and mother differ between themselves whether a seriously ill infant should be treated. In such cases, a determination must be made about who best represents the interest of the infant. A few states are now beginning to develop a hierarchy for moral decision making, but that is a legal hierarchy. It does not necessarily correspond with the moral competence of the surrogates.

The special moral status of the fetus and the infant make the substantive and the procedural issues in neo- and perinatology more complex than in adult medicine. Perinatologists and neonatologists have the same moral obligations as other physicians, complicated by the special vulnerability of their patients. They have the obligations of all physicians to understand the deeper structures in the anatomy of clinical-ethical judgments, to use an orderly approach to making the actual decision and to carry out the decision-making process itself in a morally defensible manner. The purpose of this report has been to suggest substantive and procedural schemata that can make clinical-ethical judgments more orderly and more explicit.

No matter what schema is proposed or used, the final determinant of the patient's best interests is the character of the physician, his disposition to act in a morally responsible way. Given the complexity of ethical decisions in peri- and neonatology, the type of person the physician is, is even more important than the way he structures his decisions. The best safeguard for the vulnerable patient is always a combination of a virtuous physician dedicated to the patient's welfare and an explicit and orderly system of ethical analysis.[16]

16. G.C. Meilaender, *The Theory and Practice of Virtue* (Notre Dame, Ind.: University of Notre Dame, 1984); E.D. Pellegrino, "The Virtuous Physician and the Ethics of Medicine," in *Virtue and Medicine: Explanations in the Character of Medicine*, ed. E. Shelp, Philosophy and Medicine 17 (Holland: Reidel, 1985).

II

Interventions in the Life Cycle

A. Reproduction and the Beginning of Life

8

In the Beginning: Pellegrino, Ethics, and Life Before Birth

Joseph Tham, LC, MD, PhD

Introduction

I am grateful that the editors of this volume invited me to write a chapter on beginning-of-life issues. In examining the enormous corpus of Pellegrino's writing, they could not find articles that directly address the issues of fertility control, reproduction, and abortion. It is quite a peculiar anomaly in his prolific writing career, mainly because these issues were hotly debated in the early days of bioethics during which he was a prominent protagonist.

As president of the Catholic University of America (CUA) and as director of the Kennedy Institute of Ethics, Pellegrino was directly involved in the debate. He testified in the court case against theologian Charles Curran who was fired from CUA because of his open dissent with Catholic teaching on contraception and which was therefore deemed incompatible with the university's identity. CUA won that court case against Curran.

As a doctoral student under Dr. Pellegrino, I was fortunate to enjoy long conversations with him on some of these topics since my dissertation touches on the secularization of bioethics and the debate on fertility control in the 1960s and '70s.[1] He was familiar with the stances of Joseph Fuchs, Richard McCormick, Warren Reich, Bernard Häring, Bruno Schüller, and William E. May, all prominent moral theologians and bioethicists. From these exchanges and the clues that he left in his writings, we can garner some of the reasons he shied away from addressing these issues publicly.

Perhaps one explanation is that he considered the debates on fertility control, especially contraception, to be more theological than philosophical. In his comments on my doctoral dissertation, later published as a paper, he says,

1. S. Joseph Tham, *The Secularization of Bioethics: A Critical History* (Rome: UPRA Press, 2008).

> But I was taught to discuss them in our secular society in philosophic not theological terms. Indeed keeping theology and philosophy in proper relationship with each other was part of my training by the Jesuits and Vincentians. Magisterial teachings were a guide to both right and good decisions. But in the secular world only non-theological arguments were admissible. Some of this explains my distance from the *Humanae Vitae* [HV] debate. I took HV as Magisterial teaching which I could explain to non-Catholic bioethicists. I did not try to establish HV as true except in philosophical terms which were, of course, limited. I never wrote about the issue because I did not consider myself a theologian.[2]

Strangely, he considers *Humanae Vitae*'s arguments, the Catholic Church's pronouncement in 1968 prohibiting the use of contraception in marriage, to be theological. It is strange because HV strenuously argues with natural law reasoning, which is precisely philosophy rather than theology. Pellegrino says these arguments do not convince him because they are "limited." He probably considers its reasoning unpersuasive in the secular academic environment dominated by the Anglo-Saxon philosophical tradition with pragmatic and technocratic tendencies.

Another reason for his reticence to delve into these topics could be strategic. He frequently recounted the difficulties he encountered to enroll in New York University Medical School in the 1940s because of his Italian Catholic background. To survive in those days, he mentioned the need to use reason-based arguments in his academic work, which was hostile to his personal beliefs. On several occasions, I heard him lament that his colleagues would discount his arguments merely because he was a practicing Catholic. In an environment where he wanted his ideas to be heard, he had to steer clear from specific topics that might further pigeonhole him as a conservative Catholic:

> This earned me a reputation among many clerics as a "conservative." At the same time I remained persona non grata for many non-Catholic bioethicists. . . . I was fortunate enough to speak and write as a traditional Catholic yet within the milieu of secular bioethics. It seemed to me that secular bioethicists had already made up their minds on contraception, abortion, and the related questions. I always spoke not as a theologian or philosopher but as a physician examining biomedical issues from a philosophical perspective. For me mainstream bioethics was already secularized and the task of a Catholic was to stay in the debate, give voice to the Catholic medical moral tradition and give a rational sense of the faith as consistently as possible.[3]

Finally, Pellegrino was reluctant to venture into these debates because his medical ethics theory might not be heard if he had addressed these topics. As we will see,

2. Edmund D. Pellegrino, "Some Personal Reflections on the 'Appearance' of Bioethics Today," *Studia Bioethica* 1, no. 1 (2008): 56.

3. Ibid.

medicine is based on the encounter between the physician and the patient to restore the latter's ill-health. His exclusion of social or economic considerations from this concept of health would probably apply to fertility issues such as contraception, sterilization, assisted reproductive technology (ART), and to an extent, abortion. As we can imagine, a thesis that excludes these issues from biomedical ethics could be controversial. It also explains his silence.

This chapter will predict how Pellegrino would have argued these matters based on his publications and theory of medical ethics. We will examine the issues of fertility control and abortion from the perspectives of the nature of the healthcare profession, the patient's good and self-determination, human dignity at life's beginning, and the virtue of justice.

The Healthcare Profession

Pellegrino defends the thesis that medicine's internal morality is based on the healing encounter between the physician who is professionally trained and the patient who is vulnerable when sick. As we have seen in the previous chapter, he develops from this healing relationship the ethics of virtue for the physician, the duty to act for the patient's good, and the ends of medicine and the medical profession.[4] He contrasts this vision of medicine against those who see the physician as a body mechanic, a business entrepreneur, a contractor, or a social functionary. He warns against the danger of commodification of medicine, where the patient's desire trumps the patient's good, and economic gain substitutes for professionalism. Pellegrino has firm reservations about a business model of health care, which corrupts medicine's very soul.

Relevant to this discussion is the definition of health and medical care. Pellegrino is somewhat critical of the World Health Organization's (WHO) definition of health, defined as "a state of complete physical, mental, and social well-being and not merely the absence of disease or infirmity."[5] In contrast, he prefers to restrict health care mostly to the restoration of lost functioning. While this healing is not exclusively physical, he rebuffs the intrusion of social well-being into the definition of health. Moreover, he believed that health care is the provision of assistance to persons in need of care, cure, prevention, or help related to trauma. The central feature of health care is the personal relationship between a health professional and a person seeking help (see chap. 37 below).[6]

4. Edmund D. Pellegrino and David C. Thomasma, *For the Patient's Good: The Restoration of Beneficence in Health Care* (Oxford: Oxford University Press, 1988); Edmund D. Pellegrino and David C. Thomasma, *Helping and Healing: Religious Commitment in Health Care* (Washington, D.C.: Georgetown University Press, 1997).

5. World Health Organization, Preamble to the Constitution of the World Health Organization (1946), https://www.who.int/about/governance/constitution.

6. Edmund D. Pellegrino, "The Commodification of Medical and Health Care: The Moral Consequences of a Paradigm Shift from a Professional to a Market Ethic," *The Journal of Medicine and Philosophy* 24, no. 3 (June 1999): 247, https://doi.org/10.1076/jmep.24.3.243.2523.

The addition of social concerns in the definition of health is problematic for Pellegrino because it casts the net too wide and would eventually imply that medicine ought to deal with patients' social and economic woes. Many fertility and life concerns are social without necessarily a healing dimension. In many cases, abortions are sought due more to women's social hardships and situational exigencies than for health reasons. Similarly hormonal contraceptives are taken by otherwise healthy women. Furthermore, the infertile man or woman would not need medical attention unless they want progeny.

We can foresee how WHO's redefinition may lead to the commodification of health care. Perhaps Pellegrino has abstained from writing on abortion, contraception, and artificial reproductive technologies (ART) in his medical ethics because he refuses to count these services as "health" care.

Medicalization refers to creating needs through medical procedures and products with a business mindset rather than having in mind real health needs.[7] While medicalization and commodification are present in many branches of medicine, they are prevalent in reproductive technology. However, depending on the country in which you live, many of these interventions are not covered by national health insurance. This reality implicitly confirms that many consider them not to be an essential part of medical care. At the same time, when patients need to pay out of their own pockets, doctors and other providers are turned into businesspeople.

Commodification is readily evident in ART, which has evolved into a "fertile" industry. In this medical industry, physicians become entrepreneurs, the patients become consumers who desire certain services, and a contract seals the transaction. The human body becomes an assortment of marketable components—sperm, ova, and female surrogates—to produce a child of desire. Where costs are a concern, couples seek out developing nations where prices are more affordable. Fertility tourism has become a thriving enterprise in which a child becomes a commodity.[8] According to Pellegrino, commodification destroys the profession of medicine as it deforms the healing relationship between doctors and patients: "One thing is certain: if health care is a commodity, it is for sale, and the physician is, indeed, a money maker: if it is a human good, it cannot be for sale and the physician is a healer."[9]

For the Patient's Good

For the Patient's Good is one of the significant works Pellegrino wrote with Thomasma.[10] In it he argues that a physician's primary goal is to seek the objective

7. Paul Lee and Joseph Tham, "Industrialization of Medicine and Medical Professionalism: Bioethical Critiques," *Studia Bioethica* 7, no. 3 (October 24, 2014), https://riviste.upra.org/index.php/bioethica/article/view/3018.

8. Global IVF, "Fertility Tourism in India, A Risky Prospect," *Global IVF Medical Fertility Tourism* (blog), August 18, 2011, https://globalivf.com/2011/08/18/fertility-tourism-in-india-a-risky-prospect/.

9. Pellegrino, "Commodification of Medical and Health Care," 262.

10. Pellegrino and Thomasma, *For the Patient's Good.*

good for the patient as the two enter into a healing relationship. Pellegrino may not wish to incorporate fertility-related services as a part of this relationship. Nonetheless, along with the logic of the patient's best interest, Pellegrino would further argue that the patient should not base her decisions on a distorted sense of subjective good but seek her true good.

In this work, Pellegrino recognizes the tension between medical paternalism and the patient's autonomy and between the principles of beneficence and autonomy. There has been a historic shift towards greater patient self-determination in clinical decisions due to societal, technological, and economic forces. At the same time, Pellegrino is careful not to absolutize patient autonomy with a type of "moral atomism." Despite the popularity of principlism advanced by Childress and Beauchamp, there has been much written about the tyranny of autonomy.[11] Recently, there has been further consideration of relational autonomy since a patient's decisional capacity is enhanced when he is not an island, but is in communion with his family and loved ones.[12] Pellegrino submits a hierarchical relationship to navigate between the patient's subjective value of what constitutes a perceived good and the objective moral good that medicine proposes. In our imperfect world, Pellegrino concurs with Alasdair MacIntyre that there are bound to be interminable and irresolvable conflicts (see chap. 6 above).[13] As noted, Pellegrino is quite adamant that the physician-patient relationship cannot be a legal contract where the physician does not have a say on what is "good" for the patient.

Pellegrino proposes four categories in consideration of the patient's good: the medical good, the good as perceived by the patient, the good of the patient as a human, and the good of the patient's spiritual nature.[14] The last component invites bioethics to consider the spiritual dimension, the highest good that embraces the patient's ideals, religion, destiny, meaning, and relationship: "This good is also grounded in our humanity as beings capable of commitments to ideals and beliefs beyond the needs of our material bodies. This is the realm of religious belief, or non-belief, the ultimate source of morality for most patients when confronted with their own finitude, suffering, or despair."[15] It seems that modern technology has given

11. Tom L. Beauchamp and James F. Childress, *Principles of Biomedical Ethics* (Oxford University Press, 2001); Stephen Toulmin, "The Tyranny of Principles," *Hastings Center Report* 11, no. 6 (1981): 31–39, https://doi.org/10.2307/3560542.

12. S. Joseph Tham and Marie Catherine Letendre, "Health Care Decision Making: Cross-Cultural Analysis of the Shift from the Autonomous to the Relational Self," *The New Bioethics* 20, no. 2 (2014): 174–85, https://doi.org/10.1179/2050287714Z.00000000051.

13. Alasdair Macintyre, "Why Is the Search for the Foundations of Ethics So Frustrating?" *Hastings Center Report* 9, no. 4 (1979):16–22, cited in Edmund D. Pellegrino and David C. Thomasma, "The Conflict Between Autonomy and Beneficence in Medical Ethics: Proposal for a Resolution," *The Journal of Contemporary Health Law and Policy* 3, no. 1 (1987): 31.

14. Edmund D. Pellegrino, "The Internal Morality of Clinical Medicine: A Paradigm for the Ethics of the Helping and Healing Professions," *The Journal of Medicine and Philosophy* 16, no. 6 (2001): 559–79.

15. Pellegrino, "Moral Foundations," 12–13.

humanity both a blessing and a curse. It has undeniably offered humanity solutions to many challenges like never before. Likewise, we are perplexed that the technological cure can sometimes be worse than the disease. This paradox is present in these biotechnological innovations at the beginning of life. ART could help couples overcome the suffering of infertility, and an unwanted child could be precluded with contraception or abortion. However, we wonder if, in our anxiety to control our destiny, we lose our humanity.[16] Are afflictions, anxieties, lack of control, or childlessness antithetical to a flourishing existence? Pellegrino has this to say:

> Finally, there is the illusion that a life that is free of all anxiety, suffering, or misfortune is the "good" life. But, such a life is devoid of opportunity for expression of some of our most characteristically human feelings—mercy, compassion, understanding, empathy, love, and giving of ourselves to others. . . . Miguel de Unamuno may not have been entirely wrong when he wrote: "Suffering is the substance of life and the root of personality. Only suffering makes us persons."[17]

In discussing what is good for the patient, we must account for transcendent values other than the merely pragmatic and utilitarian ones. Pellegrino has written extensively on virtues, though primarily applied to physicians. [18] The patient's good is also related to virtuous living. The Aristotelian theory of the four cardinal virtues can help patients attain the superior good on these fertility and life issues.[19] Temperance aids us in controlling gratifications and pleasures, and in the realms of sexuality, becomes the virtue of chastity.[20] Fortitude can help patients face adversities and make the right choices bravely, for instance, when confronted with the difficult circumstances of unwanted pregnancies or infertility. Justice helps us reflect on the correct use of technology, especially when dealing with early life and vulnerable populations. Lastly, prudence can help us use technology wisely, avoiding the pretense of compassion. Pellegrino explains this queen of all virtues in the cloning debate, "Compassion is not to be a moral law unto itself but must be subject to moral analysis, based on sound reasons, and carried out with forethought of consequences."[21]

16. Joseph Tham, "Suffering Technology (Part 1)," *Studia Bioethica* 7, no. 1 (2014): 55–61; Joseph Tham, "Suffering Technology (Part 2)," *Studia Bioethica* 7, no. 2 (2014): 50–56.

17. Edmund D. Pellegrino, "Doctors Must Not Kill," *The Journal of Clinical Ethics* 3, no. 2 (1992): 97.

18. Edmund D. Pellegrino and David C. Thomasma, *The Virtues in Medical Practice* (New York: Oxford University Press, 1993); Edmund D. Pellegrino and David C. Thomasma, *The Christian Virtues in Medical Practice* (Washington, D.C.: Georgetown University Press, 1996).

19. Josef Pieper, *The Four Cardinal Virtues: Human Agency, Intellectual Traditions, and Responsible Knowledge*, trans. Richard Winston et al. (Notre Dame, Ind.: University of Notre Dame Press, 1990).

20. Joseph Tham, "Lust, Shame and Bioethics," *Studia Bioethica* 1, no. 2–3 (2008): 202–10.

21. William P. Cheshire et al., "Stem Cell Research: Why Medicine Should Reject Human Cloning," *Mayo Clinic Proceedings* 78, no. 8 (2003): 1016, https://doi.org/10.4065/78.8.1010.

Last but not least, we cannot divorce the patient's spiritual good from her religion if she practices one. There has been a process of secularization of bioethics and marginalization of religion in the medical encounter. Recently, there is greater interest in this topic, especially with the discontent of overly irreligious bioethics.[22] On sex and life issues, religious mores are essential guides for many people with different creeds.[23] Commenting on the relationship between technological power and religious credence in the areas of procreation, Pellegrino writes:

> This mixture became explosive when it came into contact with the enormous powers of biotechnology to shape human existence. This power was indeed new, and it fostered illusions of a god-like humanity no longer needing a creation. . . . *Humanae vitae* does not constitute a new ethic which some hoped it would be to permit contraception. Rather it concerns the nature and purposes of procreation which imply more than simple propagation of the gene pool, or the satisfaction of personal pleasure. *Humanae vitae* simply updated Church teaching on procreation and marriage. It did not invent a new theology more suited to contemporary preferences.[24]

Human Dignity at the Beginning of Life

The thorny issues at the beginning of life concerning the human embryo and fetus in Pellegrino's thought require further discussion. The different stakeholders are the physician, the woman or parent, the unborn, and the society. We have discussed the duty of physicians above. Now, we will look at the Hippocratic ethic. Women's choices and rights are not explicitly addressed in Pellegrino's writings but are indirectly discussed regarding autonomy and a patient's true good. This section will examine the status of the embryo and fetus, the meaning of personhood, human dignity, and vulnerability. In the next section, we will look at social policies under the heading of justice.

In professing the Hippocratic Oath, new medical graduates promise a series of obligations. Even though there have been updates of the oath, it is still a public declaration that, for Pellegrino, is a promise to "use your knowledge for the sick and not primarily for your own interests." This "professing" is the essence of the medical "profession." Nevertheless, the Hippocratic tradition has been under attack and downgraded because it is considered too paternalistic. Many of the prohibitions are rendered obsolete, reformulated, or relaxed to allow for abortion, euthanasia, or sexual relationships with patients.[25] Despite these modern challenges, Pellegrino insists that Hippocrates's prime maxim, "First, do no harm," still holds true. Nonmaleficence, including not

22. Joseph Tham, "A Response to Irreligious Bioethics," *Studia Bioethica* 6, no. 2–3 (2013): 12–19; Joseph Tham, "The Secularization of Bioethics," *The National Catholic Bioethics Quarterly* 8, no. 3 (2008): 443–53.

23. Joseph Tham, John Lunstroth, and Alberto Garcia Gomez, *Multicultural and Interreligious Perspectives on the Ethics of Human Reproduction* (Cham, Switzerland: Springer, 2021).

24. Pellegrino, "Some Personal Reflections," 54–55.

25. Pellegrino, "Moral Foundations."

terminating another human life, is a stringent ethical duty of physicians.[26] Later, we will discuss the need to protect physicians from being forced to engage in these acts, through conscientious objections.

Pellegrino does not directly address the feminist demands of abortion or reproductive rights. In what we have already seen, he argues that the need for abortion is more a social need than a medical one. He is also contrary to any choice-centered rhetoric that would effectively convert the doctor-patient relationship into contract and transaction. Finally, Pellegrino wants to inquire about the authentic good of the woman seeking abortion based on the four levels of good and addressing her spirituality.

We now come to the debate on the status of the embryo and fetus regarding their humanity and personhood. We can glean from Pellegrino's writing on stem cell research and cloning the following inferences about the human embryo (see chap. 11 below):[27]

- The totipotent human embryo can develop into a complete, living human being, a member of the human species. As a result, the destruction of an embryo is the destruction of a human being.
- Philosophically speaking, the embryo is a human being in a state of active potency. It is not just potentially, but actually a human being. A gradualist view of the ontological progression of the status of the human being is philosophically unsound.
- Personhood belongs to the philosophical or metaphysical category that is not proven scientifically or empirically. Personhood does not depend on functionality or social categorization. Life precedes function, and one is first a human being before one develops and possesses functional capacities.
- Early human life is particularly vulnerable, and beneficence should be stringently applied by those who care for the unborn. Deliberate creation and sacrifice of human embryos should be rejected.
- Human dignity, or the moral status of a human being, is independent of age, size, physical characteristics or location, and emerges at the beginning of life. The equality of all human beings means that no subgroups should be discriminated against or exploited.
- Human life has a spiritual significance that transcends empirical measurements and utilitarian calculations.

Pellegrino's view on the status of human embryos can be summarized in this sentence. "In my view, they are deserving of the same respect as the infant or child" (see chap. 10 below).[28]

26. Cheshire et al., "Stem Cell Research."

27. Edmund Pellegrino, "Balancing Science, Ethics and Politics: Stem Cell Research, a Paradigm Case," *Journal of Contemporary Health Law & Policy (1985–2015)* 18, no. 3 (2002): 591–612.

28. Edmund D. Pellegrino, "Beneficence, Scientific Autonomy, and Self-Interest: Ethical Dilemmas in Clinical Research," *Cambridge Quarterly of Healthcare Ethics* 1, no. 4 (1992): 369, eweb:105936, https://doi.org/10.1017/S0963180100006551.

Pellegrino addresses another related question. Is the unborn (fetus or embryo) a patient? This query is critical for his medical ethics centered on the healing relationship between the doctor and the patient. Usually, the patient is a competent adult, or patients could be incompetent due to illness, mental incapacity, or being a minor. However, it becomes a particular challenge in perinatology and neonatology since the unborn cannot express his or her values. Due to the peculiar vulnerability of the fetus and the infant, he states, "The physician, who should be the *advocate of the patient*, has a serious responsibility to make some judgment about the moral validity of a surrogate decision maker" (see chap. 7 above).[29] Hence, we can say that both the mother and the unborn are equally patients under the physician's care.

If the unborn is a patient, and the embryo deserves the same respect as the infant or a child, then it follows that the healing relationship extends to this subgroup of vulnerable individuals. Even though Pellegrino is reluctant to include fertility treatments and abortion as a part of health care, if they were to be part of the physician's duty to care, then the patient would not only include the infertile couple or the woman but similarly the unborn fetuses and embryos. Dignity and equality of all human beings have further implications for justice.

The Virtue of Justice

Those who concur with Pellegrino's position on health care, the good of the patient, and the status of the unborn face many challenges in the current practice of medicine when it comes to the issues of abortion and ART.

Those who are conscientious about the profession of medicine and the Hippocratic ethic would be uneasy with the deformation introduced by including many of these services that will commodify human life and turn medicine into commerce or contract. Rightly so, some medical students, physicians, and healthcare institutions may wish to abstain from providing contraception, sterilization, abortion, and ART. Their objections are not unfounded, and conscientious or religious objection clauses should legally protect them. It is a fundamental human right to abstain from complicity with actions that are deemed unethical. Pellegrino cautions about Catholic hospitals' precariousness in cooperating with these practices when they receive public funding or are merged with secular institutions.[30]

While Pellegrino does not directly address the feminist concerns, his philosophy of medicine would be cautious of the rhetoric of rights as the only good of autonomously deciding women. Other justice concerns include exploitation of surrogate mothers, especially in developing countries where they are vulnerable to economic pressures, and the danger of hormonal hyperstimulation for egg donors. The

29. Edmund D. Pellegrino, "The Anatomy of Clinical-Ethical Judgments in Perinatology and Neonatology: A Substantive and Procedural Framework," *Semin Perinatol* 11, no. 3 (1987): 209 (emphasis mine), eweb:74365, https://repository.library.georgetown.edu/handle/10822/710599.

30. Pellegrino, "Balancing Science, Ethics and Politics."

practice of fertility tourism is often an unjust reduction of people, gametes, and new life to a mere means to an end.

Pellegrino appeals to the slippery slope argument, which reasons that when a moral principle is relaxed, similar practices become incrementally accepted in society both psychologically and legally. It transpired with the acceptance of contraception first for married couples, then for anyone over-the-counter; with emergency contraception, then to abortion pill RU 486; and from late-term abortion to infanticide. Similarly, in ART, we see advances moving towards "reprogenetics" that encompass cryopreservation of embryos, preimplantation genetic diagnosis (PGD), eugenics, embryo selection and reduction, and ectogenesis.[31]

Pellegrino noticed the slippery slope phenomenon in American abortion laws: "One need only recall here the history of *Roe v. Wade* to see this principle in action. First, abortion was limited to the first trimester, then allowed in the second and third, and then to late-term pregnancy. Going from the evacuation of a week-old fetus to the evacuation of the living brain of an easily identified human baby in late-term abortion is the inevitable result."[32] (Although *Dobbs*[33] removed the Constitutional support for abortion, it is less likely to reverse the societal support for abortion that grew as a result of *Roe*. To many people, legality indicates morality, and restrictions on the practice may appear too arbitrary for some.)

It is the role of the society to allocate limited health resources that will be fair and just to its citizens. Consistent with Catholic social doctrine, Pellegrino applies the common good conception of justice under the principles of solidarity and subsidiarity to health care. In beginning-of-life issues, solidarity would imply a holistic approach to support both the mother and the child both ante- and post-natally. This support should not be just material or financial, but social, psychological, and spiritual. The principle of subsidiarity would mean that the support should ideally be provided locally or bottom up rather than top down through governmental programs. In a way, these principles redirect the legal debates between life and choice toward the dignity and flourishing of the involved parties. This perspective looking out for their integral good is central to Pellegrino's approach to medical ethics. He is, however, realistic about the uneasy tension between privatized medicine and the basic need to provide health care for all. It is a complex and delicate balance at the beginning of life when ethical concerns are mixed with commercial profits, medical professionalism, science, and politics.

31. Joseph Tham, "Will to Power: A Critique of Nihilistic Tendencies in Reproductive Technology," *The New Bioethics* 18, no. 2 (2012): 115–32, https://doi.org/10.1179/2050287713Z.00000000014; Joseph Tham, "Why Should the Baby Die?" *Hong Kong Journal of Paediatrics* 17 (2012): 264–65.

32. Pellegrino, "Balancing Science, Ethics and Politics," 604.

33. *Dobbs v. Jackson Women's Health Organization*, No. 19-1392, 597 U.S. ___ (2022).

Conclusion

Despite the fact that he did not write about them, Pellegrino's position on contraception, abortion, ART, and related beginning-of-life topics conforms mostly with the Catholic position. However, it is crucial to note that the rationale and mode of arriving at these conclusions are not identical with the Catholic magisterial teaching based on natural law. Pellegrino's conclusions are based on his philosophy of medicine—the doctor-patient relationship, the nature of the medical profession, the tension between the patient's autonomy and the patient's good, and the Aristotelian virtues of a flourishing life. By arguing in this manner, he wishes to distance himself from what he identifies as Catholic "theology." Most likely, Pellegrino avoided writing about these beginning-of-life issues to strategically avert an unfair branding and possible dismissal of his philosophy of medicine as Catholic. Rather than weakening his contribution to bioethics, this analysis shows that there can be different paths to arrive at the same truth in ethics.

B. Medical Research

9

Introduction

Kevin T. FitzGerald, SJ, PhD, PhD

WHILE CLEARLY HOLDING A POSITION against human embryonic and fetal stem cell research and treatment, Pellegrino brought to this debate his *intrinsic morality of medicine* approach to the question of how best to address the ethical issues involved. The three articles he authored or co-authored represent well his approach. All three articles focus on the fundamental question of which "moral compass points" are being used to guide decisions about science, medical treatments, and policy?[1] While Pellegrino always acknowledges the diversity of perspectives that are brought to bear in bioethics on controversial issues, such as human stem cell and cloning research, he also constantly seeks—and strenuously argues for—an ethical foundation intrinsic to medicine from which all the different perspectives and arguments regarding human stem cell research can be evaluated and integrated in such a way as to achieve the common good/goals of medicine. This emphasis on the goals intrinsic to medicine, and, hence, also intrinsic to the medical profession, is a key feature that sets Pellegrino's thought and reasoning apart from many other bioethicists who addressed these issues, and is also a key source of the constructive insight he brings to this debate.

In his article on "Beneficence, Scientific Autonomy, and Self-Interest," Pellegrino chooses to focus on the personal morality of the clinical investigator because it had received much less attention than the other two crucial aspects of clinical research: acquiring new knowledge and the moral use of that knowledge (see chap. 10 below).[2] Applying his emphasis on the need for medical professionals who are committed to the intrinsic goal of medicine—the benefit of the patient—at the end of the article Pellegrino concludes: "Their best guarantee of freedom in the pursuit of medical knowledge and the good of patients lies in research that is scientifically valid and ethically responsible. To compromise either is, on the one hand, to endan-

1. Edmund D. Pellegrino, "Balancing Science, Ethics and Politics: Stem Cell Research, a Paradigm Case," *The Journal of Contemporary Health Law and Policy* 18, no. 3 (2002): 592.

2. Edmund D. Pellegrino, "Beneficence, Scientific Autonomy, and Self-Interest: Ethical Dilemmas in Clinical Research," *Cambridge Quarterly of Healthcare Ethics* 1, no. 4 (1992): 361–70.

ger patients and, on the other, to lose the privilege of freedom without which science cannot thrive."[3]

In his article, "Balancing Science, Ethics and Politics," Pellegrino reviews the entire landscape of the stem cell debate from the language used by scientists to portray their perspective of the issue to the political conundrums of policy distinctions between publicly and privately funded research. In the end, Pellegrino discerns the potential for significant damage to the medical profession if social acceptance of research or treatments, such as those using human embryonic stem cells, becomes a litmus test for how physicians are to practice and what treatments they are to choose (see chap. 11 below).[4]

An explication of the extent to which this damage might reach can be found in the article Pellegrino co-authored with several others, "Stem Cell Research: Why Medicine Should Reject Human Cloning" (see chap. 12 below). Though written by several authors with different areas of expertise, Pellegrino's influence is clear as can be seen in the section of the article titled, "The Purpose of Medicine."[5] In affirmation of Pellegrino's ethical approach, the authors conclude, "Ethical reflection always reaches, in due course, the conclusion that the least of human beings deserve the care and concern that the medical profession presumes is due all human beings."[6] Although written in concert with several colleagues addressing specifically the human cloning controversy, this statement may be one of the best summary statements of the thought of Edmund Pellegrino, M.D.

3. Ibid., 369.

4. Pellegrino, "Balancing Science, Ethics and Politics," 610.

5. William P. Cheshire et al., "Stem Cell Research: Why Medicine Should Reject Human Cloning," *Mayo Clinic Proceedings* 78, no. 8 (2003): 1010–1018. ISSN 0025-6196, https://doi.org/10.4065/78.8.1010.

6. Ibid., 1017.

10

Beneficence, Scientific Autonomy, and Self-Interest: Ethical Dilemmas in Clinical Research

Edmund D. Pellegrino, MD

THE ETHICS OF CLINICAL RESEARCH may be viewed from three different perspectives: first, the process of acquiring new knowledge; second, the moral use of the knowledge acquired; and, third, the ethics of the investigator seeking this knowledge.

The ethics of the process of human investigation has been the focus of attention ever since the Nuremberg trials exposed the atrocities perpetrated by Nazi physicians on their experimental subjects. The ethics of process emphasizes informed consent, institutional review boards, and legal regulation of human experimentation. The moral use of the knowledge gained by clinical investigation has also received wide discussion on subjects such as fetal tissue transplantation, reproductive technology, the artificial heart, and human gene therapy. Much less attention has been given to the third perspective, which is concerned with the personal morality of the clinical investigator. Yet it is the clinical investigator who is the ultimate safeguard of the safety of the experimental human subject. It is the investigator who designs the experiment, prepares the protocol for peer review, obtains the necessary consent, interprets the results, and monitors the safety of the experiment.

It is from this third perspective that I wish to examine the ethics of clinical research in this essay: first, because this perspective is so often neglected; second, because it is so central to the moral quality of human investigations; and third, because it is of such special significance in, for example, pediatric research where the vulnerability of the infant or child imposes graver responsibilities on the investigator than is the case with adults who can give informed consent or refusal.

The essence of the moral dilemma faced by all clinical investigators is the potential conflict between the investigator's roles as scientist, as physician, and as

Reprinted with permission from Edmund D. Pellegrino, "Beneficence, Scientific Autonomy, and Self-Interest: Ethical Dilemmas in Clinical Research," *Cambridge Quarterly of Healthcare Ethics* 1, no. 4 (1992): 361–69.

private individual. Each role is dominated by a different value: for science, it is truth; for medicine, it is beneficence toward the patient; and for the investigator as an individual, it is self-interest. Each value, along with its subsidiary set of standards and procedures, is legitimate in its own right. But when these values come into competition, as they may in clinical research, serious moral dilemmas result.

How investigators deal with these value conflicts ultimately determines the safety of the patient. This, in turn, determines the degree to which society will permit or restrict the scientific autonomy necessary for valid research. But no matter how rigidly human experimentation is regulated, the ethical sensitivity of the investigator remains an ineradicable factor in the moral equation.

It is this moral sensitivity I wish to examine under three headings: first, the covenantal nature of clinical investigation; second, the difference in the value systems of the investigator's concurrent roles as scientist, physician and individual; and third, the resolution of potential conflicts through application of the ordering principle of "autonomy-in-trust."

We must recognize from the outset that clinical research is an invasion of the integrity and privacy of the person who is the research subject. This invasion necessarily places the person at some risk for a result which, by the very nature of research, must be problematic. Nothing in the nature of science or its internal need for freedom and autonomy constitutes an autonomous moral right to engage in human experimentation. Rather, clinical research is a privilege accorded the investigator by social consensus. Society permits experimentation with humans because of the benefits that the whole society, as well as the experimental subject, may gain from new medical knowledge. The clinical investigator thus enters a covenant with society when he or she accepts the privilege of experimentation involving his or her fellow human beings.

In addition, the investigator is also party to a covenant with the experimental subject. For the potential good of others, of medical science in general, or of the subject him- or herself, physicians persuade a patient to become simultaneously a patient and a subject. By that act, the good of the patient, which should be primary in the purely therapeutic encounter, is now compromised by another good—the knowledge to be gained by experimental manipulation of the patient's care. The patient is, therefore, knowingly put at some risk for the sake of an uncertain good. By consenting to participate in a research protocol, the subject trustingly yields up some personal moral claims to safety in return for a potential good for him- or herself or others. The patient-subject thus has a moral claim on the investigator's fidelity to respect the implied promise that the investigator will judiciously balance the patient's interests and those of the scientific protocol.

These two covenants—one with society and one with the experimental subject—form the basis for the moral obligations specific to clinical investigation. Although these covenants arise in a trust relationship that institutional review boards and research regulations can help to safeguard, their subtle nuances cannot be spelled out in such regulations. Rather, like the therapeutic relationship, these covenants rest ultimately on the character and moral integrity of the investigator.

Society permits human experimentation under carefully circumscribed conditions because it is the only way medical knowledge can be verified for use in the amelioration and cure of human disease. In vitro studies, animal models, and computer simulations can go only so far. Ultimately, the new drug or surgical procedure must be tested in human subjects. To be faithful to both covenants—with the patient and with society—these tests must, in the first instance, be scientifically sound. To proceed with a research plan that lacks scientific integrity places human subjects at risk needlessly and irresponsibly. It does violence to society's and to the subject's legitimate expectations of science and scientists.

Even as we acknowledge the importance of the canons of good science, however, we must also appreciate that clinical medicine is not itself a science. It possesses a set of values that differ from those of science in fundamental ways. Let us look at some of these differences.

The Values of Science

The end of scientific endeavor is truth, that is, knowledge as a good for its own sake. This is the ordering value in science. To achieve this end, science must be free to examine any question and use any method appropriate to the questions it poses. It must design those questions rigorously and examine them experimentally for their verifiability and falsifiability. Observations must be accurate, honestly reported, objectively interpreted, subjected to peer review, and shared openly with colleagues and the world at large.

These values and standards imply a certain degree of objectivization of the subject under study. But in clinical investigations, the "object" of study remains a human being—either a healthy volunteer or a sick person. Here, and in the case of higher animals in general, the canons of science may conform with another set of values—those that define the endeavor of medicine.

The Values of Medicine

In contradistinction to science, whose ordering value is truth, the ordering value of medicine is beneficence. Medical knowledge is sought for a specific end: the healing, helping and curing of human beings. The good of the patient is the primary end of medicine. To achieve this end, medical knowledge must be particularized to meet the needs of this patient in this clinical situation. Its aim is not, like the aim of science, the discovery of generalized concepts, theories or laws about human physiology or disease.

In addition, medical knowledge must be applied compassionately and sensitively to the uniqueness of the predicament of illness in this person, a predicament that is not wholly penetrable by the physician. Truth-telling and confidentiality are mandatory if fidelity to the trust that patients must place in doctors is to be preserved. In those critical moments when a clinical decision is made, the character of the physician remains the patient's ultimate safeguard, for which no contract, however carefully constructed, can fully substitute. Whatever agreement is made with the patient must inevitably be channeled through, and carried out by another person—the physician.

Medicine is, therefore, not primarily science, though it uses science. In medicine, truth must be sought and applied for the good of the patient. It is not an end in itself. Medical truth, unlike scientific truth, is in service to the value of beneficence. It is this inescapable fact that not only gives a moral complexity to the activity of the physician who engages in clinical investigations but also limits the scientific autonomy of clinical investigators.

The Investigator's Self-Interest

In addition to the inherent tension between the values of science and medicine, there is a third set of values that further complicates the ethics of clinical research. These values derive from motives and impulses of self-interest that are normal to human beings but may, under the special circumstances of clinical research, present moral hazards.

Like other humans, the clinical scientist desires to advance his or her own career, improve his or her income, and experience the satisfaction of peer approval and public recognition. These motives may result in an excessive drive to compete, to be the "first" to report results, to win the Nobel Prize, or simply to enjoy the personal satisfaction of an elegant experimental design elegantly carried out. Whatever its source, unrestrained self-interest can blunt moral sensitivity. It can lead to conscious or unconscious violations of the covenants between physician and society, and between physician and research subject that permit clinical research in the first place.

For example, self-interest may erode the high standard of moral sensitivity that should guide experiments with especially vulnerable groups: the aged, the hopelessly ill, the mentally challenged, the infant with multiple defects, the fetus, the patient with HIV infection, the embryo, or that new fictive entity, the "pre-embryo." In addition, a researcher may justify her compromise of informed consent by citing the benefits to society, other patients, or the eventual cure of a serious disease. When does the investigator step over the line between a truly informed consent and telling just enough to keep the subject from being scared off? No formula can spell this out in advance. No set of regulations can fully protect even the well-educated, fully competent subject. How much more vulnerable are the poorly educated, the sick, or the parents of a desperately or hopelessly ill child or infant?

For the physician-scientist, the moral obligations of the investigator are much heavier than the legal. A legally adequate consent form may not be morally valid. A morally valid consent aims at true "consent," an "agreeing together" in which both parties enter a relationship based on mutual trust and the expectations of fidelity to that cause. Obviously, there must be written consent forms, and they must be examined and approved by peers and institutional review boards. However, it is the moral integrity of the investigator who obtains the consent that assures the moral validity of the consent procedure.

Institutional pride or hubris is a further corrosive influence. The institution's drive to be "first" is a mixed motive; it can be effective in raising institutional morale and productivity, but it is also capable of submerging moral imperatives on grounds of exigency and "survival." In competition with other hospitals or universities, institutional pride can desensitize an institutional review board to certain dubious projects. It can lead to unrealistic reliance on the character of a staff member who is nationally prominent or politically powerful.

The intellectual hubris of the investigator plays its part too. It tempts the academic physician-investigator to arrogate the privilege of scientific autonomy to him- or herself. The public, it is argued, cannot possibly comprehend the importance of scientific research or judge the details of a particular experiment. In the long run, it is further argued, the good of all will be better served if we place our trust in the integrity of the scientist's judgment.

Even if we grant that the majority of clinical investigators are honorable people, the dangers of this line of reasoning are obvious. Experience sadly teaches that there is no necessary correlation between academic prestige or intellectual acuity and moral behavior. In a matter as complex and finely balanced as clinical research, the general public may be more perceptive of the moral dilemmas than the most "enlightened" scientist.

Personal profit has, in recent years, come to rival prestige and career advancement as a motivating factor in scientific investigation. Pharmaceutical companies are understandably eager to have their products evaluated in humans by the most renowned institutions and investigators. This is understandable and even legitimate since new therapeutic agents should be evaluated by experienced and responsible investigators. To this end, for-profit companies have expanded the ways in which they can provide financial incentives. They may assume a portion of the investigator's salary, provide certain "fringe" benefits like travel and accommodations at deluxe resorts for "research" meetings, offer consulting fees, give gifts of various kinds, or promise a share in the profits from the products one is testing.

The matter is further complicated when the healthcare industry enters long-term contractual arrangements with a university to support research. The usual proviso is that a company will support research facilities and personnel in return for privileged access and a share in the patenting rights of the products developed. These industry-university compacts are especially attractive today when governmental and philanthropic sources of research funding are insufficient. Some of our most prestigious universities have entered into what may well turn out to be Faustian compacts. Eventually, many of them will be forced to pay their part of the bargain, especially if corporate profits fall to unacceptable levels or the "investment" in research turns out to be insufficiently productive financially.

Freedom of access and the sharing of research results have, traditionally, been values of science. In many of the contractual arrangements between investigators and pharmaceutical companies, there are provisos that recognize the investigator's right to publish in peer-reviewed journals. Still, it is a fact of business that one does not alert one's competitors to a new product before it is ready for marketing. It is difficult to say just how strongly such negative incentives to data sharing actually operate. It is not unfair to suppose that investigators, for fear of losing long-term research support, might consciously or unconsciously withhold information about new discoveries until the time to release it is commercially propitious.

These are a few examples of the kinds of conflicts of interest that pose a potential danger to experimental subjects. They can compromise beneficence, the central value in medicine. They can also pose a threat to the values of science. The objectivity, accuracy, and reliability of observations and data interpretation that science requires can subtly be destroyed by financial incentives as well as institutional and personal pride. The brute fact of the matter is that if investigations do not produce papers in quantity, if their results are not definitive or are contrary to the interests of those who support them, the emoluments and rewards will stop. Thus, both scientific probity and subject safety may become victims of self-interest.

Physician and Scientist—Can One Be Both?

The unique personal conflict in clinical research, and the one least soluble by simple measures, is the conflict of identity the investigator must inevitably experience between his or her role as scientist, interested in fact and data collection, and his or her role as physician, interested in care and cure. The physician-investigator must, perforce, be committed simultaneously to the values of science and the values of medicine. Even if he can avoid the temptation of self-interest noted above, he faces a difficult task of balancing and ordering the values of truth and beneficence against each other.

This problem was recognized several decades ago by Otto Guttentag who considered it serious enough to suggest that the physician caring for the patient might be other than the physician conducting the experiment. The issue Guttentag raised is far from settled. It can manifest itself at almost every point in any clinical experiment and especially in randomized clinical trials. It has never been squarely faced.

One example of this conflict is in getting consent for a patient's entry into a study. The patient's major concern is understandably for the most effective treatment of his or her disease. If the patient suffers from a chronic or fatal disease for which there is no treatment or only a dubious one, he or she is vulnerable to any promise of help. Under these circumstances, an experimental protocol raises expectations that all is not lost.

It becomes very difficult for patients to separate the physician-scientist role from that of physician-healer. For patients, these roles are personified in all the physicians taking care of them. The physician can easily obtain consent to an experimental protocol simply by emphasizing the hope of cure and downplaying the risk and experimental nature of the treatment. Even the built-in uncertainties of the randomized clinical trial can escape the patient's consciousness under these circumstances. Selective hearing is a powerful device that helps the patient sustain hope when the illness is incurable. The investigator must guard against taking advantage of this normal tendency in such people.

After consent has been obtained, the physician-investigator's role can become even more ambiguous. As the patient's physician, he or she would be impelled to intervene at the first sign of danger by modifying or discontinuing the protocol. As a scientist, he or she would be bound to persist in the protocol until the statistical evidence for harm or benefit was conclusive. If an investigator discontinues the experiment too soon, he frustrates his scientific obligations. If he waits too long, he frustrates his obligations as the patient's doctor.

When the study involves a double-blind procedure, there are further problems. If the physician does not know whether her patient is on a placebo or a potent, impotent, or toxic medication, she will have difficulty interpreting whether changes in the clinical course are a result of the disease or a therapeutic or toxic effect of the agent being tested. As the investigator, she must remain "blind" until it is statistically clear that the patient is suffering from an aberrant or beneficial effect of the treatment. As the subject's personal physician, she would need to break the code at the first suspicion of harm.

Another difficulty arises when the treatment under study is an improvement over an accepted and reasonably effective standard treatment, for example, testing the efficacy of a new drug for rheumatoid arthritis, angina pectoris, or hypertension. Here the experimental design might include a control group, a group on standard medication, and a group on the new medication. If assignment to these groups is randomized, a patient might, for a time, be deprived of standard treatment and placed on a new, untried, and possibly ineffective treatment or on a placebo. Again, the physician-as-scientist must adhere to the protocol. On the other hand, the physician-as-care-giver who gives priority to beneficence over scientific truth might not think the associated risks of a procedure justifiable for a particular patient.

There are many other circumstances that put the physician's roles as care-giver and as scientist into conflict. These role conflicts need more careful scrutiny. They do not constitute absolute impediments to clinical research; they underscore the need for the investigator to be self-critical and to recognize when the values he or she must observe as a physician are in conflict with the canons of science or self-interest.

The moral dangers inherent in clinical research can never be removed entirely. Human experimentation is a necessity if new treatments are to be discovered and validated. The regulatory measures already in force have done much to place legal, social, and institutional constraints on clinical research. But not enough emphasis has been placed on the special nature of the freedom and autonomy the clinical investigator enjoys. One way the needs of good science can be balanced against the needs of the patient-as-subject is through a consideration of the concept of "autonomy-in-trust"—the recognition of a moral order of priority that will resolve conflicts as they arise.

Scientific "Autonomy-in-Trust"

The intellectual autonomy of the scientist is autonomy held in trust. By "autonomy-in-trust" I mean the conscious acknowledgment by the investigator that he or she is allowed freedom to pursue rigorous scientific goals in human experimentation only if the welfare of the patient is always respected as primary and superior to the values of science and self-interest. This concept implies that the scientist/physician will always regard beneficence as the guiding principle even to the point of effacing his or her own self-interest or nullifying the scientific validity of his or her work. The autonomy necessary to good science is morally tenable if it is seen as held in trust for the good of the patient.

No quantitative formula can set forth explicitly the limits within which autonomy-in-trust should operate. Instead, the investigator must develop a sensitivity to the privilege he or she is permitted in doing research involving humans. That privilege is fragile and deserves constant awareness not only of the intersections among the three value systems I have been describing, but of the priority of the well-being of the experimental subject whenever there is even the potential for conflict.

Freedom is afforded the scientist to pursue problems that interest her because society believes that knowledge is a good in itself or in its application to human problems. Society recognizes that valid knowledge cannot be obtained without freedom of inquiry. For this reason, we allow freedom in the scientist's selection of problems, methods, and forms of reporting data. Society expects, in return, that the scientist be worthy of the freedom she is privileged to enjoy.

On the strength of this expectation, the scientist is thus allowed a certain latitude or "discretionary space" in the pursuit of knowledge. The dimensions of that space vary with the nature of the subjects and objects under study and with the potential effects of the study on human society. The precise dimensions within which "autonomy-in trust" can be exercised are manifestly difficult to set. Should research on the atomic bomb have been pursued? Is experimentation on the human fetus or embryo permissible? Are studies of the genetic or racial variations in intelligence warranted? Are there limits to the risks a human subject is permitted to undertake—even voluntarily—in the interests of gaining medical knowledge or helping others?

In the specific instance of clinical research, the dimensions of discretionary space must be narrowly defined. Respect for persons and the imperative of beneficence take precedence over scientific curiosity. If the investigator chooses to be physician and care-giver as well, then he or she must realize at the outset that the needs of the experimental protocol must be subservient to the needs of his or her patient. The multiplicity of possibilities for conflict between the values of science, medicine, and self-interest forces us to reconsider whether it is morally licit for one person to be scientist and physician simultaneously.

Arguments can be marshalled on both sides of this question. Those who favor unification of roles of investigator and care-giver place their trust in the binding power of the traditional ethics of medicine, the trustworthiness of the majority of investigators, and the avoidance of inter-physician conflict when one physician is clearly in charge. Those who would separate these roles hold that there is a clear advantage when one physician is unequivocally the advocate for the patient's welfare, and another for the probity of the experiment. Further, the skills, attitudes, and motivations required for physician/caregiver and physician-investigator roles are different. Different kinds of people are attracted to each role. It is unusual for one person to possess the traits of both.

There is probably no general rule that can cover all investigational situations. It is clear that better surveillance over the possibility of conflicts is in order. In some cases a separation of roles seems advisable. For example, when the investigator receives most of his support from a pharmaceutical firm, owns stock in that company, or receives financial emoluments, separation of roles seems the only morally prudent alternative. This is particularly true if the purpose of the experiment is to evaluate an improvement in an already effective treatment. The same is true when the research is clearly non-therapeutic and of no therapeutic benefit to the patient. Protection of the patient's interest should be in the hands of a physician other than the investigator, one whose prime obligation is to the patient and not the experiment.

On the other hand, when the intent of the research is to try a new agent in a patient with a serious or fatal disease, and the outcome for the patient-subject is the prime focus of the research, then it seems better for the physician giving care and the physician-investigator to be the same person. Here, the relationship between scientific knowledge and the good of the patient is so intimate that it is preferable to have one physician who can discuss and weigh both with the patient.

Clearly, a reassessment of the usually accepted fusion of roles in the physician-investigator is in order. Much depends on the contexts, purposes, and expectations of the research protocol. "Autonomy-in-trust" demands a renewed emphasis on care and precision in identifying and resolving the moral dilemmas inherent in different types of clinical investigations.

The safe rule in every case is to favor beneficence over scientific rigor when the two seem to be in conflict or when in doubt. The possible loss of knowledge cannot outweigh the possibility of harm to the subject even if the utilitarian calculus indicates great benefit to many and harm to only a few. This is the case particularly with the mentally challenged, the very young, the terminally ill, and the senile. Whenever valid consent is impossible and the experiment risky, only therapeutic research aimed at benefit to the subject seems morally defensible. It is especially dangerous to hold, as some ethicists do, that infants and the retarded, for example, owe an obligation to society or to other infants to be subjects of research. No one can presume on the willingness of another to run the risk of experimentation: we cannot presume to attribute activism to those who cannot possibly express their own views on the matter.

This applies to parents as well as guardians of infants and children below the age of competence. Parents do not have absolute rights over their children. Someone must be an advocate for the infant and the child, and this may have to be the physician. He or she may have to protect the child against even well-intended parents who might permit dubious forms of experimentation in hope that it might help others. This is another reason why it is particularly important to distinguish the role of the investigator from that of the care-giver.

Not all experimental procedures in infants and children are ruled out. Therapeutic investigations that could benefit the subject would be licit, provided all other safeguards are scrupulously observed. Non-traumatic studies of a non-therapeutic type—like examinations of body fluids, excretions, secretions, or non-invasive procedures or observations of various sorts—would be licit if they were done in association with other procedures needed in the care of the subject.

Experimentation on fetuses or embryos, spontaneously or intentionally aborted, is a special subject with its own moral complexities. It deserves more than the superficial treatment I can give here. Suffice it to say that everything depends upon what we consider to be the ontological status of the fetus and embryo. In my view, they are deserving of the same respect as the infant or child. Their vulnerability is extreme, and the rule of beneficence is particularly stringent upon those who have the unborn in their care.

But even if the subject is capable of making his or her own choices and is willing to run risks, there are limits on what ought to be permissible. A particularly dangerous recent example is so-called pay-as-you-go research. This is another manifestation of the growing belief that untrammeled competition will redound to the benefit of patients by providing incentives to investigators to try non-standard treatment.

In this kind of research, a commercial laboratory enters into a contract with a person to try out a new treatment. For example, monoclonal antibodies can be developed from the patient's own tumor cells bonded to an anti-tumor agent and then used as treatment. The patient pays all the expenses of the research, plus a profit to the laboratory. In this way, it is argued, bold new treatments will be developed that might otherwise founder in the complex review processes.

The dangers of such research are obvious. All the usual safeguards that surround human experiments are compromised. There is no peer review, no institutional review board or ethics committee, no statistically controlled trial, no monitoring of the results and the dangers. Sharing of information is seriously impeded since investigators are in competition with one another for the market. Finally, there is the serious conflict between the financial interest of the investigator, the interests of good science, and the good of the patient. A more perilous misapplication of the privilege of clinical research can hardly be imagined.

Conclusion

I have cited some examples of the ways in which the physician's role as investigator-scientist may be in conflict with his or her role as care-giver. These examples underscore the need for a more careful scrutiny of the value conflicts possible even in so well-intentioned an enterprise as therapeutic research. They are not reasons for draconian measures to regulate such research. However, there are reasons for a much greater sensitivity on the part of institutions and investigators to the fact that clinical research places restraints on the autonomy of science and self-interest.

The conflicts between the values of science, medical care, and self-interest have implications for the preparation, selection, and supervision of clinical investigations. Many unfortunate occurrences have resulted from the laxity of senior investigators who are involved in the supervision of too many projects and who are overly eager to expand their own publication lists. Some training in ethics should be mandatory for all investigators who are to be involved with human subjects. Some code or explicit set of moral commitments should be developed and subscribed to by all investigators.

Given the emphasis in Western society on autonomy, the rights of patients, and the wariness about abuses of power by science and technology, clinical investigators must recognize the value conflict inherent in their work. Their best guarantee of freedom in the pursuit of medical knowledge and the good of patients lies in research that is scientifically valid and ethically responsible. To compromise either is, on the one hand, to endanger patients and on the other, to lose the privilege of freedom without which science cannot thrive.

11

Balancing Science, Ethics and Politics: Stem Cell Research, a Paradigm Case

Edmund D. Pellegrino, MD

Introduction

Every truly effective technological advance is a challenge to culture, politics, and ethics. This is as true for the discoveries of fire, the wheel, and gunpowder, as it is for the Atom Bomb, space travel, and biotechnology. The more effective the technology, the more serious the challenges it poses.

Thomas Merton, with his acute sensitivity to the malaise of the soul of modern man, put it this way: "the problem of getting technology back into the power of man so that it may be used for man's own good is by all odds the great problem of the day."[1]

At this writing, after the horrendous assault on human life in the terrorist attacks of September 11, 2001, the "problem" of technology is easily displaced by the problem of injustice and violence of man against his fellow man. Yet, on this same day, the National Academy of Sciences released its report and recommendation regarding stem cell research.[2] We are just beginning to confront the challenge of biotechnology, a form of technology more personal, intimate, and more challenging to our concept of what it is to be human than any past technological "advances," with the possible exception of the atom bomb. How do we use our technological prowess humanely, wisely, and generously without being so overshadowed by its powers that we become its slaves?

How in essence do we place biotechnology within ethical constraints without losing its therapeutic potential? How do we do so in a democratic, morally, and pluralistically divided society, driven equally by market forces and a yearning for immortality through technology?

Reprinted with permission from Edmund D. Pellegrino, "Balancing Science, Ethics and Politics: Stem Cell Research, a Paradigm Case," *The Journal of Contemporary Health Law and Policy* 18 (2002): 591–611.

1. Thomas Merton, "The Hidden Ground of Love," in *Letters of Thomas Merton*, ed. William H. Shannon (New York: Farrar Straus Giroux, 1985), 507.

2. National Research Council and the Institute of Medicine, *Stem Cells and the Future of Regenerative Medicine* (Washington, D.C.: National Academy Press, 2002). [Updated.—Ed.]

Stem cell research is a paradigm case illustrating the complex intersections of science, ethics, and politics, which will characterize any powerful new technology. This essay seeks to outline the intersections of science, ethics, and politics through which society shall navigate in the years ahead. The stem cell research issue will serve as an example for analyzing the questions the whole citizenry must confront with the introduction of biotechnological progress—not just scientists, bioethicists, or legislators.

As a physician and a patient, the author sympathizes with the tremendous desire to make rapid progress in the therapeutic use of the fruits of our new knowledge. Persons of good will can differ on the nature, application, and source of ethical constraints on research. The search for biological knowledge—within ethical constraints—should be praised and supported. However, the ethical, political, scientific, and economic issues cannot totally be disentangled. If there is to be some moral order in the approach to policy formation and legislation, then there must be some moral compass points to guide both science and policy. No matter how much human good we envision, we cannot allow our zeal for knowledge, power, profit, or even cure to displace our humanity.

There are three fundamental questions, and sets of relationships, which recur in any legislation or policy regarding the application of biological knowledge in the public arena. This paper will discuss three sets of questions, which anchor the debate and define the moral quality of policy and legislation.

First, what scientific facts support stem cell research and how secure are they? Second, what fundamental ethical issues are at stake? Third, what are the implications of questions one and two for politico-economic policy?

I. What Scientific Facts Support Stem Cell Research and How Secure Are They?

In any situation, good policy and good ethics depend on good facts. Even so, the individuals who gather and assess the facts are not the same ones who must make public policy or legislate. These latter must nevertheless base their decisions on the testimony of experts. They must probe expert testimony critically if they are to avoid automatic capitulation to the aura of authority that follows expertise. This can be extremely difficult to accomplish since scientists may honestly disagree about the same issue. The same data may be subject to different interpretations. This factor neither accuses scientists of insincerity nor deception, but emphasizes that legislators and the public must probe scientific claims carefully if they wish to legislate and regulate wisely. This is especially true when prestige, power, and profit are so evidently at stake as they are in stem cell research.

These concerns are not confined to legislators or policy makers. Ultimately they are matters of concern for every citizen. Bioethics is no longer the restricted terrain of physicians and health professionals, nor of bioethicists. The ethical issues are pertinent for the lives of all of us in the present and in future generations.

Let us look at a few examples of current stem cell research to illustrate this point more concretely. The discussion will focus on expectations of cure, terminology, the utility of existing stem cell lines, and alternate sources.

A. Terminology

Stem cells are unspecialized cells found in the human embryo, fetus (germ cells), and adult. They are characterized by great plasticity, that is, they can, spontaneously or by manipulation, be converted into the many types of specialized cells making up the human body. Those taken from very early embryos are "totipotent," that is, capable of developing into a complete new human being. Others taken at later stages of embryonic development are "pluripotent," transformable into all other types of cells but not into a full and complete human being. Stem cells can also be found in adult human tissues, such as bone marrow, brain, muscle, umbilical cord blood, liver, and fetal tissue.s

"Embryonic stem cells" are usually harvested for experimental or therapeutic purposes from embryos five days old. "Harvesting" these cells inevitably results in the death of the embryo. Thus, whether the embryonic stem cells are toti- or pluripotent, use of them poses two ethical challenges. First, they are obtained through deliberate killing of the embryo, and second, if they are gathered at an early stage, they are totipotent and can develop into complete human beings. Therefore, the source of embryonic stem cells and their age are ethically significant in an evaluation of the legitimacy of their use.

Another manner in which to obtain embryonic stem cells is somatic cell nuclear transfer, or cloning. Through this method, a totipotential cell is produced which could, under the proper circumstances, become a complete human adult. This cell is grown in culture until it is five to seven days old, then its inner cell mass is harvested. This process, again, results in death of the embryo. This is an example of human cloning which is currently subject to a congressional ban. To circumvent this ban, it has been suggested that the term "nuclear transplantation" be used instead. This semantic device does not alter the fact that a human embryo is destroyed for its stem cells.

A new method has recently been proposed for obtaining stem cells through a process called therapeutic cloning. The goal is to clone human embryos containing genes from a patient in hopes that the clone will produce normal cells that can then be implanted in the same patient without an adverse immune system response. The first attempts using nuclear transfer failed. Another attempt using unfertilized eggs artificially stimulated to undergo cleavage produced a blastocoele cavity but no stem cells.

This is the process of parthenogenesis, and the cells thus produced are called "parthenotes." Parthenogenesis occurs naturally in simple species, but it has never been accomplished in mammals. Stem cells from parthenotes do not have the genetic make up to become a full term baby.[3]

3. Eliot Marshall and Gretchen Vogel, "Cloning Announcement Sparks Debate and Scientific Skepticism," *Science* 294, no. 5548 (November 30, 2001): 1802. These scientists include O'Rahilly, Muller, Carlson, Sher, Davis, and Stoess.

Much depends upon whether the stem cells derived from so-called excess frozen embryos are totipotent or pluripotent. This is no mere semantic quibble. If they are totipotent and each can develop into a complete human being, then one is creating, experimenting with, and killing a human embryo. If stem cells are not human embryos and not totipotent, then they can legitimately be used for experimentation and therapeutics. If there is a reasonable doubt, as any honest evaluation of the issue must conclude, the embryo deserves the benefit of that doubt and should not be used. After all, in most matters few of us would gamble what we cannot afford to lose.

Contemporary embryologists regard embryonic cells as totipotent from the zygote stage (the fertilized egg) through the blastula (four-cell stage) and possibly up to the blastocyst (five to seven days). If these dividing cells are separated from each other in these early stages they are totipotent, meaning each one can develop into a complete human being. How far beyond the blastula the embryonic cells remain totipotent or can recover their totipotency is uncertain with the embryologic evidence now in hand. If at any point embryonic cells can become complete human beings, ethical objections arise not just with the source of such cells, but with their use, regardless of the source.

Much also depends upon whether or not the cells of the frozen embryos to be used to replenish existing cell lines are young enough to be totipotent. There is some evidence that some of these cells at least may be at the four- to eight-cell stage or somewhere between the blastula and the blastocyst stage when frozen. If this turns out to be the case, they could not be used under existing Congressional guidelines, which preclude experimentation with human embryos.

This raises the question of how we define an embryo. Authoritative modern biologists[4] regard the embryo to be present from the zygote stage or right after it begins to divide. Some embryologists and bioethicists speak of the "pre-embryo," postulating a stage from the zygote until the fourteenth day before which they do not consider the developing cells an "embryo." There is no biological warrant for such an arbitrary distinction. Rather, it has the earmarks of a convenient means for justifying treatment of the developing human with less respect than one would grant it after fourteen days. In much the same way, it has been recommended that the term "ES cell" be used rather than embryonic stem cell to avoid the stigma of destroying human embryos, thus defining the embryo out of existence through a semantic ploy. The nature of the organism is not changed by changing its name.

Clearly, precision in scientific terminology is far more than an exercise in pedantry or taxonomy. Until the biological status of embryonic stem cells and their potentialities at each stage of development are clearly demonstrated, these cells should be given the benefit of the doubt and protected from destruction. Given the genuine probability that adult stem cells, in association with pharmacological and

4. O'Rahilly, Muller, Carlson, Sher, Davis, and Stoess are representative of these modern biologists.

genetic manipulation, can be as useful therapeutically as embryonic cells, it is morally prudent and mandatory to concentrate on adult cells. This is a morally and scientifically sound position whether or not one favors, or objects to, the use of embryonic stem cells.

Similar controversy surrounds the debate of whether the embryo is classified as a human person. This is an even more complicated question than the question of totipotency or what is an embryo. One need only point out that the zygote is the result of two living cells, and both are human cells. There can be no question of the earliest stages of human life being both human and living. Whether or not the embryo is considered a person is a metaphysical question not susceptible to scientific proof or falsifiability. Any attempt to define "person" in terms of a set of arbitrary biological, physiological, or sentient qualities is dangerous. Embryos, fetuses, and even impaired members of the human species may lack one or another of the arbitrarily specified properties for personhood. Persons in coma, a permanent vegetative state, the retarded, and the brain-damaged human are all at risk of being deprived of "personhood" and of social moral right to protection against harm or destruction.

Alertness to terminology is essential in the stem cell controversy. Those who favor the destruction of embryos as a source want to avoid use of the term "embryo" because they think it is too inflammatory and generates resistance. They suggest a variety of euphemisms: "nuclear transplantation" or "therapeutic cellular transfer" for cloning humans. One ethicist, who heads the ethics panel of one of the involved companies—Advanced Cell—wants to substitute "activated egg" or "cleaving egg" for the term "embryo." Another biologist offers "ovasome."[5]

These euphemisms are intended to confuse the non-scientist public and decision makers and divert their attention from the fundamental fact that human embryos are produced to be destroyed, a practice currently at least subject to congressional ban. All terms used to support or denounce embryo sacrifice should be defined clearly. Non-scientists must not be reluctant to question a scientist who, like other humans, may have a vested interest in avoiding confronting some disturbing ethical challenges.

B. Therapeutic Claims

In the minds of many, including those who have reservations about the use of human embryos, the claims for therapeutic success seem overwhelming. Most of the public's information comes from the popular media, which customarily celebrates whatever is new and sensational, or represents "progress." To be sure, responsible journalists and scientists have reported the scientific promise as well as the limitations of current research fairly and accurately. On the whole, however, the public has been exposed to a species of evangelical advertising, over promise and semantic prestidigitation overstatement that is morally and scientifically irresponsible.

5. Sheryl Gay Stolberg, "That Scientific Breakthrough Thing," *New York Times*, December 9, 2001, 3.

The fact of the matter is that no one has yet been cured by the transplantation of embryonic stem cells. In fact, the one controlled trial of fetal tissue transfer in Parkinson's disease has resulted in failure and even worsening of the disease.[6] One cannot deny the potentialities, the need, and the understandable sense of urgency for new therapeutic hope among patients and families currently afflicted with an incurable disease. Attempts at moderation of expectations, critical analysis of data and the need for time to evaluate results are unfortunately regarded as insensitivity to human suffering, or an anti-science attitude. This attitude fails to take into account the even greater harm of raising expectations unrealistically and then failing to deliver on them. The fact of the matter is that miraculous cures are a long way off in the future.

What has not been sufficiently emphasized in the popular press and the media is the simple fact that clinical application of stem cells is in its infancy and the requisite scientific knowledge is still lacking. Very little is known about the intracellular signals that initiate and direct differentiation of cells. Equally little is known about what turns genes "on" and "off," about what controls expression and silencing of genes, or about the role of chromatin, the protein-DNA complex that helps DNA enter the nucleus or the enzymes that modify these proteins.

A whole new area of investigation called "epigenetics" has emerged to modify our conception of the gene as the unit of inheritance.[7] No doubt as these new complex molecular and enzymatic entities are studied new complexities and new possibilities are certain to reveal themselves. The same can be said of the process of methylation of DNA in both healthy and cancer cells. Much remains to be learned about how the unspecialized stem cells can be made to become the specialized cells before therapy becomes a reality.

Another problem to be resolved is that of immune rejection of transplanted embryonic stem cells. This is the result of admixture with mouse cells now used as a means of accelerating growth. Obtaining stem cells by somatic cell nuclear transfer, or cloning, could prevent this, but it requires the production of a new embryo that would have to be destroyed to harvest its stem cells.

There is an enormous amount of research still to be accomplished in the fundamental cellular mechanisms if therapeutic applications are ever to become a reality. Like the Human Genome Project, the discoveries to date are only the end of the beginning. The public, policy makers, and legislators must recognize the need for increased and superior research whenever they are subjected to public pressure to "do something." That research should focus on adult sources of stem cells—not on embryonic stem cells.

6. D.D. Spencer et al., "Unilateral Transplantation of Human Fetal Mesencephalic Tissue into the Caudate Nucleus of Patients with Parkinson's Disease," *New Eng. J. Med.* 327 (1992): 1541, 1548.

7. Elizabeth Pennisi, "Behind the Scenes of Gene Expression," *Science* 293, no. 5532 (2001): 1064.

C. Limitations of Existing Cell Lines

In August 2001, President Bush ordered that federal funds be used to support stem cell research only under specified conditions. Specifically, embryonic stem cells had to have been derived prior to August 9th, the donors had to provide informed consent, the only approved source for embryonic stem cells had to be "excess" embryos created solely for reproductive purposes and no financial inducements for donors and funds would be provided for research on adult stem cells.

Since the president approved the use of federal funds for research in existing stem cell lines, new controversy has erupted about the number, viability, utility, and sufficiency of the forty to sixty cell lines. The actual number is in dispute. Recent newspaper articles provide conflicting information. A survey of fertility clinics, for example, indicates that few "parents" of frozen embryos have come forward to offer them for research.[8] Moreover, the number of cell lines actually in a viable or useful state is highly uncertain. One large source cited by the NIH is in Sweden; a spokesman from that laboratory has publicly stated that the number of available cell lines is far too high and that the number in a useful condition may be much smaller.

In addition, it is reported that perhaps a majority of the frozen embryos to be used to cultivate cell lines come from older patients and might not be of sufficient quality or viability to generate useful cell lines. Also, as cell lines age, harmful mutations occur. This is true of frozen embryos as well. Optimism about existing cell lines is also dampened by the fact that existing cell lines had to be "coaxed" to grow by incubating them with mouse cells. This raises the question of immuno-rejection problems and the possible introduction of some unknown virus into human recipients. All of these factors are coupled with a complicated and discouraging consent process. These factors have already raised a clamor for many more and newer cell lines, for fresh and better-prepared cell lines, and for relaxing the prohibition against creating new embryos as the source for those new cell lines. Progression down the slippery slope toward unrestrained research with embryonic stem cells is already underway.

Another limitation on the use of existing cell lines has evoked little discussion so far.[9] Most of the existing lines have come from small segments of the human population. In the West, most of the cells now available come from the U.S., Sweden, and Israel; in the East, they come from South and East Asia. This is a limited representation of the racial, genetic, and ethnic diversity of today's world.

It is true that humans share most of their 30,000 or so genes. However, there are some subtle yet important differences in their expression to be understood in the face of the propensity of certain populations to exhibit high prevalence of certain diseases, such as diabetes, hemoglobinopathies, hypercholesterolemia, and hypertension. Similar differences exist in the response to a variety of medications or in the immune

8. Gina Kolata, "Researchers Say Embryos in Labs Aren't Available," *New York Times*, August 26, 2001, 1.

9. Jon Entine and Sally Satel, "Race Belongs in the Stem Cell Debate," *Washington Post*, September 9, 2001, B1.

reaction to transplanted cells. A new field of pharmacogenetics has emerged to study some of these differences in drug reactions by individuals and ethnic groups.

The narrowness of the ethnic and genetic spectrum of existing cell lines means that research on fundamental mechanisms may not be readily transferable between and among ethnic groups. If there were therapeutic benefits forthcoming, they would be limited, at present, to those of European and some small number of Asian ancestries. This differentiation is also a matter of justice. For these reasons, the political pressure to expand and extend stem cell research beyond the lines approved by President Bush will increase. This in turn will generate demand for making and destroying more embryos as sources for the needed cells.

Clearly, many scientific questions remain to be investigated before stem cells from any source can fulfill the enthusiastic promissory notes made for their curative posers. What are the best conditions for growing these cells and shaping the directions their development will take? What genes control the way the stem cells grow and differentiate? What are the immediate and remote steps in differentiation? What turns the genes "on" what turns them "off"? How can immune rejection of injected cells be prevented? Even more fundamental is the series of critical questions arising from the emerging field of epigenetics, the realm of research examining the place of chromatin and histones in influencing the DNA and its functioning. This is a relatively new sector of cellular biology that may reveal it has as much, or more, influence on how genes function than DNA itself.[10]

These are all complex questions, which should be answered before acceptable treatment protocols become practicable. Answering these questions will require many stem cells and many experimental observations in order to analyze them satisfactorily. As the overestimates of the utility of existing cell lines suggest, many more new cell lines will be demanded.[11] Pressure to relax the president's conditions, particularly his restrictions on the production of new embryos, will be extremely difficult to resist. If promising therapeutic reports come from experimental animal models of disease, then the "slippery slope" will be further greased.

These difficulties make it necessary to focus new attention on alternate sources for stem cells, chiefly those not involving the death of human embryos. It is providential that President Bush has set aside federal funds to support research on alternative sources of stem cells derived from tissues as varied as bone marrow, placental cord blood, muscle, bone, and brain. Animal studies indicate success in converting mouse bone marrow cells into brain cells (glia), mouse skeletal muscle into all major blood cell types even on second transplantation, adult marrow cells into brain cells, human stromal cells into rat neurons, and neural cells into lymphoid and myeloid cells. The scientific community should vigorously pursue all of these possibilities.

The possibilities and limitations of adult stem cells must be appreciated as well. Some of their advantages based on recent investigations include (1) stem cells of several

10. Pennisi, "Behind the Scenes of Gene Expression," 1064–70.

11. National Academy of Sciences, "Stem Cells," report, recommendations l and 2.

types are already predisposed biologically to generate a particular tissue and thus might be able to produce that tissue or its components more readily than less specialized stem cells; (2) they secrete growth factors, which might facilitate growth of other cells as well in a transplanted tissue site; (3) some adult stem cells have been shown to migrate to an injured tissue or other sites, for example, neutral stem cells to brain tumor sites; and (4) they might be genetically engineered to produce other compounds normally needed, but deficient, or even to secrete a drug for therapeutic purposes.

Along with these advantages are certain current disadvantages when using adult stem cells: (1) they are difficult to isolate; (2) their plasticity might be minimal; (3) it is not known how the culture environment affects their functions; and (4) they may not remain undifferentiated or even fail to become differentiated into functional cells. These questions can be resolved by intensive research on adult stem cells, which would be ethically permissible since such research does not involve the death of embryos.

There is also the possibility that embryonic stem cells are not experimentally more beneficial than cells from these adult sources with regard to their plasticity or the possibility of transformation into useful cell types. The emphasis placed on the supposed superiority of embryonic stem cells may even be misplaced on strictly scientific grounds. Attention must be turned to other sources that lack the moral objections that accompany the use of embryonic stem cells.

II. What Ethical Issues Are at Stake?

As difficult as the assessment of scientific facts may be, stem cell research is further complicated by the need to place those facts within an ethical framework. The mere fact that a therapeutic benefit may be scientifically demonstrable is insufficient to justify its adoption. The challenge in a technologically oriented society is to use capabilities wisely, humanely, and within ethical constraints. In a morally pluralist society like ours, the identification, validation and application of moral norms is laborious and controversial. Often, no ethically satisfactory compromise is possible. Regardless, every effort must be made to confront the ethical questions first.

Certain foundational ethical norms and principles must be examined to make this analysis. Admittedly, they may be defined, interpreted, or applied in different ways. Nevertheless, a few moral principles will make their appearance recurrently in the debates. These principles have been prominent in the paradigm case of stem cell research. The ordering of science in relation to politics must be done in an ethical framework, or its decisions will simply be reduced to expediency or exigency.

A. The Ethics of Science

The first ethical norm is the norm of good science, or science that fulfills the obligations of the scientist as scientist. Much of what has been discussed thus far in this essay relates to the ethics of good science, establishing certainty with the factual foundations of the information supplied by one's colleagues, the public, and the policy makers. This means accuracy and honesty in both reporting and sharing those results so that they can be falsified or validated by other investigators. It also means providing negative as well as positive evidence for one's hypothesis or experimental results.

In the case of therapeutic research, special attention must be placed on these requirements for several reasons. First, the anxiety and eagerness for a cure for those afflicted with any illness for which a new treatment is described must be considered. Second, the implication for public policy if a new technique or treatment will be made widely available is another factor. Third, the whole realm of ethical implications to society when the data focuses on control of some aspect of the beginning and ending of human life is also significant. Economy of pretension is a particular virtue for the responsible scientist.

B. Respect for the Dignity of the Human Being

The moral use of biotechnology must recognize the inherent worth of the human being as a human, and as a being capable of a rational, self-determining existence. Most of the rights in the United Nations Declaration are expressions of this principle, such as the right to life, liberty, freedom from oppression, equality of opportunity, and access to health care.[12]

In the stem cell controversy, the focal issue is the moral status of the embryo. At what stage of human development do we impute respect for the dignity of that embryo? It does not help to evade the issue by the invention of euphemisms like the "pre-embryo" or the "ES cell" for embryonic stem cells. Nor does it help to reduce a metaphysical concept like personhood to a biological construction. Personhood is not definable in terms of length of gestation, development of certain anatomical structures, or the capacity for sentiency or social relationships.

Such definitions are far too often semantic contrivances designed to escape the stigma of destroying a human life in its earliest and most vulnerable stages. Moreover, these definitions would also exclude from personhood those persons afflicted by mental retardation, senility, or lack some modicum prescribed of physiological or other sociological characteristics arbitrarily defined.

From the philosophical point of view, the embryo is a human being in the state of active potency. That is to say that from its inception, the embryo is set on a course of development, which will actualize its potentialities. Unless interrupted by man or nature, the embryo possesses the characteristics essential to a human being at the earliest stages of its development. It is actually, not just potentially, a human being. Consequently, destruction of an embryo is the destruction of a human being. Stem cell research, any method of which relies on the death of the embryo as a step in stem cell retrieval, is morally wrong. This contravenes use of fertilized zygotes, embryos, or clones human ova. Use of adult stem cells would be licit unless they were made first into embryos.

12. Universal Declaration of Human Rights, 1948–1998.

C. Avoiding Evil Even if Good Comes of It

In a highly pragmatic society like ours, the probability of a good result for many, or for a noble cause, may well override the respect owed humans. Harm to some humans is thus justifiable if it will help others or benefit the common good. Human decisions are made with some good in mind, therefore what is evil is always done under the appearance of good. This is the argument of those who justify destruction of human embryos on utilitarian grounds to obtain cells for experimental or therapeutic purposes. The "special" nature of the embryo will be "recognized," but it will be set aside if its death can bring life to others. This is the argument now being used to justify relaxation of the congressional and presidential ban on producing new embryos or establishing new cell lines. It is also essentially the argument used by the president's National Bioethics Commission to justify embryonic stem cell research.

What is at issue here is a choice between two theories of ethics—consequentialism and deontology. The former argues that the morality of an act depends upon whether its consequences are so good that they outweigh the wrong necessary to bring those consequences about. The deontological view holds to the contrary, that certain acts are so wrong in themselves that they may never be done even if good may come of it. Of course, if one holds that there is no wrong done in destroying very young embryos in the first place, then the debate becomes moot. In any case, the question must be faced that some choice must be made between truly respecting the human embryo and overriding that respect when the embryo's death provides significant benefits for others.

D. Avoiding Complicity

Even if one does not do harm oneself, one may share in the guilt of an ethically improper act or decision. The degree of moral complicity in a morally wrong act will depend upon whether or not the wrong intent is shared and how closely one facilitates the wrong act. The resulting "moral distance" is crucial in separating morally licit from illicit cooperation. This is not the place to review the well-established distinctions between different degrees of moral complicity.[13] Better familiarity with the ethics of cooperation is requisite if moral judgment is to be made about the use of embryonic stem cells obtained by the destruction of embryos, even if one does not do the actual destroying oneself.

This is necessary because some investigators have tried to escape association with the killing of embryos to obtain their stem cells or avoid the congressional sanction against such use with federal funds. They do so by using cells obtained in this manner under private auspices. Nevertheless, using cells obtained from the death of human embryos is an instance of moral complicity. An individual shares the intent for which these cells were harvested when one uses them for experimental or

13. Orville N. Griese, *Catholic Identity in Health Care: Principles and Practice* (Braintree, Mass.: The Pope John Center, 1987), 373–419.

therapeutic purposes. One also encourages further harvesting by providing an outlet for the cells thus obtained.

E. The Reality of the Slippery Slope

Once a major moral principle is compromised, it becomes incrementally easy to extend that compromise to analogous or similar cases. Logically, one finds it difficult not to assent to extension of the compromise to similar related cases. With each incremental compromise the principle becomes progressively diluted until it becomes all but non-existent. Little is left that can be defended.

Once a principle is relaxed, it becomes easier psychologically to accept compromises even of morally robust norms. It is always the first infraction that is most difficult. Moral revulsion, once some individuals are able to overcome it, loses its power to restrain others. Moral revulsion is easier to overcome with each subsequent compromise. Although many ethicists revile the slippery slope and label it a myth, they fly in the face of logic, psychology, history, and fact when they do so.

One need only recall here the history of *Roe v. Wade* to see this principle in action. First, abortion was limited to the first trimester, then allowed in the second and third, and then to late-term pregnancy. Going from the evacuation of a week old fetus to the evacuation of the living brain of an easily identified human baby in late-term abortion is the inevitable result. Similarly, assisted suicide and euthanasia in the Netherlands began with certain restrictions for physical illness and "unbearable" suffering. Today it has been extended to children, infants, and psychological suffering as well. All this has occurred despite regulations and laws designed to control abuse.[14]

Already, the National Academy of Sciences has stated that research will be impeded if it is confined to currently available cell lines, as the president's decision requires. The National Academy emphasizes that mutations occur as cell lines age, and become less useful as a result. In addition, current cell lines have been admixed with animal cells, raising the probability of immunogenetic responses if they are transplanted into humans. Furthermore, the academy report asserts that private entrepreneurs are unlikely to develop new lines of cells until the profit potential is more apparent.[15]

In the stem cell controversy, there is already mounting pressure to use federal funds to support the use of cells obtained from killing embryos under private auspices not supported by private funds. Already the quality, quantity, and variety of existing stem cell lines is insufficient, and more embryos must be created to make needed research possible. In the same way the actual totipotentiality of embryonic stem cells is being re-interpreted as pluripotentiality—ostensibly to avoid the stigma of destroying cells from early embryos each of which is capable of, on its own, becoming

14. E.D. Pellegrino, *The False Promise of Beneficent Killing in Regulating How We Die*, ed. Linda Emanuel (Cambridge, Mass.: Harvard University Press, 1998).

15. Rick Weiss, "Stem Cell Research and the Future of Regenerative Medicine, Broader Stem Cell Research Backed," *Washington Post*, September 11, 2001, 1.

a complete new human being. The slippery slope is as applicable to compromise in the meaning and use of words as it is with respect to principles and actions.

This is not, nor is it intended to be, the sum total of relevant ethical principles requisite to an analysis of policy formation and legislation regarding biotechnology. The goal has been to illustrate some of the ethical issues in assessing the ethical aspects of legislation and regulation.

III. What Are Political and Economic Issues?

Any effective technological innovation applied over a wide range of population must, in a democratic society, be subjected to some societal regulatory mechanism. In the ordinary course of events, this will mean legislation and regulation. In the end this will involve some limitation on the use of the technology in order to protect the common good of all society. Deciding what should be legislated, and how, will mean that public policy must be derived from a careful balancing of scientific knowledge, ethical principle, and constitutional rights.

In this respect, stem cell research is a paradigm case. The difficulties of striking a reasonable balance were illustrated on August 6, 2001, when President George W. Bush announced his decision on the use of federal funds for stem cell research. The president attempted, with great sensitivity, to balance the putative therapeutic benefits of stem cell research with the respect owed the life of the human embryo. He also tried to consider the divided opinions of Americans on these issues. As a result, he struck a politically attractive compromise of the ethical and scientific issues.

On the one hand, the president affirmed the sanctity of the life of the embryo by forbidding further destruction of frozen embryos. On the other, he permitted use of existing cell lines, which had been harvested after the destruction of human embryos. He established a commission to examine ethical issues as they arise, authorized funding for research in adult stem cells not involving embryo destruction, and forbade human cloning for any reason. The president's rulings applied only to the use of federal and not private funding.

As the debate resumes, the focus will be on the ethical principles mentioned earlier in this essay and, in the public arena. It will also involve the principles of freedom of choice and social justice. State and federal legislators must confront the practical problem of protecting the rights of freedom, personal choice, privacy, and conscience. They must also consider the life of the embryo, while avoiding exploitation in the use of the knowledge and power of this new technology.

There is no way such power can be left entirely to the individual choice of investigators, entrepreneurs, or private persons. To argue that the pursuit of knowledge, even if ostensibly for human good, should be unrestricted is to submit the public good to the fortuitous will of a small group of experts and entrepreneurs. Some undoubtedly will be ethically responsive, while others will not. History and the frequency curve of the distribution of the moral sense among people do not justify such naive confidence.

In any case, personal freedom and autonomy are neither ethical nor legal absolutes. To live in society peaceably and safely, certain restrictions on both have always been necessary in civilized societies. However, once public policy has been decided, the choice between using a new technology or not using the results of stem cell research will reside with the individual. Those who find a policy morally reprehensible should be free, within the methods available in a democratic society, to object and to work for changes in the policy.

A clear distinction must be preserved between what is legislated and legal and what is moral. Legality does not confer moral status upon stem cell research any more than it confers moral status upon abortion or euthanasia. Given the moral and religious diversity of present-day America and the greater diversity of its future, we must expect to live with such differences peaceably. We must depend on persuasion, debate, and clarification of differences when public policy violates our sense of what is morally permissible.

This surely seems to be the future for such technological innovations as stem cell research, genetic engineering and therapy, and perhaps even cloning. On the other hand, if a particular technology is deemed ethically wrong, then it should be forbidden in the private as well as the publicly supported sector. Surely, the morality of stem cell research cannot depend on the source of its funding. In any case, each of us will be challenged to determine whether we can in good conscience make use of knowledge and techniques obtained from morally improper sources of stem cells.

Whatever decision is taken as public policy, it must be arrived at and administered justly. This means conforming to the requirements of both commutative and distributive justice. Even when the decision regarding what to legislate is taken, the decision must also be made to legislate justly. This means conforming to the principle of justice—rendering to each his due, equally to equals and unequally to unequals. This principle operates at the level of individual relationships where it is called commutative justice and at the societal level as distributive justice. Only a few examples of each will be mentioned.

In the realm of commutative justice lie the moral obligations of individual investigators to individual patients and experimental subjects and of physicians to individual patients. Investigators are, in justice, obliged to obtain properly informed consent; to be objective in reporting their results; to share results that are crucial to the health of the population with the public and other investigations. Premature and inflated reports of success in order to gain a competitive advantage are not morally admissible. These obligations are derived from the trust society and persons must place in the expertise of physicians and scientific investigators. Ensuring these and other aspects of commutative justice is a fit subject for regulatory consideration in stem cell research legislation and policy.

In the realm of distributive justice lie the broader political and economic questions of availability, access, distribution, quality, cost, and ownership of any cell lines that may eventually prove useful. The commercial possibilities already promoted by biotechnology companies will attract interest among investors, who will be entitled to some return on their investment. The costs of research and development will have to be returned through the price of cell transfer treatment. If stem cells should prove to be effective, the high cost of treatment will become a moral issue. Disproportionate access based only on ability to pay cannot be just in the case of a life-saving or curative treatment. Government subsidies for the economically less advantaged will be necessary. In the current commercialized climate of American health care, assurance of justice in distribution will surely become a matter for legislative action.

This question of distributive justice is especially acute in stem cell research since the cells available for federal funding come from predominantly European and a small number of South Asian sources. This is a small representation of the ethnic diversity of our world. The research currently permitted may not be transferable and consequently may be unavailable to many ethnic groups in the U.S. and the world. Yet on the basis of equity and justice as well as humanitarian grounds, it would be unjust not to make these benefits available to all.

The problem of ability to pay is linked to the limited sources from which existing cell lines have been drawn. They derive from "excess" embryos resulting from in vitro fertilization, a costly process denied to the poorer members of our society. If it turns out that stem cells can be engineered to reflect the unique qualities of the humans who donate them, then the economically less fortunate will again be left out. The same problem arises when embryos are cloned to meet the specification of the donors for cells of a particular type. Custom-tailored stem cells will likely be costly and well beyond the means of the poor.

These questions will become particularly acute should there be a shift in our healthcare system to one of universal entitlement. Legislators will be heavily involved in establishing principles of fair distribution and access. Justice and fairness are not intrinsically matters of expediency, and some concept of social justice must be evolved for access and payment when truly effective biotechnologies appear.

The question of monopoly already looms as a genuine issue. One corporation, for example, controls the rights over many of the stem cell lines now eligible under President Bush's plan for federal funds.[16] The same corporation, Geron of Menlo Park, California, owns commercial rights to the technology that produced Dolly, the first cloned sheep, as well as rights to telomerase. Telomerase is an enzyme that can restore the telomere to full length in vitro and seems to extend the age of cells. Inhibiting telomerase might be useful in inhibiting growth in cancer cells. Geron at present is making little or no profit on these possibilities, but its CEO, Thomas Okarma, says his company aims "to dominate the market."[17]

Further, the patent on embryonic stem cells is held by the University of Wisconsin Research Foundation, and Geron Corporation holds commercial rights to six cell types and seeks to exercise its options over twelve more. Okarma expresses

16. Andre Pollack, "The Promise in Selling Stem Cells," *New York Times*, August 26, 2001, 1.

17. Ibid., 11.

a view not rare among investigators in this field, "I am not apologetic for our intellectual property. We paid for it, we earned it, and we deserve it."[18] To what extent does this frank assertion of property rights over biological material conflict with what is owed in justice to individuals and society? Who really "owns" the cell lines from a "donated" embryo?

Geron also has plans for "therapeutic cloning," creating new embryos from a patient's own cells to obtain stem cells specific to the donor, thus avoiding the dangers of immuno-rejection. Another company, Advanced Technology of San Francisco, has been formed to compete with Okarma. Clearly, the question of how much can be left to the market and how much is to be controlled in the public interest already presents itself. In a nation already commodifying health care, the likelihood of extension of commodification to human cells is unfortunately high. The fact that Geron Corporation also has its own "ethics committee" is not at all reassuring. Money, profit, and ethics are not natural allies.

All of this is now in the private sector, which is not covered by President Bush's recent decision. Ethical and legal constraints are at the moment not available to protect either the embryo or the public interest. Indeed, those who work within President Bush's constraints, like NIH and university investigators with federal grants, look to private industry and research to supply the cell lines they believe they need to continue their research.[19] One need only mention these facts to appreciate how complex and intermingled are the ethical, scientific, and socio-political issues we as a nation must face.

In a recent update, the National Institute of Health expressed confidence that patenting of new discoveries would not affect research. However, it mentioned neither the ethical status of such patenting nor of its impact on the public interest. Are stem cell research and the products it generates simply commodities? Or are there humanitarian and human rights dimensions that are being ignored in pursuit of the principles of utility and market economics?

Important as these considerations may be, they leave aside the question of whether there can be two standards of morality: one for the private sector and one for the federally funded sector. If it is morally wrong to create new embryos to obtain stem cells on federal grants, how can it be morally permissible to do so with private funds? This is a rather extreme example of the difficulties created when morality is left entirely to social convention or convenience.

Very serious questions will arise in the realm of professional ethics. Currently, this field, which deals with the obligations of physicians and other health professions, is already in flux with many calling for a "new" ethic. What will the professional obligations be of physicians who see a grave moral problem in providing treatments obtained from embryonic stem cell research dependent on destruction of the embryo?

18. Ibid.

19. Ceci Connolly and Justin Gillis, "Thompson: Stem Cell Work Viable," *Washington Post*, September 6, 2001, 1.

This question will become intensely debated if stem cell treatment becomes genuinely effective. The pressure on physicians to make stem cell treatment available will be enormous. Must physicians be "value neutral" as some contend and submerge their personal ethical beliefs and provide treatment simply because they are physicians? Is it even possible to be "value neutral" in this arena? Will compliance with a socially accepted treatment be a condition of licensure regardless of the physician's personal beliefs? What will be the impact on this question of universal healthcare entitlement if this becomes a policy? Will physicians morally opposed to the destruction of embryos be barred from practice or even entry into medical school?

The same questions will apply to religiously sponsored hospitals and other healthcare institutions. Will Catholic hospitals, for example, be allowed to receive public funds or reimbursements if they do not provide a full range of services? Will a choice have to be made between religious sponsorship, institutional integrity, and survival? Will dissenting hospitals and physicians be limited to caring for those who share their beliefs only?

These issues touch on the much deeper question any policy must face in a morally pluralist society. How are we to live peaceably and protect the human rights of all citizens when we are "moral strangers," that is, members of communities whose ethical and or religious beliefs do not permit certain compromises? Stem cell research is a paradigm case of the kind of questions that will also rise with the actualization of some of the promise of the genome project, human cloning, and biological innovations as yet unimaginable.

These are some of the inter-related and intermingled questions which confront our society and which will generate demands for public policies, regulations, legislation, and accountability. They are inter-related in the nexus of science, ethics, and politics. Clearly, in the real world, these issues are not so easily dissectible. Yet if they are to be examined, the nature of each must be recognized and a proper balance struck between them.

In striking a balance, the ethical considerations should guide both the use of science and the economic and political uses of that science. This means that decisions may move more slowly than interested parties, patients, biotech companies, and scientists desire. If anything is clear in the example of stem cell research, it is that haste, overselling of results, and political pressure can lead to more conflict rather than less. The debate has only begun. All of the general public will be either victims or beneficiaries of our collective deliberations. Let us hope that they are morally sound, scientifically correct, and politically just.

12

Stem Cell Research: Why Medicine Should Reject Human Cloning

William P. Cheshire, Jr., MD; Edmund D. Pellegrino, MD; Linda K. Bevington, MA; C. Ben Mitchell, PhD; Nancy L. Jones, PhD; Kevin T. FitzGerald, PhD; C. Everett Koop, MD; and John F. Kilner, PhD

SPECULATION THAT THE WORLD'S FIRST HUMAN CLONE may have been born,[1] combined with reports that human embryos have been cloned for research purposes,[2] calls for careful public and professional scrutiny of the critically important matter of human cloning.

Human cloning is the asexual production of a human being whose genetic makeup is nearly identical to that of a currently or previously existing individual.[3] Whereas the deliberations of international, national, and state regulatory bodies have, in most cases, favored the prohibition of what has been called reproductive cloning—in which a cloned human embryo is created with the intent that a human clone will be born—they have differed considerably over what has been termed

---

Reprinted with permission from Elsevier from William P. Cheshire, Jr. MD, Edmund D. Pellegrino, MD, Linda K. Bevington, C. Ben Mitchell, PhD, Nancy L. Jones, PhD, Kevin T. FitzGerald, PhD, C. Everett Koop, MD, and John F. Kilner, PhD, "Stem Cell Research: Why Medicine Should Reject Human Cloning," *Mayo Clinic Proc* 78 (2003): 6–14. © 2003 Mayo Foundation for Medical Education and Research. Reprinted with permission from Elsevier.

1. R. Roth, "U.S. Doc Makes Human Cloning Claim," *CBS News,* January 20, 2004, https://www.cbsnews.com/news/us-doc-makes-human-cloning-claim/; M. Ritter, "Company Claims Birth of Human Clone: Experts Are Skeptical," *The Middletown Press,* December 28, 2002, https://www.middletownpress.com/news/article/Company-claims-birth-of-human-clone-experts-are-11941297.php.

2. J.B. Cibelli, R.P. Lanza, M.D. West, and C. Ezzell, "The First Human Cloned Embryo," *Sci Am.* 286 (2002): 44–51; J.B. Cibelli, A.A. Kiessling, K. Cunniff, C. Richards, R.P. Lanza, and M.D. West, "Rapid Communication: Somatic Cell Nuclear Transfer in Humans: Pronuclear and Early Embryonic Development," *E-biomed J Regenerative Med.* 2 (2001): 25–31.

3. National Bioethics Advisory Commission, *Cloning Human Beings* (Rockville, Md.: The President's Council on Bioethics, June 1997), 17–18; President's Council on Bioethics, *Human Cloning and Human Dignity* (New York: PublicAffairs, 2002).

research cloning. Research cloning involves the creation of a cloned human embryo for the purpose of scientific investigation of early human development or for medical research aimed at developing treatments for disease. Because embryonic stem cells are pluripotent, having the capacity to differentiate into the full range of human tissues, some believe that these cells hold the potential to revolutionize medicine by providing a source of replacement tissue that might one day restore the health of persons suffering from a variety of debilitating conditions.[4] Transplanted embryonic stem cells derived from a patient's clone may be compatible (as would adult-derived stem cells from the patient) with that patient's immune system and hence, in principle, be resistant to immune rejection.

Our contention is that human cloning should not be permitted, whether for research or reproductive purposes. While we enthusiastically affirm the importance of medical research and ardently support the goal of healing people, we believe that the harm human cloning would bring to medicine would exceed the anticipated benefits.

Reproductive Cloning

An overwhelming majority of scientists, healthcare professionals, policymakers, bioethicists, theologians, and the general public have indicated their opposition to the birth of cloned human beings.[5] The following concerns have been advanced.

Human cloning would be hazardous to the gestating clone and the surrogate mother. The current state of nonhuman animal cloning technology is so rudimentary that the procedure has resulted in a staggeringly high occurrence of severe physical and genetic defects and premature aging in cloned offspring.[6] Embryologists estimate that a single successful human cloning might come at the cost of hundreds of failed attempts.[7] Even if issues of safety were overcome, which is unlikely apart from unethical human experimentation, compelling ethical objections remain.

Human cloning would signify an egregious disrespect for personal autonomy. In forcing on the human clone a selected identity bound to certain, perhaps unfulfilled, expectations placed on the genetic original, cloning would frame that person's life and limit that person's autonomy permanently. Cloning would also encumber that person with profound emotional burdens. The cloned individual would not be born with the special privilege of having a unique genetic identity, but rather would

4. Committee on the Biological and Biomedical Applications of Stem Cell Research, Board on Life Sciences, National Research Council, Board on Neuroscience and Behavioral Health, Institute of Medicine, *Stem Cells and the Future of Regenerative Medicine* (Washington, D.C.: National Academy Press; 2001).

5. Kass, *Human Cloning;* R. Jaenisch and I. Wilmut, "Developmental Biology: Don't Clone Humans!" *Science* 291 (2001): 2552; Lydia Saad, "Cloning Humans Is a Turn Off to Most Americans: Embryonic Cloning for Research Is Also Opposed," Gallup poll analyses, *Gallop News Service,* May 16, 2002.

6. S.P. Westphal, "Cloned Monkey Embryos Are a Gallery of Horrors," *New Scientist,* December 12, 2001; I. Wilmut, "Are There Any Normal Cloned Mammals?" *Nat Med.* 8 (2002): 215–16.

7. Jaenisch and Wilmut, "Developmental Biology."

always live in the shadow of the other person whom he or she was intended to duplicate genetically. The social stigma of being known as a clone, combined with confused parentage and expectations of measuring up to the achievements of the genetic original or of "replacing" a deceased loved one, could result in unimaginable psychological turmoil.[8] If cloning became common practice, its deviation from the traditional design and accompanying moral responsibilities of the human family might well disrupt social stability.

Moreover, cloning brings to mind images of assembly line manufacture more suited for the making of replaceable appliances than unique human beings. Deeply held public intuition thus regards the prospect of human cloning to be a repugnant departure from the intimate and richly meaningful process of natural procreation. Unlike other reproductive technologies that assist procreation, cloning seeks to produce a human being with a particular genetic code. It is not technology that we oppose, but rather the misuse of technology that enables some people to exert nearly absolute control over the genetic makeup of others. This substitution of human genetic replication for procreation would constitute a serious affront to human dignity.[9] If cloning proceeds along its current path of development, it will foster a grave devaluation of humanity. Whether cloning were to become a widespread or an occasional practice, its acceptance would shift societal attitudes away from appreciating people as distinct individuals and toward a new way of sizing up people as useful or attractive commodities of technology assembled to satisfy others' expectations.

Some will defend human cloning as a right of reproductive liberty that ought never be restricted. However, there exists no inalienable right to engage in human cloning as a means of realizing one's desire for a child, regardless of the particular motivation behind such a desire.[10] Furthermore, although reproduction is a private matter, development and implementation of genetic technology on which reproductive decisions will be based are matters of definite public interest. No reason has been advanced that is weighty enough to justify overlooking the considerable hazards described herein and resorting to cloning as a means of human reproduction. The disturbing dangers of human cloning to public health and well-being should be of concern to physicians in particular because the menacing key to this Pandora's box is a medical procedure.

Implications of a Partial Ban

Proposals to ban human cloning for purposes of reproduction have attracted broad support. However, enacting a ban solely on reproductive cloning, while simultaneously permitting research cloning, would almost certainly fail to achieve its stated objective. For the following reasons, we contend that a partial ban could well

8. Kass, *Human Cloning.*

9. W.P. Cheshire, "Toward a Common Language of Human Dignity," *Ethics Med.* 18, no. 2 (2002): 7–10.

10. C.D. Forsythe, "Human Cloning and the Constitution," *Valparaiso Univ Law Rev.* 32 (1998): 469–542.

result in instances of both types of cloning, leading to a society that most Americans would deem undesirable.[11]

Unenforceability.—First, a partial ban would be unenforceable. If a ban on reproductive cloning only were adopted, enforcement would necessarily entail the legally mandated destruction of human embryos created for research cloning. Such required destruction would not only constitute a form of clear discrimination against a class of human beings based on the means of conception but also would likely be objected to or wholly disregarded by many, particularly by those who desire to implant the embryos. Because the legality of terminating the lives of unborn human beings by abortion is frequently defended as a matter of personal choice, it is difficult to imagine that most Americans would welcome a governmental policy that mandated the destruction of embryonic human life and the punishment of those who defied the law (either by knowingly implanting a cloned human embryo or by giving birth to a human clone).[12] Such acts of defiance would be viewed by many as the private exercise of a reproductive option entitled to certain protections, effectively circumventing a partial ban. Although we do not believe that people have a right (rooted in reproductive liberty) to create human beings via cloning, we nevertheless maintain that parents should never be forced to destroy their offspring, *once created,* regardless of their method of origin.

Currently, the parents of embryos created via in vitro fertilization (IVF), for example, are given a great measure of decision-making power regarding the fate of their embryos. Some fertility clinics exceed clinical policy requirements in their efforts to determine parents' wishes regarding stored embryos, and all clinics are obligated to honor decisions both for and against implantation.[13] Our autonomy-steeped culture would surely have difficulty accepting a policy that would deny people the same choice simply because their embryos were created through cloning.

Regardless of their legality, both IVF and reproductive cloning are technologies that lie within the realm of reproduction. Because of the private context of reproduction and the under-regulation of the US fertility industry, prohibiting the implantation of cloned human embryos would be a formidable task met with considerable resistance, regardless of whether reproductive cloning is legally permissible. Of importance, although public consensus favors a law prohibiting the reproductive cloning of human beings, continued legislative stalemate on proposals to adopt a comprehensive cloning ban prohibiting both research and reproductive cloning might mean that even reproductive cloning would remain legal.

11. J.F. Kilner, "Human Cloning," in *The Reproduction Revolution: A Christian Appraisal of Sexuality, Reproductive Technologies, and the Family,* ed. J.F. Kilner, P.C. Cunningham, and W.D. Hager (Grand Rapids, Mich.: WB Eerdmans Publishing Co., 2000), 124–39.

12. L.R. Kass, "Preventing a Brave New World: Why We Should Ban Human Cloning Now," *New Repub.* 224 (2001): 30–39; L.B. Andrews, *The Clone Age: Adventures in the New World of Reproductive Technology* (New York: Henry Holt & Co., 1999), 74.

13. ASRM Ethics Committee, *Disposition of Abandoned Embryos* (Birmingham, Ala.: American Society of Reproductive Medicine; 1997).

If cloned human embryos were created in the laboratory for research purposes only, the mandate that they not be implanted or otherwise allowed to progress toward birth would prove extremely difficult to uphold. Therefore, the birth of cloned human beings—the very thing that a ban on reproductive cloning should prevent—would likely result.

Compassionate Transgressions.—Second, if cloned human embryos were available for research, appeals to compassion within the privacy of the physician-patient relationship would likely lead to their implantation. Consider the following hypothetical scenarios.

A cloned human embryo is created with the intent of producing tissue needed to save the life of a seriously ill child. Before the tissue can be obtained, the ill child dies. Her grieving parents, distraught over their tragic loss, request that the embryo be implanted so that they may have another child who is a near genetic duplicate of the daughter whom they so desperately miss.

A man agrees to be cloned with the intent of donating the resultant embryo to research. Subsequent to creation of the cloned embryo, he learns that both he and his wife are infertile. Realizing that their prospect for having a genetically related child suddenly appears to be compromised, the man changes his mind and requests that his clone be implanted in his wife instead of donated to research.

In such cases, it would be difficult for many physicians to deny the wishes of those desiring to implant a cloned embryo.

Ineffective Detection.—Third, violations of a partial ban would often go unnoticed. If laboratory creation of cloned human embryos was permitted but implantation of such embryos was banned, it would be infeasible to monitor the fate of each and every cloned embryo. The somatic cell nuclear transfer procedure typically results in the creation of multiple embryos. To prevent a single embryo from being implanted within the private context of the physician-patient relationship would surely prove to be impossible. Moreover, policies that would require genetic testing of every neonate at birth to ensure that he or she is not a clone (and that would penalize the parties responsible for implanting a cloned embryo) would likely be regarded as a violation of privacy. Even if such testing were allowed, it would fail to ensure that reproductive cloning had not occurred because the baby could be a clone of an unknown or unrevealed person, rather than being a near genetic duplicate of one of the parents.

As a result, threats to levy fines or inflict other punishments would not always deter those wishing to engage in technology they perceived to be undetectable. Policies that prohibit and penalize those who request or assist in, the implantation of cloned human embryos would therefore, ultimately fail to prevent reproductive cloning once cloned human embryos were produced for research purposes.

Facilitation via Technological Advance.—Fourth, a policy that prohibited cloning for reproduction while permitting cloning for research would actually facilitate the means to achieving the activity it intended to prevent. To permit research cloning as a legitimate activity of science would undoubtedly result in an increased number of human clone births.

Ongoing embryological research is poised to overcome many of the remaining technical obstacles to human cloning. Some experts estimate that the successful and efficient production of healthy cloned human embryos suitable for implantation might be months or at most a few years away.[14] If this methodology is perfected and IVF practitioners are trained in its use, the implantation of cloned human embryos would no longer be the distant prospect of a few laboratories in possession of specialized resources but could become a simple and brief procedure within reach of most fertility clinics that perform intracytoplasmic sperm injection or other labor-intensive forms of fertilization.

If the goal is to avoid instances of human reproductive cloning, as advocates of a partial ban fervently assert, then a law designed to prevent a requisite activity occurring over a period of months or years should be regarded as preferable to a law forbidding the implantation of cloned human embryos that could be accomplished in only minutes. Many existing US laws offer precedents. For example, to prevent private citizens from developing nuclear weapons, current laws ban the unlicensed possession of plutonium and enriched uranium.[15] It would be foolhardy to distribute plutonium widely and then expect people not to engage in the production and use of nuclear weaponry because it is easier to withhold the means to produce a weapon than it is to prevent its production and use.

Even persons who do not find research cloning to be morally objectionable may nevertheless oppose cultivating the industry due to the inevitability that cloned human beings would be born if the development of technology for research cloning were not banned.[16]

Evasive Language.—Fifth, a partial ban would eliminate only the language of reproductive cloning. To speak of a distinction between "reproductive" cloning and "research" cloning is to neglect an important commonality between both forms of cloning. Regardless of intent, both generate in the same manner a human embryo. Therefore, both methods of human cloning are reproductive in that they give rise to new individual human lives.[17] A partial ban clearly understood would not truly be a ban against cloning but against the implantation—and hence the survival—of human clones.

The choice of language applied to cloning should recognize that, on biological grounds alone, the human embryo is a living human organism. Structurally, the embryo is genetically complete. What is necessary for continued growth is suitable nurture and environment, two conditions that live human beings need as much in their adult stage as in their embryonic stage. Metabolically, at every cell division the

14. M. Boiani, S. Eckardt, H.R. Schöler, and K.J. McLaughlin, "Oct4 Distribution and Level in Mouse Clones: Consequences for Pluripotency," *Genes Dev.* 16 (2002): 1209–19; D. Solter, "Cloning v. Clowning," *Genes Dev.* 16 (2002): 1163–66.

15. Atomic Energy Act, 42 USC §2077 and 42 USC §2014 (aa) (1954).

16. Associated Press, "Sen. Smith Changes Stance on Stem Cells," *Northwest News Channel 8* Website, May 6, 2002.

17. J. Langman, *Medical Embryology,* 4th ed. (Baltimore, Md.: Williams & Wilkins, 1981), 1.

embryo copies the complete human genome with nearly perfect fidelity and, in transcribing his or her genetic code, has begun the journey toward actualization of all the functional capacities that uniquely typify a being of the species *Homo sapiens.*

Some well-intentioned thinkers will defend research cloning and human embryo research in general on the grounds that, rather than being fully present at conception, human worth develops gradually as the nervous system reaches a stage of maturation when certain functional capacities are demonstrable. We consider such a gradualist view to be an inadequate account of the value of human life. To suppose that human life consists only in functional capacities is to mistake the detection of life for its existence. Life ontologically precedes biological function, and one must first be a human being to develop and possess human capacities. Similarly, although some have argued that the embryo fertilized in vitro must enter the womb to count as human, we maintain that the moral status of a human being is independent of age or geographical location.

A gradualist view can also run counter to the widely accepted belief that diminished or less developed capacities may obligate increased care or protection. Some of the very people who would gain from the alleged benefits of research cloning are themselves in a state of functional decline due to degenerative disease. If one accepts the gradualist criterion that moral worth depends on one's stage of development or function, then, by the same logic, individuals who are ill or disabled (e.g., those with Alzheimer disease, Parkinson disease, or spinal cord injury) would have an uncertain claim to full human worth because of their loss of function. From the gradualist perspective, the widely held belief in human equality grounded in a common basis for dignity vanishes.

Because material traits alone are an unsatisfactory guide to assessing human moral status, no empirical description of a human being, no matter how exact, can fully grasp the magnificent complexity of the individual within. Drawing from their experience in responding to human frailty and suffering, physicians understand this well. The practice of medicine teaches that the meaning of human joy cannot be fully explained by a map of molecules in motion, or the tears of human suffering by the trickling of neurotransmitters. Thus, human dignity, which medicine recognizes as being irreducible to mere physical characteristics, cannot be denied on material grounds to a portion of the biological life span. To discount the emergence of human dignity at the very beginning of life—when life, although just barely measurable yet has distinctly begun—would be a serious error. Language that undermines the humanity of early human life inevitably exposes other vulnerable classes of humanity to the risk of similar devaluation. Regrettably, medicine has witnessed throughout history the tragic consequences that ensue when a certain subgroup of human beings is denied the full status of humanity for the purpose of research or economic gain.

Summary.—A ban on human cloning for both research and reproductive purposes would be the most effective and ethically responsible safeguard against the birth of human beings via cloning. Once human embryos were developed to the stage at which stem cells are present, a primary objective of research cloning, they would also be suitable for implantation.[18] Then, as we have illustrated, the birth of cloned human embryos would be only a short step away; once a cloned human embryo was implanted in a woman's body, no responsible public policy would mandate the termination of pregnancy.

Advocates of a less than comprehensive ban may respond to the preceding arguments by pointing out that all legal bans function imperfectly. Although it is of course true that no law functions perfectly (e.g., people continue to murder even though homicide is illegal), a society serious about prohibiting a certain act should adopt laws that will reduce, to the greatest extent possible, the likelihood of that act occurring. Although a comprehensive ban on human cloning may indeed fail to prevent all instances of reproductive cloning, prohibiting not only the implantation but also the creation of cloned human embryos would prove to be a far more effective mechanism for securing a society free from reproductive cloning. A comprehensive cloning ban is also the only policy consistent with the priority medicine should place on the value of human life.

Legal and Ethical Precedent

Historically, extracorporeal human embryos have been afforded certain protections[19] that have received broad support. For example, many present-day proponents of embryo and stem cell research dependent on the destruction of already existing human embryos created during IVF procedures recoil at the prospect of deliberate creation and sacrifice of human embryos.[20] A substantial proportion of the public is also opposed to the creation and destruction of cloned human embryos for research purposes.[21]

Policy Timeline.—In 1994, the National Institutes of Health proposed that the federal government begin funding research in which human embryos were created for destructive experimentation. This proposal was greeted with nearly universal condemnation from the public, various professional communities, and the major news media.[22] President William J. Clinton appropriately chose not to grant federal fund-

18. W.J. Smith, "Cloning and Congress: No Ban Is Better than a Phony Ban," *The Weekly Standard,* July 1, 2002.

19. Consolidated Appropriations Act, §510, Pub L No. 108-7 (2003).

20. C. Krauthammer. "Research Cloning? No." *Washington Post,* May 10, 2002, A37, https://www.washingtonpost.com/archive/opinions/2002/05/10/research-cloning-no/29211114-8967-4e73-8861-f9f944313e4d/.

21. Poll on American support of human cloning, "Stop Human Cloning," April 22, 2002, https://www.cloninginformation.org/new-poll-american-people-oppose-all-human-cloning-support-president-bushs-call-for-a-comprehensive-ban/; Cloning opposed, stem cell research narrowly supported: The Pew Research Center for the People & the Press and the Pew Forum on Religion & Public Life, "Public Makes Distinctions on Genetic Research," April 9, 2002, https://www.pewresearch.org/politics/2002/04/09/public-makes-distinctions-on-genetic-research/.

22. "Embryos: Drawing the Line," editorial, *Washington Post,* October 2, 1994, C6; "Embryo Research Is Inhuman," editorial, *Chicago Sun-Times,* October 10, 1994, 25.

ing for the creation of embryos for research, and Congress went one step further in passing an amendment prohibiting funding for all research harmful to human embryos.[23] When opposition to this amendment has been voiced by members of Congress, bipartisan support for its prohibition against creating human embryos for destructive research has remained undiminished.[24]

The National Bioethics Advisory Commission's report on cloning human beings,[25] the National Institutes of Health guidelines for embryonic stem cell research,[26] and the Stem Cell Research Act of 2001[27] all explicitly proscribed the special creation of embryos for research purposes. More recently, the President's Council on Bioethics unanimously rejected human cloning for purposes of reproduction, while pronouncing a majority recommendation for a moratorium on human cloning for biomedical research.[28]

Ethical Tradition.—In addition to constituting a break with US legal tradition and much of public sentiment, research cloning violates existing ethical guidelines designed to protect human subjects.

The Nuremberg Code (1945) stipulates that, "No experiment should be conducted where there is *an a priori* reason to believe that death or disabling injury will occur."[29] This fundamental principle of nonmaleficence is reflected also in the Declaration of Helsinki (1964, latest revision 2000), the Belmont Report (1979), and the Council of Europe's Convention on Human Rights and Biomedicine (1996), which represent the accumulated wisdom of a half century of reflection on the grave consequences of conscripting nonconsenting human beings for destructive research. Of importance, these moral codes do not exempt human beings at the beginning or end of life as somehow not being under the protection due all human subjects. To the contrary, these codes demand even greater degrees of protection for the most vulnerable of human subjects, that is, children. We maintain that cloned human embryos, as human beings, likewise are vulnerable human subjects and are worthy of such protections.[30]

Although a strong lobby to legalize research cloning has been formed by certain scientists, biotechnology firms, and patient advocacy groups, the fact remains that the prospect of creating and destroying human embryos for research purposes has,

23. Consolidated Appropriations Act.

24. N. Lowey, *Congressional Record,* 142 (July 11, 1996): H7343, quoted in www.nrlc.org/Killing_Embryos/factsheetcloning.html.

25. National Bioethics Advisory Commission, *Cloning Human Beings.*

26. National Institutes of Health, "National Institutes of Health Guidelines for Research Using Human Pluripotent Stem Cells," *Federal Register* 65 (2000): 51976–81.

27. Amendment to the Stem Cell Research Act of 2001: Referred to the Committee on Health, Education, Labor, and Pensions, 107th Cong, 1st Session (April 5, 2001) (statement of Arlen Specter, senator).

28. Kass, *Human Cloning.*

29. G.J. Annas and M.A. Grodin, eds., *The Nazi Doctors and the Nuremberg Code: Human Rights in Human Experimentation* (New York: Oxford University Press, 1992), 2.

30. T. Smith, *Ethics in Medical Research: A Handbook of Good Practice* (Cambridge, England: Cambridge University Press, 1999), 278; Ramsey Colloquium, "The Inhuman Use of Human Beings: A Statement on Embryo Research," *First Things* 49 (1995): 17–21.

for valid reasons, been consistently opposed in both the legal and the ethical arenas. It is incumbent on advocates of research cloning who wish to overturn well-established ethical standards to make and defend a sufficient and compelling case. No convincing case has been presented that provides substantive arguments for rejecting the existing set of governing principles that has been carefully formulated and deeply etched into the prevailing ethos of our culture.

A policy allowing research cloning would therefore run counter to US jurisprudence regarding the treatment of human embryos and to the intent of ethical codes designed to protect human subjects in research. Of note, a non-comprehensive ban permitting research cloning would establish, for the first time in US history, a class of human beings created for the sole purpose of experiments that will destroy them and whom, ironically, it is a crime not to destroy. In recognition of this fact, and of research cloning's inherent potential for reproductive applications, the burden of proof must lie on those who wish to justify such a momentous break with legal and ethical precedent. We assert that no such justification has been offered, as defended in the subsequent section.

Appeals to Medical Benefit

In justifying a policy promoting the creation and subsequent destruction of cloned human embryos, advocates of research cloning have frequently turned to the rationale of utility. Utilitarianism in its classic form advocates acting in whatever ways will result in the "greatest good for the greatest number." (Although other forms such as rule utilitarianism have emerged in ethical theory, they are not nearly as influential in public discussion.) Proponents who invoke a utilitarian rationale for research cloning maintain that the lives of nascent human beings should be regarded as having less value than the anticipated health benefits for others and the increased medical knowledge that research cloning would allegedly offer.

Some scientists and biotechnology companies seeking to close in on medical breakthroughs, as well as some patients who hope that such advances will result in a treatment or cure for their particular affliction, have been among the most vocal advocates of research cloning, and the pressure they apply to those who seek to prohibit such a practice is immense.[31] Although the goal of medical breakthroughs is laudable and the hopes of patients are certainly understandable, the praiseworthy endeavor to alleviate human suffering does not justify the use of all possible means.

We affirm the importance of weighing consequences as an empirically grounded method for deciding many difficult issues in medical science, but we also recognize that utilitarianism as an ethical theory is incomplete. It has distinct flaws, which even its most sophisticated articulations are incapable of resolving. This is because utili-

31. C. Ezzell, "Stem Cell Showstopper?" *Sci Am.* 285 (2001): 27; D. Solter and J. Gearhart, "Putting Stem Cells to Work," *Science* 293 (1999): 1468–70; R.P. Lanza, J.B. Cibelli, and M.D. West, "Human Therapeutic Cloning," *Nat Med.* 5 (1999): 975–77; R.P. Lanza, J.B. Cibelli, and M.D. West, "Prospects for the Use of Nuclear Transfer in Human Transplantation," *Nat Biotechnol.* 17 (1999): 1171–74.

tarian philosophies attempt to sever ethical reasoning from any prior commitment to moral norms other than maximizing utility. Immaterial realities that on independent grounds we know to be valid—such as love, justice, and human dignity—have no validity per se (i.e., apart from their consequences in particular situations).

The utility defense of human cloning fails because utilitarianism alone is incompetent to give due regard to human life, which in so many ways (as the practice of medicine reminds us) transcends calculation. Just as there is more to right and wrong than the utilitarian calculus, so there is more to humanity, even wondrous embryonic humanity, than empirical measurements can apprehend. Therefore, human life-and-death decisions should not be left solely to the impersonal assessment of utilitarian analysis. The skillful surgeon knows to operate with the greatest care near vital structures, and there are some things that the scalpel of utilitarian logic should never be allowed to cut.

In summary, we find appeals to utility, even the utility of medical benefit, to be an insufficient defense of research cloning. In further support of this contention, we offer the following arguments.

Utilitarian Justifications Are Self-Defeating.—The utilitarian rationale for research cloning is self-defeating. According to utilitarian theory, inflicting some harm in the pursuit of benefit is justifiable only if there is no other less harmful way to secure that benefit. Research cloning fails to meet this criterion because an increasing body of evidence suggests that stem cell research from nonembryonic sources (as well as other methods of tissue repair or regeneration) may hold equal or even greater promise for treating human infirmities.[32]

How much utility research cloning would afford is simply unknown. With human embryonic stem cell research still in its infancy and published research on stem cells obtained from cloned human embryos lacking, the claim that the utility of such experimentation justifies a practice long held to be unethical and illegal is extremely weak. Although a few advances to date suggest that such research might yield various therapies in a number of years,[33] embryonic stem cell research has been

32. "Treating Disease with Adult Stem Cells and Embryonic Stem Cells: Adult Stem Cells More Promising, More Successful," http://www.stemcellresearch.org/facts/quotes2.htm; W. Lillge, "The Case for Adult Stem Cell Research," *21st Century Science & Technology Magazine* (Winter 2001–2002); D. Orlic, J. Kajstura, S. Chimenti, et al., "Bone Marrow Cells Regenerate Infarcted Myocardium," *Nature* 410 (2001): 701–705; C.M. Verfaillie, "Adult Stem Cells: Assessing the Case for Pluripotency," *Trends Cell Biol.* 12 (2002): 502–508; Y. Jiang, B. Vaessen, T. Lenvik, M. Blackstad, M. Reyes, and C.M. Verfaillie, "Multipotent Progenitor Cells Can Be Isolated from Postnatal Murine Bone Marrow, Muscle, and Brain," *Exp Hematol.* 30 (2002): 896–904; D. Lu, P.R. Sanberg, A. Mahmood, et al., "Intravenous Administration of Human Umbilical Cord Blood Reduces Neurological Deficit in the Rat after Traumatic Brain Injury," *Cell Transplant.* 11 (2002): 275–81; J. Chen, P.R. Sanberg, Y. Li, et al., "Intravenous Administration of Human Umbilical Cord Blood Reduces Behavioral Deficits after Stroke in Rats," *Stroke* 32 (2001): 2682–88.

33. Committee on the Biological and Biomedical Applications et al., *Stem Cells and the Future of Regenerative Medicine;* US Department of Health and Human Services, National Institutes of Health, "Stem Cells: Scientific Progress and Future Research Directions," http://www.nih.gov/news/stemcell/scireport.htm.

fraught with so many obstacles[34] that currently not a single therapy has clearly benefited a patient.[35]

Utility justifications must establish that any harms resulting from a certain practice will be outweighed by the resultant benefits. Furthermore, as utilitarian philosophy has traditionally acknowledged, distant or uncertain benefits should be given less weight than present or certain harms.[36] Research cloning would undoubtedly inflict substantial harms without the certainty of near-term benefit. Human embryos sacrificed for their stem cells would immediately suffer definite harm in that the harvesting of these cells would necessitate their death. By contrast, the speculative benefits of research cloning would likely not occur until decades later, if ever. As such, multitudes of human embryos would likely need to be destroyed before a single patient could benefit from this research.

In the broadest sense, utilitarian justifications of research cloning are self-defeating because they are unduly persuaded by potential good consequences that appear at first glance. However, attempts to justify research cloning must also consider the full range of bad consequences that may not be as immediately apparent but are just as real. For example, women donating their eggs for use in the cloning process would be required to take superovulatory drugs and receive numerous hormone treatments before undergoing the invasive extraction procedure. Such a procedure carries rare yet serious health risks, including ovarian rupture, severe pelvic pain, bleeding into the abdominal cavity, acute respiratory distress, pulmonary embolism, possible increased risk of ovarian cysts and cancers, and potentially infertility.[37] Thomas Okarma, president, CEO, and director of Geron Corporation, Menlo Park, California, has no ethical qualms about destroying embryos for research but now believes that the potential of research cloning for medical benefit is "vanishingly small"; estimates are that treatment of a single patient via research cloning would require "thousands of [human] eggs on an assembly line."[38] Others have calculated similarly overwhelming estimates.[39] As a result, certain

34. G. Vogel, "Cell Biology: Stem Cells: New Excitement, Persistent Questions," *Science* 290 (2000): 1672–74. "Embryonic Stem Cell Research: A Reality Check," March 2002, http://www.stemcellresearch.org/facts/quotes3.htm.

35. University of Wisconsin-Madison, "Clinical Application Still Years Away," http://www.news.wisc.edu/packages/stem cells/index.html?get=patients.

36. H. Sidgwick, *The Methods of Ethics,* 6th ed. (New York: Macmillan Co., 1901).

37. International Center for Technology Assessment, "Six Reasons Why Progressives Should Not Support the Harkin-Specter Bill (S. 1893) on Human Cloning," http://www.cloninginformation.org/info/icta-why_oppose_harkin-spector.htm; J. Norsigian, Statement before the US Senate Health, Education, Labor and Pensions Committee, March 5, 2002; L. Andrews and D. Nelkin, *Body Bazaar: The Market for Human Tissue in the Biotechnology Age* (New York: Crown Publishers, 2001); A. Venn, L. Watson, F. Bruinsma, G. Giles, and D. Healy, "Risk of Cancer after Use of Fertility Drugs with In-vitro Fertilisation," *Lancet* 354 (1999): 1586–90; A. Delvigne and S. Rozenberg, "Epidemiology and Prevention of Ovarian Hyperstimulation Syndrome (OHSS): A Review," *Hum Reprod Update* 8 (2002): 559–77.

38. D. Gellene, "Clone Profit? Unlikely," *LA Times,* May 10, 2002, Business section.

39. W.J. Smith, "Practical Council: Important Stuff from the Kass Commission," *National Review Online,* August 13, 2002.

groups of women, such as those economically disadvantaged, would be at great risk of exploitation, a danger of biomedical research that existing ethical codes are designed to prevent.

In addition, if cloned embryos created for the purpose of research were instead implanted into a woman (which we have argued is a highly likely prospect), the probable deformities in, and even death of, the cloned children[40] would constitute great harm for them and result in unspeakable grief for their parents and families.

Moreover, some experimental evidence suggests that embryonic stem cells (especially those obtained from cloned embryos) might actually constitute harm to patients who receive therapies derived from such cells.[41] Of note, some ethically questionable acts previously justified in the name of utility, for example, transplantation of embryonic cells and fetal tissue into patients, have actually resulted in grave harm, rather than therapeutic benefit.[42] For the preceding reasons, an increasing number of scientists doubt that research cloning will ever yield the balance of benefit over harm that some anticipate.[43]

Such doubt is buttressed by two further observations. First, even if research cloning were to yield the promised cornucopia of medical therapies, some patients who are morally opposed to the destruction of embryonic human life would likely refuse these treatments, unless they are willing to abandon their moral convictions. Thus, conscientious abstention would diminish the utility of this technology. Other patients, desperate for treatment, might compromise their principles using their own form of utilitarian reasoning but at the cost of a troubled conscience. Second, the allocation of funding and other resources to human cloning and other human embryonic technologies would divert precious resources from the development of morally noncontroversial technologies. Because Americans are so divided on this important issue, the most prudential policies will be those that avoid conflict in society by rejecting contentious human cloning agendas and embracing the promising avenues of stem cell research from nonembryonic sources and other therapies that do not depend on the destruction of human embryos. Moreover, supporting research on the development of therapies acceptable to patients with moral convictions opposed to research cloning would benefit both patients and researchers alike because such research programs would be less liable to become entangled in public controversy.

40. Jaenisch and Wilmut, *Developmental Biology*.

41. "Treating Disease with Adult Stem Cells"; J.S. Odorico, D.S. Kaufman, and J.A. Thomson, "Multilineage Differentiation from Human Embryonic Stem Cell Lines," *Stem Cells* 19 (2001): 193–204. "Scientific Problems with Using Embryonic Stem Cells," November 2001, http://www.stemcellresearch.org/facts/escproblems.htm.

42. R.D. Folkerth and R. Durso, "Survival and Proliferation of Nonneural Tissues, with Obstruction of Cerebral Ventricles, in a Parkinsonian Patient Treated with Fetal Allografts," *Neurology* 46 (1996): 1219–25; C.R. Freed, P.E. Greene, R.E. Breeze, et al., "Transplantation of Embryonic Dopamine Neurons for Severe Parkinson's Disease," *N Engl J Med.* 344 (2001): 710–19.

43. "Treating Disease with Adult Stem Cells"; P. Aldhous, "Can They Rebuild Us?" *Nature* 410 (2001): 622–25.

To assert that the utility of research cloning justifies the certain destruction of cloned human embryos created for this purpose is to make a claim that is highly dubious even on utilitarian grounds. Utilitarian calculation, if done correctly, argues against pursuing research cloning. Invoked as a justification for research cloning, it is self-defeating.

Utilitarian Justifications Are Dangerous.—Not only is the utilitarian claim to produce the most beneficial outcome inaccurate in the case of research cloning, the attempt to do so is itself unacceptably dangerous.[44] As long as a favorable balance of good over bad consequences ensues for society as a whole, utilitarian logic allows for literally any harm to be inflicted on an individual or a minority group regardless of whether doing so is widely regarded as unethical or even evil.

For example, scientists would acquire some extremely valuable information about the body's response to ionizing radiation by injecting radioactive plutonium into a group of people and then studying them over time. Although a few people might suffer harm, an untold number of persons at risk for occupational exposure to radiation might receive great benefit from the knowledge gained. However, conducting an experiment of this type would definitely be unethical. Lest anyone think that such an experiment would be inconceivable in the United States, it is worth noting that this very protocol was carried out, in the absence of informed consent, at the Los Alamos Scientific Laboratory, New Mexico, in the late 1940s.[45] History provides numerous other examples of unethical medical research, including the experiments conducted at Tuskegee, Alabama; at the Willowbrook State School, Staten Island, New York; and during some foreign totalitarian regimes. All these research protocols were defended at the time by utilitarian reasoning but were later recognized to have been unethical. In each case, it became clear that the potential for obtaining benefits from the research, regardless of how important such benefits may have been or who would have gained from them, did not render the research justifiable.

The Purpose of Medicine

The physician, out of concern for the patient who desires a child or a stem cell transplant, may initially be attracted to the argument that a ban on human cloning should allow certain exceptions based on compassion. Of note, however, a medical procedure is not judged to be good simply because a patient may desire it. What the patient perceives as his or her own good must be brought into proper relationship with other levels of good—what is good for the patient, for humanity, and for the cloned human being. Otherwise, an act motivated by unrestrained compassion may become harmful.

Although emotionally compelling, compassion is by itself an insufficient guide to ethical behavior. Rather, wisdom looks both to compassion and to understanding.

44. Kilner, "Human Cloning."

45. Advisory Committee on Human Radiation Experiments, *Final Report,* GPO Stock No. 061-000-00-848-9 (Washington, D.C.: US Government Printing Office).

Compassion is not to be a moral law unto itself but must be subject to moral analysis, based on sound reasons, and carried out with forethought of consequences.

If human cloning for biomedical research were to become a reality, physicians would be called on to perform the procedures to extract oocytes from healthy women for the purpose of generating cloned embryos. Physicians would face the temptation to implant some of those embryos as treatment for infertility (with possible risk to their careers if a partial ban on cloning were enacted). If it proves impossible to grow replacement organs from embryonic stem cells in vitro, physicians could be called on to implant into women cloned embryos to be grown to the fetal stage and later terminated to serve as transplant donors. Physicians, in prescribing drugs, vaccines, or cellular therapies developed from research on cloned embryos or their stem cells, would occasionally incur complicity with the prerequisite destruction of early human life. Physicians would inherit the strange tasks of counseling patients about the option of receiving treatments derived from clones of themselves and of informing patients morally opposed to receiving such treatments that more expensive non-cloning alternatives might not be covered by their insurance plan.

Considered realistically, human cloning fails to qualify as a healing act. The physician's integrity as a healer requires that he or she always act to preserve life, and never by means of human death. The intention to heal can in no way justify the act of terminating the life of another human being. Accordingly, physicians consider nonmaleficence to be a more stringent ethical obligation than beneficence.[46] Such ordering of ethical priorities in medicine has long placed the profession of medicine on higher moral ground than that reachable by pragmatic rationalizations. This is why Hippocrates' prime maxim, "First, do no harm,"[47] has endured since antiquity as a guiding principle among physicians who, by honoring it, have earned their patients' trust.

To rewrite medical ethics to permit human cloning would ensnare physicians in a perilous compromise of professional standards. To acquiesce to human embryonic cloning would be to disregard, to an unprecedented degree, the value of new human life. Human cloning would also represent a decided step toward the devaluing of humanity universally because justifications of human cloning research disturbingly imagine a category of dismissible human life. Such a designation is utterly foreign to the Hippocratic ethic, which respects human beings at all stages of life.

Although the potential for eventual health benefits from research cloning has been vigorously advanced, its violation of human dignity by treating nascent human life as no more valuable than an expendable means to others' ends falls short of the purpose for which medicine exists. The medical good aims not only to improve physiologic function and to secure the good as perceived by the patient but also strives to

46. T.L. Beauchamp and J.F. Childress, *Principles of Biomedical Ethics,* 4th ed. (New York: Oxford University Press, 1994): 189–258, 271–87.

47. Hippocrates, *Epidemics,* Book I, sect XI.

safeguard and promote the higher good for the patient.[48] This higher good includes the preservation of the dignity of humans as humans, for which the physician is obligated to guard the welfare of the most vulnerable of human beings, including the cloned human embryo.

Conclusion

Human cloning, for whatever purpose, represents an abuse of scientific freedom, not its realization. This new technology should adhere to the standard that science should always serve humanity, never that a segment of humanity would be created to serve science. As history has conspicuously recorded, no program sacrificing those at the margins of humanity to science has ever stood the test of time. Ethical reflection always reaches, in due course, the conclusion that the least of human beings deserve the care and concern that the medical profession presumes is due all human beings. Whether the ethical cinder of human cloning will lodge in the eye of society's conscience is an issue still within the reach of sensible preventive intervention.

For the sake of their patients, as well as the future of humanity, we urge healthcare professionals to oppose all forms of human cloning. In keeping with the Hippocratic ethic, we recommend that biomedical research on non-embryonic stem cells be pursued and funded aggressively. We also commend legislation and policies at all levels that will protect people from the unfavorable outcomes of human cloning, both now and for generations to come. Only a ban prohibiting both research and reproductive cloning will offer such protection.

We thank Mr. Russell C. DiSilvestro for his critical review of the submitted manuscript.

48. E.D. Pellegrino, "The Internal Morality of Clinical Medicine: A Paradigm for the Ethics of the Helping and Healing Professions," *J Med Philos.* 26 (2001): 559–79.

C. Genetics and Genomics

13

Genomic Medicine: Using Pellegrino's Appreciation of Ancient Wisdom to Achieve Twenty-First-Century Healthcare Benefits

Kevin T. FitzGerald, SJ, PhD, PhD

AS THE READER MOVES THROUGH THIS VOLUME containing the contributions of Dr. Edmund Pellegrino to the world of clinical ethics, two fundamental aspects of Dr. Pellegrino's work will become abundantly clear. The first is his extensive knowledge of ancient insights into the meaning of human existence, in particular issues of health and well-being as well as the human experience of illness and suffering. The second, and more challenging, aspect—more challenging both for those in health care and for Dr. Pellegrino to defend convincingly in the current clinical ethics arena—is Dr. Pellegrino's argument that these ancient insights remain relevant, and even crucial, for how healthcare professionals should determine the best treatments for their patients in light of the recent and continuing rapid expansion of medical information and technology. If Dr. Pellegrino was correct in his analysis and evaluation of contemporary health care, then applying his insights and approach to clinical ethics to one of the most rapidly developing, and even revolutionary, areas of twenty-first-century health care should give us clearer evidence of the real worth and usefulness of his contributions to our current healthcare landscape. It is the goal of this chapter to evaluate the contemporary worth and usefulness of the Pellegrino approach by examining the promises and perils of genomic medicine through the lens of Dr. Edmund Pellegrino's argument regarding the need to train healthcare professionals according to the virtues and relationships inherent to the internal morality of medicine.

To map more precisely the healthcare terrain that will be the focus of this chapter a brief description of several key characteristics of genomic medicine should be useful. First on this list should always be the all too common misunderstanding that a person's genome determines who the person is and will become. Though there are some genetic conditions that cause relatively certain and significant impacts on an individual's health, the vast majority of genetic characteristics linked to disease can vary in both likelihood and severity of impact. In fact, one of the most rapidly expanding areas of genomic medicine involves epigenetics, which investigates how genes are turned on and off to varying degrees in one's genome by behavioral or environmental effects rather than changes in the genetic sequence.[1] This understanding of genetics as an important but partial contributor to one's overall health is critical to the successful development of genomic medicine. Hence, applying genomic medicine information well to a patient's care will require an approach that is able to integrate genetic and genomic information appropriately into the entire scope of the patient's situation.

The challenge this integration will be for a given patient is increased due to the second key feature of genomic medicine—the enormous amount and complexity of the data now available due to recent technological advances in genetic testing and sequencing. Determining which parts of this massive and complex information set is relevant to a patient's care will require both professional expertise and detailed knowledge of a patient's goals and desires.[2] This knowledge of the patient's goals for care will also be crucial for the third key characteristic of genomic medicine—direct genetic interventions to treat disease. This transition in genomic medicine from a focus on diagnosis and prognosis to genetic-based treatments will force healthcare professionals to address definitively the array of complex ethical issues surrounding the intentional alteration of patient genomes. This issue of gene treatments has been debated for decades but still today remains unclear in its ethical application as demonstrated by the recent broadly condemned research project of Dr. He wherein he attempted to genetically engineer female embryos to make them resistant to HIV infection.[3]

While they clearly denounced Dr. He's experimental intervention, members of the research community also acknowledged the need for broad public engagement to help guide future decisions regarding the appropriateness of human genetic interventions.[4] This acknowledgement by the scientific community of the need for public input in order to determine how healthcare professionals and patients should move forward, or not, with human genetic interventions—especially heritable germline changes—raises the central question of the appropriate role for the public in determining what clinical research is beneficial to pursue and what is not. We can begin here to investigate how the approach of Pellegrino may both diverge from others in the current biomedical ethics arena and at the same time be more likely to provide insights and foster decisions beneficial to the public that the scientific community now so eagerly wishes to engage.

1. https://www.genome.gov/genetics-glossary/Epigenetics.

2. John Marshall et al., "From Genomics to Multiomics," *JCO Precision Oncology,* in press (2021).

3. Antonio Regalado, "China's CRISPR Babies: Read Exclusive Excerpts from the Unseen Original Research," *MIT Technology Review* (December 3, 2019), https://www.technologyreview.com/2019/12/03/131752/chinas-crispr-babies-read-exclusive-excerpts-he-jiankui-paper/.

4. Second International Summit on Human Gene Editing, http://www.nationalacademies.org/gene-editing/2nd_summit/index.htm.

Though the research community has added its voice to the general outcry for broad public engagement to determine the trajectories and goals that human genome editing and genomic medicine should pursue—and which trajectories should be avoided—many question the motives and goals researchers and biotechnology companies have regarding this public engagement. Is the goal of the engagement merely to inform and educate the public in a way that encourages agreement with the values and goals of researchers and developers? Or is the goal of this public engagement to stimulate and facilitate substantive dialogue and relationship-building among the various global communities to provide a comprehensive and nuanced mutual understanding of the diverse hopes and fears different communities bring to this rapidly expanding area of biotechnology? Whatever the motives behind this call for public engagement from the research and biotechnology communities, one can still insist on pursuing the more inclusive goals of substantive dialogue and relationship building by referencing the Statement by the Organizing Committee of the Second International Summit on Human Genome Editing from November 29, 2018, in their call for "an ongoing international forum to foster broad public dialogue."[5]

A useful framework for grounding and guiding a global public engagement regarding the research and development of human genome editing can be found in Pellegrino's application of the ancient concept of a covenant between two groups or individuals to clinical research (see chap. 10 above).[6] Pellegrino outlines the framing of a covenant for biomedical research when he states: "Society permits experimentation with humans because of the benefits that the whole society, as well as the experimental subject, may gain from new medical knowledge. The clinical investigator thus enters a covenant with society when he or she accepts the privilege of experimentation involving fellow human beings."[7] And, determination of the benefits to be gained by society requires input from society as to the needs society wants addressed which may or may not fit well with the desires of the research and technology development communities.[8]

If significant discrepancies arise between the scientific and public communities as to how genetic medicine should be developed, who should determine what should and should not be done? Pellegrino's approach to this question would combine both the covenantal relationship between these groups and the fiduciary responsibilities of clinical researchers to research participants and society to formulate an answer: "Although these covenants arise in a trust relationship that institutional review boards and research regulations can help safeguard, their subtler nuances cannot be spelled out in such regulations. Rather, like the therapeutic relationship, they rest ultimately

5. Second International Summit on Human Genome Editing, *Continuing the Global Discussion: Proceedings of a Workshop—in Brief* (Washington, D.C.: National Academies Press, 2019), https://www.nap.edu/read/25343/chapter/1#8.

6. Edmund D. Pellegrino, "Beneficence, Scientific Autonomy, and Self-Interest: Ethical Dilemmas in Clinical Research," *Cambridge Quarterly of Healthcare Ethics* 1, no. 4 (Fall 1992): 361–70.

7. Ibid., 362.

8. Ibid., 363.

on the character and moral integrity of the investigator."[9] This insight from Pellegrino regarding the ultimate ethical responsibility of medical researchers to foster and develop their own integrity and virtue is not one often emphasized in recent discussions about how medical research should be regulated and safeguarded. Most often the discussion focuses on structures such as oversight committees and policies that should be put into place to protect participants and society while the research moves forward. Perhaps this lack of attention paid to training virtuous researchers as a key part of the research community's relationship with the public is one reason the Pew Research Center found in 2019 that only 35 percent of the public thought that medical researchers care about the best interests of the public all or most of the time.[10]

Hence, if the research and biotechnology communities are truly serious about responding to the public's concerns about the development and implementation of genomic medicine and technology, then embracing a Pellegrino covenantal approach to their relationship with the public along with incorporating Pellegrino's emphasis on training virtuous professionals might provide the foundation for achieving the levels of trust and honest exchange of ideas that everyone purports to desire. Employing this Pellegrino approach could also facilitate a mutual understanding between the biomedical and patient communities that would include the fact that any development in genomic medicine comes with risks of creating inadvertent harms and insufficient benefits. In addition, in light of these risks healthcare professionals will promise to continue to care for patients as best they can no matter the treatment setbacks and disappointments. As Pellegrino recognized, such an understanding between the medical and patient communities is necessary to achieve the real benefits that both communities claim to desire from the application of new technologies such as genomic medicine.

Pellegrino's emphasis on covenant and virtuous professionals also provides advantages for addressing the other two key characteristics of genomic medicine—its challenging complexity and its often-limited overall impact on people's health. Again, the significant advantage of Pellegrino's approach is the focus on the importance of a relationship of trust between the patient and the healthcare professional, in part grounded in the integrity and virtue of the professional. Rather than acting merely as an unbiased, simple conduit of information for patient decision making for fear of infringing on patient autonomy, Pellegrino's healthcare professional will recognize both the vulnerability of the patient to genomic information overload and misinterpretation, and the responsibility of the healthcare professional to acknowledge honestly the current vast uncertainty that surrounds most of the genomic data that can be acquired through testing and sequencing.[11] Then, after listening to the

9. Ibid., 362.

10. Cary Funk, Meg Hefferon, Brian Kennedy, and Courtney Johnson, "Trust and Mistrust in Americans' Views of Scientific Experts," *Pew Research Center*, August 2, 2019, 15, https://www.pewresearch.org/science/2019/08/02/trust-and-mistrust-in-americans-views-of-scientific-experts/.

11. Edmund D. Pellegrino and David C. Thomasma, *The Virtues in Medical Practice* (Oxford: Oxford University Press, 1993), 56.

goals and concerns of the patient and incorporating those into the current capacities and limitations of genomic medicine, Pellegrino's healthcare professional will work with the patient, and family and friends if so desired by the patient, to arrive at the best treatment plan for the patient considering both the patient's values and the patient's current circumstances.[12]

This highly interactive and forthright relationship with the patient or research participant a Pellegrino clinician would pursue is particularly well-suited to the complex information challenge genomic medicine presents. Both the protocols for acquiring genomic information and the interpretation of the information acquired can be challenging even for those practiced in these techniques. Next Generation Sequencing of a patient's genome involves several major steps which include DNA fragmentation, library preparation, massive parallel sequencing, bioinformatics analysis, and variant/mutation annotation and interpretation.[13] This last step of interpretation can be particularly difficult especially considering the fact that each human being is genetically unique—even identical twins, though they would have far fewer differences than regular siblings. All this complexity adds to the importance of the relationship the patient has with the patient's primary healthcare professional. If the patient can trust that professional to be virtuous—for example, competent, compassionate, honest, self-effacing, etc.—then the patient can be assured that the healthcare professional will have the patient's best interest as the focus of any decision regarding the use or not of genomic medicine in the patient's treatment plan—or participation in clinical research if that is an option that fits with a patient's overall values and goals.

The richness and complexity of genomic information is being pursued because it can lead to more individualized patient treatment that may help healthcare professionals provide genetically targeted treatment benefits to the patient while reducing harms by avoiding patient genetic susceptibilities to adverse effects of certain treatments. Ironically, some of the advances in clinical knowledge obtained through genomic research are actually causing more confusion than help when it comes to healthcare treatment benefits and harms. One example of this situation is the discovery from genetic research that the current growth hormone treatment used to stimulate the growth of children predicted to be shorter than 5' tall for girls and 5'3" for boys may also be significantly increasing the risk of these treated children having type 2 diabetes or cancer when they become adults.[14] Hence, parents today now face the difficult dilemma of having to weigh the risks of their children facing the social

12. Edmund D. Pellegrino and David C. Thomasma, *A Philosophical Basis of Medical Practice: Toward a Philosophy and Ethic of the Healing Professions* (Oxford: Oxford University Press, 1981).

13. Dahui Qin, "Next-Generation Sequencing and Its Clinical Application," *Cancer Biology & Medicine* 16, no. 1 (2019): 4–10.

14. Jaime Guevara-Aguirre et al, "GH Receptor Deficiency in Ecuadorian Adults Is Associated with Obesity and Enhanced Insulin Sensitivity," *The Journal of Clinical Endocrinology & Metabolism* 100, no. 7 (July 2015): 2589–96.

stigma of being shorter than most people against the increased risks of disease their slightly taller treated children would face as adults. Those who knew Dr. Pellegrino personally will recognize the additional irony of this situation because Dr. Pellegrino was less than average height and so was well aware of the additional challenges people of short stature can face. Perhaps his own challenges growing up helped shape his approach to clinical ethics and the remarkable care and concern he had for his patients and for anyone who came to him for help or advice.

Pellegrino's concern for each person in the person's unique circumstance is fundamental to his approach to the proper practice of medicine. His emphasis on professional responsibility, integrity, and virtue, and his focus on the covenantal relationship between patients and their healthcare providers, provide an excellent framework for helping people also address the non-genetic causes of illness and injury. As mentioned above, because we are entering an era of individualized, genomic medicine, it will be all the more important to help a patient to be aware of the limitations of genetic information and interventions and to be aware of all the social determinants of health that may be impacting the patient's health much more than any aberrant genetic sequences the patient has. Hence, all healthcare professionals can benefit from embracing the Pellegrino approach to clinical practice. Employing Pellegrino's approach will help to ensure that all patients will be justified in presuming that their clinicians have their patients' best interest as the goal of their interaction. This clinical relationship of trust and understanding will be all the more crucial when genomic medicine develops to the point where it becomes standard of care for those who can afford it, but perhaps not for those in less well-resourced communities. Healthcare professionals will be challenged then to advocate for equal access for all who are in need—as Dr. Pellegrino did during the length of his career—because, as he argued, that is what virtuous healthcare professionals do when they are in a covenantal relationship with their patients.

Considering the key aspects of genomic medicine reviewed above that need to be addressed in order to integrate genetic technology well into global health care, engendering a broad public presumption of trust in clinicians and their research will obviously be a significant challenge in light of the Pew survey results and tragedies like the experiments of Dr. He. Dr. Pellegrino experienced the challenges of public trust raised by biomedical research over his 60 plus years in medical practice and education. In response to those challenges, he reflected extensively upon the ethical options that would best serve the goals of medicine. His conclusions included the need for a return to the ancient focus on a covenantal relationship between healthcare professionals and patients, and the need for healthcare professionals to practice medicine virtuously. If the goal of both the research community and the global public is to reap the potential benefits of genomic medicine and minimize the potential harms, then we would do well to explore seriously the approach of Dr. Pellegrino to once again mine the riches of ancient wisdom for insights into how best to proceed in the twenty-first century.

III

Controversies at the End of Life

A. End of Life

14

Introduction: End of Life

Daniel P. Sulmasy, MD, PhD

ONE OF THE MOST IMPORTANT TOPICS in clinical medical ethics is the care of patients approaching the end of life. As Pellegrino observes at the beginning of "Withholding and Withdrawing Treatments," physicians have recognized the limits of medicine since the time of Hippocrates (see chap. 15 below). All patients are mortal. Medicine is a finite craft. The ethically best care is always cognizant of the fact that medicine must be humble, recognizing its limits. In this article, Pellegrino provides an expanded version of the approach to caring for dying patients who cannot speak for themselves that was briefly described in chapter 1. He provides guidance on how to assess decision-making capacity and how to decide who speaks for the patient when he or she lacks capacity. He then introduces a simple and sound approach to deciding about when it is ethically appropriate to withhold or withdraw life-sustaining treatments for dying patients. The first question to ask is whether the treatment is effective. If it will not appreciably alter the natural history of the disease then there seems to be no ethical reason to provide it. He argues that this is fundamentally a judgment made by physicians, who have the expertise to determine whether an intervention will work in a particular set of clinical circumstances. The next question to ask is whether it will be beneficial—that is to say, whether the benefits of the treatment outweigh the burdens for a particular patient. Pellegrino argues that this judgment is largely one that the patient must make as he or she is the best judge of how valuable more days, weeks, or months might be compared with the burdens of the underlying illness and the proposed treatment. Of course, this judgment must be informed by the physician's expert knowledge, but assessment of the possible benefits and burdens and their probability rest mostly with the patient.

In "Withholding Treatment, Surrogate Decisions, and Rationing," Pellegrino provides further insights into decision making on behalf of those who cannot speak for themselves, including a useful set of criteria for deciding whether a surrogate is morally valid (see chap. 16 below). He argues vigorously against permitting the surrogate or the healthcare treatment team to decide to withhold care on the basis of the economic needs of society, urging us, except in extraordinary circumstances, to keep our focus on the fundamental purpose of clinical medicine—promoting the good of the individual patient.

Importantly, care at the end of life is not only about deciding what ought *not* to be done. In "Ethical Issues in Palliative Care" (see chap. 17 below), Pellegrino argues vigorously for the moral imperatives of treating pain and telling the truth (gently if necessary). He takes on the vexatious question of whether to continue hydration and nutrition in patients who are not terminal, but suffer from the permanent vegetative state. He quotes the eighteenth-century physician-ethicist John Gregory as saying that pain relief and efforts to "smooth the avenues of death, when inevitable," are central ethical duties for physicians.

The final two articles in this section address the question of futility in care at the end of life (see chaps. 18 and 19). Pellegrino notes that while the concept is centuries old and historically has denoted what he has dubbed "ineffective" care, the connotations of the term have rendered it much more ambiguous and subject to abuse. Futility ought not, he argues, be reduced to third party determinations of quality of life or economic considerations.

These articles equip the reader with essential tools for navigating the roiling ethical waters of care at the end of life. Almost all healthcare professionals will face these issues, making this preparation vital.

15

Withholding and Withdrawing Treatments: Ethics at the Bedside

Edmund D. Pellegrino, MD

Introduction

In his treatise *The Art*, Hippocrates defined the purposes of medicine this way:

> to do away with the sufferings of the sick, to lessen the violence of their diseases, and to refuse to treat those who are over-mastered by their diseases realizing in such cases that medicine is powerless.[1]

Further on in the same treatise, he adds:

> Whenever therefore a man suffers an illness which is too strong for the means at the disposal of medicine he surely must not expect that it can be overcome by medicine.[2]

In these words, the putative Father of Medicine recognized the limits of medicine and gave moral sanction to decisions to desist with treatment when it becomes futile. For most of medicine's history, physicians followed this Hippocratic dictum.[3] Only in the modern era, when medicine's capabilities expounded prodigiously, did the tendency arise to treat against nil odds. Physicians, understandably, wanted to offer their patients the full benefits of the miracles which medical science had conferred on medical practice.

But in the last two decades, it has become clear that medicine's miracles are not unmixed blessings. Mechanical respirators, artificial hearts, dialysis machines, and

---

Reprinted with permission from Edmund D. Pellegrino, "Withholding and Withdrawing Treatments: Ethics at the Bedside," *Clinical Neurosurgery* 35 (1989): 164–84.

1. Hippocrates, "The Art," in *Hippocrates*, vol. 2, trans. W.H.S. Jones (Cambridge, Mass.: Harvard University Press, 1967), 193.

2. Ibid., 203–205.

3. D.W. Amundsen, "The Physician's Obligation to Prolong Life: Medical Duty without Classical Roots," *Hastings Cent. Rep.* 8 (1978): 23–30.

resuscitation techniques can prolong the act of dying and at great financial, social, and emotional costs to individuals and society. Now the central ethical question is: When is it morally permissible or even mandatory to withhold or withdraw life-sustaining treatments? How is Hippocrates's moral dictum to be implemented amid the technical complexities of contemporary medicine?

This is the most frequent ethical dilemma in clinical medicine today. It is one which most of us will be forced to face not only in the treatment of our patients but in our own lives and in the lives of those for whom we act as surrogates. It is also a dilemma of special significance for neurological specialists because their expertise is the basis on which other physicians depend for those assessments of brain function that are so critical to the moral quality of the final decision.

In the last two decades moral and local sanction for withholding and withdrawing life-sustaining treatments has come from a wide variety of sources.[4] As a result, consensus is emerging on a moral algorithm to guide these decisions. My purpose in this essay is to examine the ethical issues and difficulties in applying this algorithm at the bedside.

My focus will be the gap between the seeming simplicity of the general schema for ethical decisions and its use in particular cases. Clinicians have the unique task of translating moral principles and rules into concrete decisions despite the uncertainties and uniqueness of each patient's experience of illness. This is what makes clinical ethics a more strenuous exercise than its classroom analog.

Despite patient and surrogate participation, ethics committees, and court opinions, the physician is still bound in a covenant of trust to the good of his or her patient. It is through the physician that any moral algorithm finally is expressed in a morally right or wrong action. The physician must, therefore, have a framework for ethical decisions, but he or she cannot accept any framework uncritically. Physicians must be able to make both technical and moral decisions to fulfill the obligation of trust inherent in the healing relationship.

For these reasons, every clinician must understand the substructure of any proposed moral algorithm. I will examine this substructure by looking at six questions crucial to its practical application. (1) Who should decide? (2) By what criteria? (3) How should conflicts among decision makers be resolved? (4) What

4. Council on Ethical and Judicial Affairs of the American Medical Association, "Withholding or Withdrawing Life Prolonging Medical Treatment," March 15, 1986 (unpublished opinion); Hastings Center, *Guidelines on the Termination of Life-Sustaining Treatment and the Care of the Dying* (Briarcliff Manor, N.Y.: Hastings Center, 1987); New York State Task Force on Life and the Law, *Life Sustaining Treatment: Making Occasions and Appointing a Health Care Agent* (New York: New York State Task Force, 1987); Office of Technology Assessment, *Life Sustaining Technologies and the Elderly* (Washington, D.C.: U.S. Government Printing Office, 1987); C. Pulhs, "Defining Death," *Br. Med. J.* 291 (1985): 666–67; President's Commission for the Study of Ethical Problems in Medicine and Biomedical and Behavioral Research, *Deciding to Forego Life Sustaining Treatment* (Washington, D.C.: U.S. Government Printing Office, 1983); Sacred Congregation for the Doctrine of Faith, *Declaration on Euthanasia* (Vatican City: May 5, 1980).

dangers inhere in decisions to desist? (5) Is there a morally prudent way to proceed in the face of uncertainty? (6) What fundamental issues remain unresolved?

The Moral Algorithm

The moral algorithm gaining acceptance today runs this way: Is the patient competent? If so, the patient has a moral and legal right to make his or her own decisions about acceptance or rejection of treatments of all kinds. These decisions take precedence over the wishes of physicians or family. If the patient was once competent but is now incompetent then we must seek some way to come as close as we can to what the patient would have wanted were he or she able to make the decision. The source of this judgment can be some advance directive or instruction (e.g., living will or durable power of attorney). In the absence of these, we seek the decision of a valid surrogate. If the patient has never been competent, (e.g., infants or those with intellectual disabilities or mental illnesses that would incapacitate them), a valid surrogate makes the decision.

The criteria to be used by the decision maker are not as easily decided upon as who makes the decision. Several criteria are in common use: diagnosis, prognosis, benefit and effectiveness of treatment, futility or burdensomeness of the treatment, brain death or permanent brain dysfunction, costs of care, quality of life, and age. If conflicts among decision makers occur, they must be resolved by conferences among healthcare team members and families, pastoral counseling, psychiatric assistance, ethics committees, or as a last resort, the courts.

Who Shall Decide?

The Question of Competence

The patient's competence to make his or her own decisions is the first and perhaps the most crucial decision in the whole algorithm. How we assess competence will propel the patient down two morally divergent pathways toward autonomy or toward paternalism. Clinicians must make as precise an assessment of competence as possible.[5] Even if, eventually, a psychiatrist is consulted, the decision to seek consultation already implies some judgment of competence.

Agreement on what constitutes competence, however, is not easy to come by. Usually, it is defined as a capacity to make a reasoned judgment about a particular clinical choice. 'This involves the capacities to receive information, recognize its relevance, understand the gravity of each option, make a choice consistent with one's own value system, and communicate it. Competence is a limited capacity. It does not entail the capacity to make all decisions or handle all of one's affairs. It is not a

5. T. Engelhardt Jr., E.E. Shelp, and M.A. Gardell, *When Are Competent Patients Incompetent? A Study of Informed Consent Determinations in Primary Care* (Dordrecht: D. Reidel, in press). [published as M.A. Gardell Cutter and E.E. Shelp, eds., *Competency: A Study of Informed Consent Determinations in Primary Care* (Dordrecht: Kluwer Academic Publishers, 1991)—Ed.]

legal determination. Indeed, legal incompetence does not preclude competence for ethical decisions. Competence does not require that the choices be agreeable to the doctor, family, or society, or a "reasonable" person. A person may be intellectually challenged, depressed, or psychotic in other spheres and still have the capacity to choose according to personal values. Nor is competence age-linked. Legal minors and also the very elderly may be competent.

For all these reasons, we should presume that patients can make their own decisions. The burden of proof is on those who deny the patient's capacity. Yet there are still too many physicians who consider patients "irrational" when they reject indicated medical treatment. Others genuinely doubt that any ill patient, particularly one acutely ill, can ever make truly competent decisions. The emotional impact of illness and the complexity of medical technology they say conspires against the possibility of truly informed decisions. Those who hold these views are the "strong" paternalists.[6] For them, the current dominance of patient autonomy is a violation of the dictum "first, do no harm."

The Competent Patient

The majority of medical moralists today, however, argue for the autonomy of the competent patient. I agree since I think there are few greater violations of beneficence than to override the patient's moral right to decide what is in his or her own best interests. To respect autonomy is to act beneficently; to violate it is maleficent.[7]

In actuality, the strong paternalists do not treat the patient by the use of force or a court order. More usually they violate autonomy indirectly by manipulating consent through the selective presentation or withholding of information. Even though the intent is the good of the patient, deception and coercion of this kind are morally inadmissible. Particularly reprehensible is the boast of some physicians, "I can get any decision I want by the way I present the facts."

In very acute situations, for example in the early hours of a severe burn or severe trauma, there may be some justification for erring on the side of treatment. This is a weaker form of paternalism. The same is true when competence is doubtful because of reversible disturbances of brain function resulting from shock, toxemia, or fever. The physician has an obligation first to treat these reversible causes and restore competence. As soon as this is accomplished, the wishes of the patient should be ascertained and followed.[8]

6. J. Childress, *Who Should Decide Paternalism in Healthcare* (New York: Oxford University Press, 1982).

7. E.D. Pellegrino and D.C. Thomasma, *For the Patient's Good: The Restoration of Beneficence in Healthcare* (New York: Oxford University Press, 1998).

8. E.D. Pellegrino, "Informal Judgement of Competence and Incompetence," in Engelhardt, Shelp, and Gardell, *When Are Competent Patients Incompetent* [in Cutter and Shelp, *Competency*, 29–48—Ed.]

When competence waxes and wanes, the last competent decision should prevail. The physician ought not to speculate that the patient may have changed his or her mind. On the other hand, competent patients should be permitted to change their minds whenever they wish and are competent mentally to do so.

An unsettled question is whether competence is an all-or-none or a threshold phenomenon. From the threshold view competence and incompetence are seen as on a continuum in which some combination of criteria is deemed sufficient to require compliance with the patient's wishes. Others insist on a more precise cut-off point. Still, others measure competence in relationship to the gravity of the decision. They are less likely to consider a patient competent if he or she refuses an effective lifesaving treatment than if he or she refuses a marginal or questionable one. On this view, the degree of autonomy permitted varies inversely with the gravity of the decision.

If a patient is adjudged competent, he or she is morally entitled to autonomous decisions. But how absolute is this autonomy? Is the physician merely an instrument of the patient's wishes, as some are beginning to suggest and even insist? There are good moral reasons for limitations on the patient's autonomy.

One limitation is when the patient asks the physician or institution to do something contrary to their religious or moral values, for example, the issue of asking a Catholic physician to assist in euthanasia or abortion. The physician is as much an accountable moral agent as the patient. The physician's values cannot be subverted any more than the patient's.

A second limitation is based on the principle of justice. If the patient's wishes result in serious, palpable harm to another person (e.g., a human immunodeficiency virus (HIV) seropositive male who wishes this knowledge to be kept from his pregnant wife), then autonomy is limited. More difficult is the question of how much liberty patients and families should have in demanding very expensive treatments.

The Incompetent Patient

If the patient is incompetent, then the decision is made through some surrogate mechanism. The moral requirement here is to come as close to what the patient would wish were he or she able to decide, not what the physician or surrogate would wish if he or she were the patient. When an advance directive is at hand a valid living will, or other verifiable statements, it "substitutes" for the patient's will. In the absence of advance directives, the autonomy of the patient is transferred, first to his or her chosen surrogate and then to others if the patient has not made a choice.

Surrogates must meet several tests of moral validity whether they are family members, friends, or court-appointed guardians: First, they must meet the same tests of competence already discussed for the patient's decisions; second, they must be free of conflict of interest, financial or emotional; and third, they must know the patient's values well enough to make a so-called substituted judgment for the patient, that is, they should provide evidence that their decision reflects the patient's values.

Families do not always, or necessarily, meet these requirements. A biological relationship does not confer automatic moral authority. Often family members have not seen the patient for a long time or have been estranged from him or her for years. Also, they may favor overtreatment consciously or unconsciously to assuage guilt for their own animosity to the patient or favor undertreatment in order to accelerate an inheritance or to take revenge for actual or imaginary injuries inflicted on them by the patient.

The physician has a special obligation to be the advocate for the patient's best interests. He or she must therefore make some effort to ascertain the moral validity of surrogate decisions, to document the fact that a surrogate knows the patient and his or her values, and to determine that the surrogate is free of obvious conflicts of interest. The surrogate's decision must be in the best interests of the patient. This does not call for a full-scale investigation of surrogate motives, etc. but it does entail the duty to make a reasonable assessment of possible conflicts of interest. It is necessary also to comprehend and take into account the normal anxiety and guilt a surrogate might experience in making serious decisions about another person's life.

The physician is the steward of the interests of the incompetent adult or infant. He or she cannot escape moral accountability by a "Pontius Pilate" act—blaming others for decisions which he or she carries out. His or her primary covenant is with the most vulnerable person—the sick, comatose adult, or infant—not with society, the family, or the HMO.

There will be times, especially in emergencies, when an advance directive or surrogate decision maker is not available. Then the physician's moral obligations are sharpened because of the potential conflicts with his or her own values. Appointment of a legal guardian may be advisable. Consultation with others is mandatory. Whenever possible, a surrogate ought to be appointed to relieve the physician of any actual or presumed conflict of interest. Better mechanisms are needed for the assignment of responsible surrogates under these circumstances.

When there is doubt about what the patient would wish, the patient should be treated. The moral onus rests on anyone who chooses to shorten life. The supposition is that most patients would wish to live. Physicians must be especially careful to avoid decisions not to treat that are based on their own value systems or on their evaluation of the quality or burden of the patient's life or the value of the patient to society. If the treatment is medically indicated, it should be instituted, at least until valid surrogates are available or the patient recovers sufficiently to act on his or her own behalf.

Up to this point we have dealt with procedural ethics, the ethics of the process of decision making. We now turn to substantive ethics, the moral quality of the decision itself.

By What Criteria?

Whoever makes the decision, that decision itself must be grounded in morally valid criteria. Here the clinician has grave obligations as the technical expert upon whose judgments and clinical knowledge so much of the ethical decisions must depend. Substantive issues in clinical medicine will intersect with ethics, economics, religion, and even politics. The physician's irreplaceable expertise is in his or her knowledge of the technical facts. If they are shaky, the whole process of ethical decision making will be distorted.

Diagnosis and Prognosis

In every case, diagnosis and prognosis are the first and indispensable criteria. They are essential to deciding whether a treatment is futile or, to use Hippocrates's phrase, "beyond the means at the disposal of medicine." It is the clinician's responsibility to make as accurate an assessment as possible of the chances for recovery, its completeness, and its residual disabilities for each option in management of the patient's problems. But as every experienced clinician knows, the prognosis is the weak sister of the clinical arts.

Even such a frequently used prognostic criterion as a terminal or preterminal state is fraught with difficulties. Such designations are notoriously variable and arbitrary. Yet, many physicians in good conscience will not entertain withdrawal of treatment unless they are morally certain the patient is in a "terminal state."

But how is this state defined? At the one extreme, we may all be "preterminal" in that we shall all die. Some legal jurisdictions define "terminal" as any prediction of death up to one year. I find it safer to consider a patient terminal when death, to the best of our limited prognostic abilities, is foreseeable within hours, days, weeks, up to one month. This is admittedly arbitrary but some practical limit must be set if decisions are to be made. A more debatable alternative is to apply the criteria for withdrawal of treatment whether the patient is terminal or not.[9]

The term preterminal is too vague to carry much weight in ethical decision making. I would suggest eliminating it. The same can be said of the distinction between withholding and withdrawing treatment. Most physicians have greater psychological difficulty withdrawing treatment they have started than withholding them. Actually, a stronger case can be made in favor of starting treatment. This at least affords time for a better grasp of the clinical and moral context. Nonetheless, the distinction, from a moral point of view, seems to serve little purpose.

One of the responsibilities of the profession is to gather the kind of data needed to support reliable prognostic statements. Currently, attempts to "score" the prognosis of patients in intensive care units with respect to ethically significant decisions are promising.[10] Equally crucial are the data of properly controlled clinical trials and consensus conferences. We still have a long way to go before diagnosis and prognosis become quantifiable scientific enterprises. We ought to admit these uncertainties to patients, families, and ourselves.

9. *Bouvia v. Superior Court* [of Los Angeles—Ed.], 179 Cal. App. 3d 1127, 225 *Cal. Rptr.* 297 (Cl. App., Apr. 16, 1986).

10. Pulhs, "Defining Death."

Brain "Death" as a Criterion

Nothing illustrates the prognostic dilemmas more acutely than the knotty questions that surround the use of brain death and dysfunction of the person as criteria for withdrawal of treatment. Here ethics is very much dependent upon clinical and physiological data, much of it still incomplete and still debatable. It would be presumptuous indeed for an internist to lecture neurologists and neurosurgeons on such matters as total brain death, brainstem death, neocortical death, or persistent vegetative states. Indeed, brain death criteria have been carefully set forth.[11] But there is still debate on when the "person" is dead.

From the ethical point of view, one can ask: Which of these criteria is indicative of death of the person? This question is essential in establishing a moral foundation for terminating life-support systems, artificial feeding and hydration, removing organs for transplantation, or writing do not resuscitate (DNR) orders.

Some neurologists like Pallis have made a strong case for equating death of the person with death of the brainstem, particularly the ascending reticular activating system.[12] Others define the "point of no return" as death of the neocortex, in which the brainstem is spared but patients remain in a persistent vegetative state.[13] Such patients are considered incapable of ever regaining conscious or sentient states. Some consider them dead as "persons," if not biologically dead, because they will never be able to regain the faculties that distinguish us as humans. Others disagree strongly and require "total" brain death to consider the person dead.[14]

These questions are at the intersections of physiology, metaphysics, theology, and ethics—a very busy and vexed crossroads indeed. The debate is, however, not a mere academic charade. It makes a very great difference to the practical decisions about what we consider ethically permissible. Like it or not, we cannot avoid implicit assumptions about what we mean by personal death. When we deny meaning to this question we have already taken a position. The intensity of the debate is obvious in both the medical and philosophical literature, and it stems from sharp differences in philosophical presuppositions.

The neuroscience community has the expertise and thus the obligation to define the physiological dimensions of brain dysfunction as precisely as possible. Ethicists, theologians, and philosophers for their part must be familiar with the relevant findings of neurobiology. It seems unlikely that consensus will be forthcoming soon, either in the scientific or the ethical communities. But at a minimum, these communities must be in a continuing dialogue with each other.

11. President's Commission, *Deciding to Forego Life-Sustaining Treatment.*

12. Pulhs, "Defining Death."

13. P.A. Byrne, S. O'Reilly, P. Quay, and P. Salsich, "Brain Death—the Patient, the Physician and Society," *Gonzaga Law Rev.* 18 (1982–1983): 429–516; D.A. Shewmon, "The Metaphysics of Brain Death, Persistent Vegetative State and Dementia," *The Thomist* 49 (1985): 24–80.

14. R.L. Barry, "Ethics and Brain Death," *The New Scholasticism* 61 (1987): 62–98.

For the moment there is some moral uncertainty in what is becoming the most common ethical decision—withholding or withdrawing treatment from brain-damaged persons. There is little question that total brain death is sufficient warrant for discontinuing all life-sustaining measures since the chances of recovery are nil. Personal death and brain death are synonymous here.

As to partial brain death, I incline to Pallis's view. As long as there is evidence of brainstem function, I feel on surer moral ground to continue whatever measures meet the criteria of benefit and effectiveness, exercising particular caution to require futility and burdensomeness for the removal of food and fluid. The death of the person is uncertain, then the decision to desist is also uncertain.

With presumed neocortical death and persistent vegetative states, I am inclined to caution erring on the side of treatment until the prognosis becomes morally certain. As studies on the prognosis of patients with persistent vegetative states become more secure we should be able to refine these decisions further.

Effectiveness and Benefit

Two important criteria are the effectiveness and benefit of proposed treatments. The two are not synonymous. Effective treatments are those which demonstrably alter the natural history of an illness or alleviate an important symptom. Beneficial treatments are those which bring some good for the patient, not simply medical benefit, but benefit in terms of his or her value system. Antibiotic treatment of pneumonia in a patient dying of metastatic malignancy is effective, but not beneficial if it merely postpones the moment of dying when neither patient nor surrogate wishes to prolong the dying process. There are times when patients might want to postpone death, for example, to see a child born or a grandchild graduate. Then the treatment for pneumonia would be both effective and beneficial. Another example is in the use of analgesics. They are effective for pain relief in terminal cancer and therefore beneficial, but not effective so far as the natural history of the disease is concerned.

Ordinarily, treatments ought to be both effective and beneficial to warrant their use. This applies to life-support measures like respirators, artificial hearts, dialysis, or cardiopulmonary resuscitation (CPR) as well.

Artificial feeding and hydration are in a special category because of their deep symbolic meaning and their close identification with care. There is substantial debate about whether they should be classified like any other medical treatment or regarded as care which would always be continued even when other life-sustaining measures can validly be withdrawn. The Council on Ethical and Judicial Affairs of the American Medical Association and the recent Hastings Center task force take the former view while the Pontifical Academy of Sciences classifies artificial feedings as care, which should never be withdrawn.[15]

15. Council on Ethical and Judicial Affairs, "Withholding or Withdrawing"; Hastings Center, *Guidelines on the Termination*; Pontifical Academy of the Sciences, "Medical and Legal Questions of

Clinicians differ on this point depending upon their fundamental beliefs about life and its meaning. Personally, I am very cautious about removing artificial feedings. The intent is too often simply to accelerate death. The possibilities of abuse and laxity of standards are great. I believe withdrawal is permissible under two conditions: (1) the patient is competent and terminal and refuses food and fluid; and (2) the patient is incompetent and terminal and artificial feedings are both futile and excessively burdensome. With modern methods these criteria should be met infrequently. However, in a terminally ill patient hyperalimentation that is neither effective nor beneficial is morally unwarranted.

Futile and Burdensome Treatment

Most moralists agree that a treatment that is futile or excessively burdensome ought to be discontinued. But once again the problem is how to define the terms "futile" and "burden."

Ordinarily a treatment with little chance of altering the natural history of the primary disease process can be considered futile. But how poor should those chances be? Allowances must be made for differences in values among physicians, families, or patients. Some like to defy the odds; others accommodate them. Subtle differences of this kind are often a source of conflict at the bedside.

The same ambiguities accompany assessments of burdensomeness. No clear-cut definition is possible. What is a burden to one is to another a challenge to be overcome. Competent patients can make these determinations for themselves. But it is difficult to tell what is burdensome for a comatose or otherwise incompetent patient. Externally, signs of distress may be reflexes below the level of consciousness or they may signal genuine suffering.

Opinions vary about whether patients in coma or with other manifestations of brain dysfunction suffer when food and fluids are withdrawn or even how much they record of the manipulations they are put through. Experienced clinicians are familiar with the patient who recovers and describes in detail the events around her bed while she was in a "coma." Such phenomena are deterrents to facile judgments about a patient's capacity for suffering.

Often the burden is more on the family and the medical care team who must carry out the nursing care, pass the nasogastric tube repeatedly, do the feedings, dress the bedsores, and come in day-by-day to see no palpable result to their efforts.

Burden can also be interpreted in terms of justice to third parties, for example, the diversion of resources in time, energy, personnel, and equipment from persons

---

Artificial Prolongation of Life," *L'Osservatore Romano*, November 11, 1985. [R.W. Chang, S. Jacobs, and D. Lee, "Use of APACHE II Severity of Disease Classification to Identify Intensive Care Unit Patients Who Could Not Benefit from Parenteral Nutrition," *Lancet* 8496 (1986): 1483–87, doi: 10.1016/s0140-6736(86)91511-4, PMID: 2873287. This note is listed in the references of the original but missing a citation in the text. It is included here in a hopefully likely place so as not to leave it out altogether.—Ed.]

who might profit to those who might not. This is a genuine problem in some hospitals and nursing homes. It promises to increase as more patients are kept alive in impaired states of functioning. Its solution rests with the ethics of resource allocation, a topic of its own.

Attractive as it sounds, the criterion of benefit/burden ratio is also fraught with difficulties. There is no formal calculus of benefits and burdens that is generally applicable because both benefit and burden are value-laden judgments. There is a glimmer of hope that benefits and burdens—which ought to be couched in the patient's terms—may become semi-quantifiable with improvement in the analytical methodology of clinimetrics.[16] This has yet to be tested in ethical decision making but any attempt to measure the intricacies of physician-patient interactions in decision making would be welcome.

I mention these difficulties in definition not to protest the use of the criteria of futility and burden but to warn against their too easy acceptance. As with quality of life, which I shall discuss below, no one can decide what is a "burdensome" life for another. I am particularly wary of presuming that the "burden of living" would be too much for a handicapped child, or even one with Down's syndrome, and thus justifying the withholding of indicated treatment.

When we move from clinical and physiological criteria to the psychosocial and economic, the issues are even more problematic. Should quality of life or economic costs be factored into the decision?

There is no question that many clinicians, families, and even courts take "quality" of life as a valid criterion for desisting from treatment, especially in the aged or in disabled and handicapped infants. Quality of life is a morally defensible criterion only for the competent patient. Only the competent patient can judge what quality of life means in terms of personal values, religious beliefs, or life plans within the limitations on autonomy. Only the patient can decide when life is so burdensome that it is not worth living.

With the incompetent patient—and especially with the never competent (those with mental disabilities or illnesses, or infants)—we have no idea what constitutes a quality of life from the patient's point of view. It is impossible to decide what is a quality life for anyone else. The opportunities for abuse, by imposing one's own values, by devaluating certain categories of persons, or by making them objects of some social philosophy are genuine.

Economics

The same can be said for economics as a criterion as for quality of life. Competent patients have a moral right to decide not to mortgage or deplete their family's resources and can reject costly treatments. They can stipulate this in a living will or through a durable power of attorney to become effective should they become incompetent.

16. A.H. Feinstein, *Clinimetrics* (New Haven, Conn.: Yale University Press, 1987).

But with an incompetent patient who had made no such stipulation or with infants who cannot make one, economic factors cannot be given moral weight. This is because the physician's covenant requires that he or she act in the patient's interests. This means that the physician must offer his or her patient the treatments that meet the criteria of benefit and effectiveness, as defined above. Patients expect physicians to be their advocates, not the "gatekeeper" of society's resources.[17]

Unless we change the nature of the covenant with the patient, the physician is morally bound to place the sick person first. This is not to deny the reality of the economic impact of individual decisions on others. Nor is it to sanction a "blank check" approach to medical care. Rather, it emphasizes that questions of social and public policy cannot be resolved at the bedside of individual patients. They must be the subject of public debate. Their focus is on categories of patients and services, not individual patients. Any decision to limit access to indicated or effective treatments must be made publicly and must apply equally to all members of society falling within specific disease categories.

There are other third-party interests that some moralists defend as legitimate criteria for removal of treatment. Among them we can mention the emotional and physical strain on families; the heavy demands on institutions and personnel; and the diversion of resources from the young to the old, from the healthy to the handicapped, and from other social goods to medical care in hopelessly ill patients. These are undeniable human dilemmas associated with the continuing care over months or years of a person in a persistent vegetative state or a child with multiple handicaps and mental retardation.

While these are genuine moral issues, the clinician must nonetheless focus on the interests of the person who is ill. The physician is the steward of the interests of the patient. 'The patient's vulnerability, dependence, and inability to defend his or her own interests generate the moral obligations of that stewardship.

Age as a Criterion

There is a growing tendency among responsible medical moralists to suggest, either through voluntary actions or public policy, that limits ought to be placed on the amount and kinds of care given to the elderly.[18] Some suggest that when competition for the same scarce resource occurs, for example, an organ for transplantation, an intensive care unit (ICU) bed, etc., preference should automatically go to the young. It is certainly true that a disproportionate fraction of the costs of health care arises from the increasing longevity and expanded medical needs of the aged. As a result, if age is not explicitly used as a criterion, it is often implicitly a decisive factor.

17. E. Pellegrino, "Rationing in Healthcare: The Ethics of Medical Gatekeeping," *J. Contemp Health Law Policy* 2 (1986): 23–45.

18. D. Callahan, *Selling Limits: Medical Goals in an Aging Society* (New York: Simon and Schuster, 1987); Office of Technology Assessment, *Life-Sustaining Technologies.*

Yet, reflection on the moral validity of such a criterion raises serious questions. What is the cut-off age to be? Who will decide? Will the ability to pay make a difference? Does each human life have the same intrinsic value? Are the aged less worthy of care simply because they are aged? Is a 29-year-old less worthy of care than an 18-year-old?

It would be an act of admirable civic responsibility and charity if the elderly were to voluntarily arrange that their care be discontinued when it becomes fruitless. This should be the aim of a widespread educational program. But for clinicians or families to decide against effective and beneficial treatments on the basis of age alone violates the trust that the sick person must place in the physician or those who act as surrogates.

Age alone is a poor indicator for moral decisions. Chronology, physiology, and quality of life do not parallel each other in any precise way. Such an arbitrary criterion as age cannot be trusted to anyone—even physicians who, after all, have no special predisposition for compassion or for acuity in moral reasoning. The morally defensible way to use age as a criterion is to weigh it along with other clinical factors in deciding whether the treatment will be effective and/or beneficial. Age will on physiological grounds rule out certain procedures like cardiac transplantation because the likelihood of success is low. But this is a factor inherent in the procedure, not external to it. If techniques improve, older patients can become medically eligible.

From the ethical point of view, the aged are entitled to more rather than less concern. They are far more vulnerable to usurpation of their rights of self-determination than the young. Their competence to make decisions is more often questioned. Those who care for them and even their families are prone to paternalism under the guise of beneficence.

Finally, distinction must be made between age as a factor in social policy and age as a factor in individual clinical decisions. If the aged participated in the decision, it is conceivable, but to my mind still perilous, that morally licit policies could be fashioned to include age as a factor. But such policies would have to bear the most critical scrutiny and be motivated by the direct necessity.

Conflict Resolution

Even when the clinician has negotiated the ethical shoals of "who shall decide?" and "by what criteria?" he or she faces the dilemmas of resolving conflicts between, and among, him- or herself and the other participants in the decision. This is inevitable in all democratic and morally pluralistic societies where moral consensus is disappearing rapidly. As a consequence, deeply held convictions may clash at every level from the meaning of the physician-patient relationship and what constitutes beneficence, justice, or autonomy, to the ethical theory that should prevail and whether the ultimate source of all ethics should be man, God, or society. At whatever level they arise, conflicts must be managed in a morally defensible way. I can only suggest a schema that attempts to meet this requirement.

First of all, it is essential to locate the source of the conflict. The least negotiable conflicts are based in religious or philosophical convictions. Examples are the conflicts about abortion, sterilization, voluntary euthanasia, discontinuing life-support measures (especially food and fluid in nonterminal patients), organ donation, etc. When a clinician encounters personal conflicts at this level, two courses are open. If a compromise does not violate an important moral principle, cooperation with the request is possible. If, however, the conflict docs compromise such a principle, the health professional should withdraw after arranging for someone else to care for the patient. If the health professional feels grave harm is being done, he or she may feel morally impelled to intervene by objecting, reporting the incident to the responsible authority, or even seeking court action—courts as a last resort, they are not qualified to make moral decisions. A court decision leaves the moral issue unresolved but at least allows some action to be taken.

Courts, however, are essential when we perceive that certain fundamental rights may be violated. Examples are court orders to permit transfusion of infants born to Jehovah's Witnesses, to withdraw feeding tubes from totally brain-dead patients, to restrain an institution's decision to withdraw feeding tubes or treatment deemed beneficial and effective in newborn infants. Obviously the proponents of diametrically opposed views may feel equally impelled to resort to law to fulfill what they perceive as their moral duty to prevent harm to others.

If the conflict is at a less fundamental level it may be resolvable by negotiation, by consultation with chaplains, psychiatrists, social workers, and other health team members, or by use of an ethics committee. When conflicts arise, clinicians have a responsibility to employ all measures necessary to assure that the moral values of all parties are heard, respected, and given due weight. The fundamental principle is that no one has moral authority over another and that the moral rights of all must be respected. How best to protect each person's moral beliefs in a morally divided society is a problem of the greatest public importance today.

All too often conflict occurs among family members or other valid surrogates. In a few states a hierarchy of surrogates has been established by statute.[19] This does not resolve the moral dilemmas of who is best able to make a substituted judgment. What is morally right and what is legally defined may be in conflict. The physician, as part of the covenant with an incompetent sick person, must work with the surrogate chosen by the patient or, lacking this, the one who most closely satisfies the criteria for moral validity set forth above.

When law and ethics conflict, the dilemma is particularly vexing. No one can tell another person how much pain or penalty he or she must suffer to defend a moral principle even against the law. In a democratic society those who feel morally obliged to disobey a law must expect to suffer the consequences (e.g., conscientious objection in wartime). This can become a genuine issue in cases in which courts may decide on an action that violates the philosophical or religious beliefs of a health professional or institution.

19. J. Areen, "The Legal Status of Consent Obtained from Families of Adult Patients to Withhold or Withdraw Treatment," *JAMA* 258 (1987): 229–35.

So far, resolution of such conflict has been achieved by transferring the patient to another physician or hospital. But opinions like those of Judge Compton in the *Bouvia v. Superior Court* case are worrisome. In that case, the judge implied that the physicians had a duty to assist the patient to fulfill her wish to die by removing a nasogastric tube. Even more direct are the cases of court-mandated obstetrical interventions.[20]

As the participation of team members becomes more explicit, the conflicts among them become more serious. No member of the health team has moral authority over the others, despite differences in technical and legal authority. The physician, for example, cannot order a nurse to turn off a respirator if she or he thinks this is an immoral act.

When intra-team conflicts occur, the usual mechanisms must be invoked—negotiation, consultation, ethics committees, or ultimately, and always reluctantly, the courts. The central ethical fact is that every competent person is a moral agent, responsible and accountable for his or her own acts. We cannot escape responsibility for violations of conscience by taking refuge in fear of malpractice suits, hospital or public policies, financial exigency, or even the dictates of law.

Institutional mechanisms are badly needed that recognize the fact that conflicts of conscience may occur in managing the decisions to withhold or withdraw treatments. Opportunities for airing the grounds for conflict should be provided, and it should be possible to withdraw from a case when one's conscience requires it without disrupting patient care or imperiling one's career. The same thing applies to the sensitive question of responsible "whistle-blowing" when impaired physicians, nurses, or administrators are the issue.

In short, a pluralistic society must encourage moral behavior, recognize that differences in what constitutes morality exist in a democratic society, and provide ways to reconcile conflict in a peaceable manner.

Inherent Dangers

Even if morally qualified people are making the decisions and using morally defensible criteria, there are dangers inherent in decisions to withhold or withdraw treatment. They must always be kept in mind since every indication is that the criteria for withholding and withdrawing treatments will be further liberalized.

I have already alluded to the precarious state of clinical prognostics. If prognosis is in error the whole train of ethical decisions is switched in the wrong direction. But there are other dangers. One is the tendency gradually to devalue the lives of certain types of patients who require prolonged, costly care—the old, the retarded, the social misfits, the chronically insane, the street people, and the badly handicapped infant and child. The "slippery slope" need not become a reality. But it is always a danger, as the all-too-fresh memory of German medicine in the Hitler era so graphically reminds us.[21]

20. V.E. Koldcr, J. Gallagher, and M.T. Parsons, "Court-Ordered Obstetrical Interventions," *N. Engl. J. Med.* 316 (1987): 1192–96.

21. I. Alexander, "Medical Science under Dictatorship," *N. Engl. J. Med.* 241 (1949): 39–47.

Another danger is to discourage innovative treatments and research in groups of patients who, consciously or not, have already been written off. We do not know what technological advances may produce in even the most hopeless situations. If we write off certain types of patients, their treatment ceases to be a challenge.

Moreover, the economic pressures that seem to weigh heavily, even for sensitive moralists, may be overdrawn. There are ways to care for patients more economically than at present. Also, if we examine our expenditures for things like alcohol, tobacco, advertising, gambling, cosmetics, and even potato chips, as well as the money wasted on unnecessary workups and treatments that do not meet the criteria of effectiveness or benefit, we will find many billions that could be redirected to health care had we the will to do so.[22]

The Neonate

I have not mentioned here the special difficulties in withholding and withdrawing treatments in infants, especially the newborn. There is a large literature here that I cannot hope to summarize. Suffice to say that infants are in a special state of vulnerability. They have no opportunity to express values. They are totally at the mercy of physicians, parents, and guardians. The emotional, fiscal, and social burdens of keeping handicapped infants alive are all too obvious. The tendency to decide that "quality of life" or "burdens of living" are too great to be visited on the innocent is strong.

Equally distressing are the difficulties of diagnosis, prognosis, and assessment of brain function in neonates, upon which so much depends. If there is caution to be exercised in the care of incompetent adults, those cautions must be exercised with even greater sensitivity in infants. This is too complex a topic to be covered in this paper except to say that the same principles can apply but with much greater sensitivities to the uncertainties.[23]

Moral Prudence in Emergency Medicine

All of the dangers and complexities I have been outlining take on special urgency in emergency care. Usually, there is little time for prolonged ethical discussion. Actions must be taken quickly and under stress. The question of withholding treatment often is more urgent than withdrawing treatment. How vigorously does one treat the severe third-degree burn, the barely alive suicide victim who had plunged ten stories, the severely brain-injured auto accident victim, when the prognosis in such cases is so dismal? Why start what must almost inevitably be discontinued? When should such patients be resuscitated?

22. Pellegrino, "Rationing in Healthcare."

23. E.D. Pellegrino, "The Anatomy or Clinical Ethical Judgments in Perinatology and Neonatology: A Substantive and Procedural Framework," *Semin. Perinatal.* 11 (1987): 202–9.

Who should get the remaining ICU bed? Who should be transferred out? What criteria should be used to make these decisions? Should it be first come, first served? medical need? value to society? merit? equality? ability to pay? lottery? age of the patient?

The criteria and principles I have outlined in the body of this paper are as applicable in emergency as in chronic care. What is different is the telescoping of time intervals, the uncertainties of patient or surrogate intent, and the greater difficulties of accurate prognostication. To deal with this telescoping, I suggest a few guidelines that try to assure morally prudent decision making:

(1) The moral onus is on those who move to discontinue treatment. When in serious doubt about prognosis, competence, or the wishes of the patient or surrogate, the patient should be treated.
(2) Withhold emergency treatment rarely unless prognosis and other criteria are indisputable.
(3) Starting treatment does not entail a commitment to continue to the "bitter end."
(4) Allow time for the clinical picture to emerge. Frequent assessments of prognosis, competence, and patient or surrogate wishes are essential. "Vector in" on the decision; don't "jump in."
(5) Moral decisions based on diagnoses of brain dysfunction and destruction must be carefully scrutinized, especially in infants and old people. Consultation and confirmation must be sought.
(6) Beware of categorizing patients too early or changing their categorization too late.
(7) Consult frequently with all parties to a decision.
(8) Ask not "Is this what I would want done to me?" but "Is this what I think the patient would want?"
(9) Withholding or withdrawing treatment must always be associated with continued care for the patient.
(10) Perform ethical postmortems, rethinking your decisions, especially those that had to be made in haste.
(11) Examine all formulae and algorithms (including this one) with a critical eye.

Some Unresolved Issues

The emerging consensus on discontinuance of life-sustaining treatment is largely in the procedural realm. It deals primarily with the ethics of the process and less with the substantive morality of the decisions themselves. But we must keep in mind that, because philosophical premises may be incorrect, it is possible to follow a morally correct procedure and arrive at a substantively immoral decision.

Thus there are still unresolved fundamental philosophical problems in the current decision-making schema. We should continue to examine and clarify them even though they may seem abstruse to practical people. Is there a real difference between withholding and withdrawing treatment? Is there a distinction of kind or only of degree between killing and letting die, between active and passive euthanasia? Is personal death synonymous with total, neocortical, or brainstem death? Is passive euthanasia the same as assisted suicide or homicide? Is there a difference between withdrawing treatments because they are burdensome and futile and doing so because of quality of life considerations? Is not the intent the same in each case—hastening the death of the patient? If it is permissible to "allow" a patient to die because an illness is incurable or burdensome, why is it wrong to "assist" dying?

These questions still recur in discussion of withdrawal or withholding of life-sustaining measures. They reflect differences in our concept of the purpose, destiny, and meaning of human life. While the moral algorithms emerging from various groups are providing some answers, the deeper questions still remain for many people because of differences in deeply held religious and philosophical beliefs. A growing issue is the extent to which these beliefs; should be respected when they run counter to law or public or institutional policy.

These fundamental questions are not the kind that can be settled by either plebiscites, mandates, or negotiation. They demand a continuing dialogue among ethicists, theologians, clinicians, and policymakers. Out of that dialogue can come those refinements and fine-tunings needed to make our clinical decisions complete.

We can agree with Hippocrates that there should be limits to medicine. But deciding when to desist is far more complex for us than for him. He did not face the prodigious powers of today's medicine and the difficulty of balancing their benefits and harms. Yet, paradoxically, we have the same tool for making our decisions that he had: the discipline of ethics, a discipline born in his era. The more technologically advanced we become, the more the physician must temper technical prowess with ethical sensitivity. "Doing" ethics has become as crucial as "doing" science for anyone who aspires to be a competent clinician.

16

Withholding Treatment, Surrogate Decisions and Rationing: Some Selected Aspects

Edmund D. Pellegrino, MD

THE REMARKS OF MY COLLEAGUES WERE CLEAR, concise, and authoritative. They set a standard which I will find hard to emulate. Rather than summarizing each paper, I will select one or two issues in each and offer some comments on their ethical implications. [NB: The following remarks refer to the manuscripts collected in *Health Care, Law, and Ethics: The Harold and Jane Hirsh Symposium,* edited by Richard F. Southby and Harold L. Hirsh (Washington, D.C.: The George Washington University, 1989).]

Surrogate Hierarchy

I will start with Professor Areen, whose focus is on the decision-making process in the matter of withholding or withdrawing medical treatments. Dr. Areen's focus was the surrogate or proxy decision and the role of the family. She described the current trend in the courts in several jurisdictions to establish an order of preference or legal hierarchy among surrogates or proxies. Dr. Areen focused on the decision-making process in which the matter of withholding treatment or not starting treatment or withdrawing treatment that's been started evolved. I think there are a series of questions that boil down to fundamentally three: Who decides? Second, how, under what circumstances, what procedures? And third, by what criteria? Dr. Areen spoke primarily of the question of who decides, and she talked about surrogate and proxy decision making and the role of the family.

Reprinted with permission from Edmund D. Pellegrino, "Withholding Treatment, Surrogate Decisions and Rationing: Some Selected Aspects," ed. Richard F. Southby, PhD, and Harold L. Hirsh, MD, JD (The Harold and Jane Hirsh Symposium, The George Washington University Health Policy Series) *Health Care, Law, and Ethics* 3 (1987): 151–72. Courtesy of The George Washington University.

The aspect I would like to illustrate first is what I consider to be a division between the legal criteria for a valid surrogate or proxy decision, which is what the courts are moving toward in making out a hierarchy of decision-making persons, and giving certain powers to proxies and surrogates. However, there are also a morally valid set of criteria or criteria for a morally valid surrogate. And these may not necessarily be the same as a legally valid surrogate.

Moral Criteria for Surrogates

To be considered a morally legitimate surrogate, a person must meet the following criteria:

First, the person must be competent, just as the patient who makes his own decisions must be competent. In the case of a decision by proxy or surrogate, the decision-making power of one person is assumed by another. Because the moral right of autonomy is, in a sense, transferred to the surrogate, he or she becomes culpable of making a reasoned judgment on behalf of the incompetent patient.

To meet the second criterion, the surrogate must know and reflect the values of the person whom he or she is representing, otherwise there would be no possibility of a true substituted judgment. This presents us with the very difficult task of ascertaining whether the surrogate has had recent and intimate enough contact with the person whom they represent to, in fact, be considered a genuine "representative." This is a real problem in today's mobile society where families are not necessarily the ones who know the patient's values best. Moreover, a biological relationship does not automatically confer moral competence, nor does it assure a true substituted judgment—a judgment that the patient, if able, would have made. I hope that the courts will not ignore these facts when establishing their hierarchies of decision makers. Someone not included in the statutory hierarchy of decision makers may have a much greater knowledge of the patient's values than legally designated surrogates.

That means that in some way we must ascertain whether in fact, they've had contact with this person. This is a real problem in today's society which is so mobile. Families may not know what the decisions would have been of those to whom they are biologically related.

Moral Aspects of Surrogateship

So as in the first instance when a biological relationship does not confer moral competence, it also does not confer familiarity with the decision that would have been made. And therefore I hope some of these statutes begin to not ask the moral questions themselves but to take into account the moral question that I'm raising of how do we, in this hierarchy of designated persons, account for the fact that someone with no biological connection whatsoever may have a greater familiarity. A third criterion for morally valid surrogateship is that there should be no conflict of interest between the surrogate and the patient whom he or she represents. Unfortunately, this is not always the case. Families or potential heirs may be consciously or unconsciously motivated to under-treatment in order to hasten an inheritance, to relieve themselves of the burdens of caring for an elderly or chronically ill parent, or to prevent erosion of family finances by prolongation of the life of a seriously ill patient. If they are to be accepted as morally legitimate substitute decision makers, they ought to be free of the more obvious or dangerous conflicts of interest that can compromise their capacity to act on behalf of the other.

We know unfortunately in this messy world of ours that one may be impelled to under-treat if one is waiting for the settlement of an estate and one may convince oneself rather easily that "dear old dad" really wouldn't have wanted to be treated this way, he wouldn't have wanted to "suffer this way." But of course that may mean that the Rolls Royce may stay in the garage longer than you wanted so perhaps it would be a good idea to move a little bit faster or to assume that he would not. We're not imputing evil motives but simply saying that, humans being human, it is entirely possible that that may be the motivating factor.

As for the patient for whom they are responsible, healthcare professionals and institutions have responsibilities that transcend whatever statutory requirements may be established for legally valid surrogate decision making. They have an obligation based on this duty of beneficence to ascertain whether or not the surrogate meets these criteria of morally valid decision making. Although this does not require exhaustive investigation into the motives of surrogates, it does call for sensitivity to the presence of conflicts of interest. The health professional should find and work with those who give the best evidence of really knowing the patient's mind even if they have no legal relationship with the patient. This means too, in instances where the moral credibility of a surrogate is in question, that the health professional has an obligation to protect the patient and to invoke legal procedures to challenge the morally suspect surrogate.

Emotional Factors: Biological vs. Non-Biological Surrogate

Another situation we see not too infrequently is the emotional problem between biological or non-biological relatives, those of kin, those next to, those representing other human beings, in which they try to work out their guilt feelings for never having treated "dear old dad" properly for most of his life. And now, by George, they're going to be able to say for the rest of your life that you did everything for him and therefore kept him alive for two-and-a-half years in a permanent vegetative state so that they can say "we did everything."

Or the other point of view, the surrogate knows what a terrible person he is today as a consequence of the maltreatment he had as a child and the sooner that son-of-a-bitch dies, the better. I hate to use such vulgar language but this is the notion I get in dealing with patients sometimes, and therefore under-treating.

It is the morally valid surrogate that we as the health professional need to be concerned about. This puts the health professional squarely in the middle of a very serious dilemma. Because while the instrumentalities that have been talked about, the ethics committees and the courts and the guardians and so on have a role. I don't mean to depreciate them by any means nonetheless the physician or the nurse, the health professionals on the front line, are having to deal with that decision as to raise the question in the first instance of whether or not this is a morally valid, no matter whether it's legally valid or not, it's a morally valid substituted judgment.

The moral competence, the moral motivation, the ethical status of the decision-making surrogate is an important and crucial issue and one of those places where law and ethics come into intersection with each other.

Other Potential Decision Makers

These same criteria apply to the other instrumentalities of ethical decision making—durable power of attorney, ethics committees, legal guardians, and even court orders. Although this presents the health professional with some very serious moral and legal dilemmas, his or her inescapable prime responsibility to the patient may sometimes leave no other way out of such dilemmas. At the least, the health professional cannot feel excused from moral responsibility just because a surrogate fits a statutory definition or decision-making hierarchy. This is another instance in which the expanding inquiry into medical ethics can make the work and life of health professionals more complicated and difficult than it has been in the recent past.

Physician's Role

Under certain circumstances, the physician may become the surrogate. This is apt to be the case in overwhelming emergencies before valid surrogates can be found or when the patient is without close acquaintances or friends.

In passing from a moral point of view the physician has no greater competence in value judgments, in moral judgments, not legal, than anyone else. Therefore, we ought to be very careful about placing that in the hands of a physician and ought to perhaps set up some other mechanism, at least to provide a check on his own values which are easy to impose upon the other human being because the sick person is dependent, anxious, vulnerable, and exploitable. All of those need to be taken into account.

Under these circumstances, the physician may be compelled to decide for an incompetent or comatose patient. Here our concern must be for the possibility of conflict of interest on the physician's part. He or she may want to be faithful to his or her own moral values or unconscious desire to keep the patient alive or to hasten death for selfish reasons. More generally, the danger is the imposition of the physician's values, his or her judgment about the quality of life of the patient, or his or her biases or prejudices about the social, moral, or personal worth or merit of a particular patient whose medical care is consuming valuable resources.

While he or she can generally be assumed to be motivated by benevolence, the physician has no greater capacity for ascertaining the larger good of the patient—his or her welfare (other than his or her medical good)—than anyone else. The physician has no special competence in value judgments. Indeed, he or she is often impelled only by medical values and indications rather than the other interests of the patient that may be just as significant.

Except under the most urgent situations of grave emergency, the physician should act with caution as surrogate decision maker. Even in emergencies physicians are obliged to take measures to assure that their decisions were morally defensible. Consultation with other physicians, appropriate clergy, hospital ethicists, ethics committees, and the appointment of a legal guardian are all measures which the physician might take to this end. No competent or morally responsible physician should fear these as intrusions. The mature and honest doctor will recognize that help of this kind does not displace her as primary advocate for the patient but rather enables her to exercise that advocacy better.

What About Legal Guardians?

Parenthetically, the criteria for morally valid surrogateship apply to legal guardians as well as other surrogates. Mere appointment by a court and mere legal status do not confer moral competence. Because the physician or the institution must, in all cases, guard the welfare of the patient, this may sometimes mean seeking court relief if a legal guardian acts in what is perceived as other than the patient's best interests.

I agree with Dr. Areen's admonition that the courts must be used judiciously and not invoked carelessly or where other means for settling ethical differences among decision makers are available. Injudicious use of courts to settle disputes over ethical issues is inappropriate, costly, and time-consuming. The court's authority lends an unjustified weight to its moral opinions, which are often the result of superficial moral reasoning.[1]

Law, Conscience, and Moral Convictions

Let me turn now to a few selective comments about Professor Wadlington's paper. I would like to start with his reference to the intersection of law and statutes with the conscience and moral convictions of the participants in medical-ethical decisions. He talked about the right of conscience and the possibility of law and statutes taking this into account.

I want to make this point because the moral limitations—I'm not talking about the legal—the moral limitations on the competent patient as well as the surrogate decision is when it asks the physician or other health professional to do something which runs counter to their belief system.

Physician's Conscience Moral Values

Now, no matter what the legal criteria may be for decision making, you cannot ask that physician to violate his conscience by participating in or cooperating in an

1. E.D. Pellegrino, "Medicine and the Courts: Ethics at the Intersection," Seventh Annual Thomas Ryan Lecture, Georgetown University School of Law, November 18, 1986.

act which he considers to be intrinsically evil. Now that's a limitation. How do we deal with that? Well, obviously here I would say he ought to withdraw from the case. But let us make it more complicated, namely, what happens when he feels not only that this is a violation of his conscience but it is of such grave assault on a human being that he feels impelled to stand and interfere with the decision of the surrogates out of the principle of his own duty and obligation to the patient.

Now clearly, again, he must not in my view, he has no moral right to interpose that view; nonetheless, he feels a moral wrong is being done, he has a moral obligation to do what's available to him. Sadly and unfortunately he has to have recourse to what we have in this society, perhaps the ethics committee is a buffer but then the courts are available. Dr. Areen has on other occasions very, very wisely cautioned us as physicians against using the courts too much. And I quite agree with her. Nonetheless, put yourself in the position of someone who does disagree.

Let me take it a step further, in the most recent decision, Judge Compton in the *Bouvia* case, in his concurring opinion made the point, implied I think, rather strongly that it was almost a duty. In fact he didn't use the word "duty" but the implication was there of the health professional to assist a competent person who had decided to end his or her life. That's the way it was put, not discontinue because of the burden of suffering, but their decision directly to end their lives.

Again, if one disagrees, what is one's moral obligation? It's a limitation surely on the competent patient or the surrogate if the physician's moral conscience is violated. On the other hand, he doesn't have any moral right to impose on the other person and yet by not imposing a grave evil may be done.

The Role of Third Parties

What does the third party do when the decision not to treat offends his ethics or moral convictions? It was the third party who started the trouble if you want to see it as trouble in the Baby Jane Doe case, her right-to-life lawyer.

The minute we say that, of course we get into the question of categorization. Leave that aside for the moment. Let us say someone well-motivated who says, "A great harm is being done." Now we feel morally, if not by law at least, to assist someone in danger. The law doesn't require it of us, it's a minimalistic ethic. But we require that one intervene when someone is being assaulted. It's a shame if you walk by and don't say something.

What do we do under those circumstances? I think this question will come up increasingly even though we lay out procedures—and I'm not opposed to this. I think one of the great virtues of the President's Commission was that it set up a set of procedures on which there is a general consensus: I don't want to lose the substantive moral issue in our elation over the procedure on which we have an agreement.

So when a third party raises that question, we have a larger question of general ethics: What is the responsibility of each of us to the other? What is our moral obligation in terms of a general principle of benevolence or at least not harming other human beings? So that's a very difficult complexity in some of these issues. As to the right to conscience, I'm not so much concerned with it as a legal issue but as a moral question: how do we deal with that?

How do we deal then with the differences of points of view among the surrogates in the family? Let us grant that they're morally valid and all satisfy the criteria. The hierarchy established by the courts does not establish a moral hierarchy. Some may feel that what is being done is in fact wrong and that they have a responsibility to the person for whom the decision is being made. How do we resolve those issues? Those seem to me to be crucial.

Another point I'd like to make is to say a word or two about the question of the dangers of withholding treatment: the ethical and moral dangers. This also comes out of Professor Wadlington's discussion when he talked about the issue of the definition of the terminal patient. What do we mean by that? If we define that, we may get into the question of withdrawing life support measures sooner and sooner. Someone said at one point, I believe, that one state defines "terminal" as any patient who in the best medical judgment will die within a year. I don't know what that jurisdiction is.

Physician as Patient Advocate

But again falling back on the moral obligation of the health professional, remember the prime moral obligation of the health professional is to stand for the good or the interests of the patient. So that even when the surrogates come to a decision, even if it fits the criteria or moral validity, A, or legal validity, B, but they arrive at a decision which the healthcare professional thinks is wrong, what is his obligation? Is it to the family? Is it to the courts? Or is it to the patient?

Procedures Are Not Everything

Now I'm making things more difficult when my colleagues tried to make them simpler, which is of course the way we like to go. But I think these issues cannot be settled only by procedure. But certainly, if we begin to stretch that definition of terminology too much, then we run into possibilities.

Every clinician knows that the weakest part of medicine is prognosis; that every clinical statement is a statement of probabilities. Yes, our statements of clinical probabilities are that a universe of persons with these characteristics has this distribution of possible futures. But where does "X" fit one person in that distribution? We don't know if we cannot eradicate the fact that those have to be probability statements. But if you will just read the newspaper, and by the way these are not arguments for not withholding treatment, so please do not jump to that conclusion, but there are arguments for extraordinary care in making those decisions.

You all saw the case of the woman in our own area who was comatose for two-and-a-half months, whose husband, a Presbyterian minister, and the physician agreed to withhold treatment as she had made a valid anticipatory declaration. She had said, "If I get in this state, don't continue it." So, they made a morally right decision in my view—prognostically, the physicians were not wrong—yet, she fell outside the frequency curve, and she recovered. She defended their actions. But, keep in mind, our prognosis is weak. Second, it tends to force the clinician, or at least drive the clinician, away from the heroic measures which sometimes can work. That causes us once again to re-think probabilities. But having been a clinician for forty-five years and having done some heroic things that had no real justification in scientific medicine and yet redounded to the benefit of the patient, I don't want to exclude that possibility. I'm not suggesting we do it every time or it justifies keeping people alive like the woman you reported about, that I think is ridiculous. But there are situations in which we can move too fast. What I'm saying is that we need some braking, and that means critical reflection. There are errors.

Here I must emphasize the moral limitations on the autonomy of competent patients and of surrogate decision makers. The nub of the question is what we are to do when what is legally permissible or legally prescribed runs counter to the personal beliefs of the health professional. In a morally pluralistic society, this can become a matter of serious conflict which is not resolvable by simple reference to what law requires.

Patient vs. Surrogate Autonomy

While I believe we ought, on moral grounds, to respect the autonomy of both the competent patient and the surrogate in the case of incompetent patients, there are also at least two moral restrictions to place on the exercise of that autonomy.

One restriction applies when the autonomous decision poses a serious, direct, discernible danger to an innocent third party—as for example when a person who is seropositive for HIV (with confirmatory tests also positive) demands that the information be withheld from his pregnant wife who may or may not yet test seropositive. Here the danger of transmission of a fatal disease to the unsuspecting wife and the potential of irretrievable harm to her fetus overrides, on the grounds of justice, any obligation we may have to protect the patient's autonomy. Other examples would be the refusal of a mother with cavitary sputum-positive tuberculosis to be treated or hospitalized to protect her infant or family, or refusal of institutionalization by a homicidal paranoid schizophrenic.

The second limitation is placed on patient or surrogate autonomy when the physician or other health worker is asked or ordered (by the patient or surrogate) to do something he or she considers a grave violation of conscience. Here we might think of the Orthodox Jew who holds to the strictest interpretation of his tradition that it is not permissible to shorten human life, and the Roman Catholic who can never sanction abortion under any circumstances. Both actions are under certain specific circumstances legal, but for the Orthodox Jew or Roman Catholic, they are seriously immoral.

The moral restraint on autonomy in these cases derives from two things: First, the physician or health professional is the instrument necessary to carrying out the patient's request. She or he thus is asked to cooperate directly in what she or he considers an immoral act. Second, the health professional is a moral agent, and like the patient, is accountable for his or her moral choices. If he ought not impose his values on the patient, neither ought the patient impose her values on the physician or other health worker.

The kind of impasse that can result is illustrated in the well-publicized Bloomington Baby case in Indiana.[2] The issue centered on whether or not a baby with Down's syndrome and an easily correctable surgical defect should be treated. The parents and the obstetrician thought not. They argued that the baby's quality of life would not be satisfactory enough to warrant its living. The pediatrician felt that the baby should be treated. Recourse was had to the courts which upheld the parents and the decision was made to starve the infant.

Without arguing the merits of the opposing positions into this case at this moment, I want to consider the dilemma of the pediatrician who is asked to participate in withholding treatment which he thinks is morally required. On what I have already said it is clear that as a moral agent himself he must be free to refuse to cooperate. He then has two choices, one is to withdraw respectfully from the case giving his reasons. The other is more difficult. Should he take steps to intervene—despite what the parents wish—if he thinks that the decision to withhold treatment from the infant is morally wrong?

The Law Is the Last Resource

The fundamental question here is whether or not in a morally pluralistic society such decisions are to be left solely to parents and consenting physicians, or whether those who witness what they regard as morally wrong actions ought to take such steps to intervene with such means as a democratic society makes available?

I would argue from the principle of beneficence—in medical or ordinary human affairs—that we have a duty to prevent serious harm to others. Therefore, I would uphold the recourse to law. Although this will not settle the moral question, it is the only measure available to prevent serious harm being done.

This was the issue in the famous contretemps of a year or two ago over the regulations of the Department of Health and Human Services regarding the care of seriously defective infants. The story is a complex one that I cannot detail here. Suffice it to say that the issue is the degree of freedom and discretion that can legally be afforded parents and their physicians in deciding to withhold or withdraw life-sustaining treatments from seriously ill infants.

Leaving aside the legal question of what the state's interests ought to be, I would argue on ethical grounds that the agreement of parents and physicians cannot be absolute, and that health professionals and others may be morally bound to intervene through legal means to contravene the decisions of parents and/or physicians.

2. *Doe vs. Bloomington, Indiana* Ct. App. (Feb. 3, 1983), cert. denied, 104 S. Ct. 394 C 1983.

This point assumes considerable importance in the light of some recent court decisions that seem to threaten the moral conscience of the health professional. Judge Compton in the most recent appellate court decision in the famous *Bouvia* case went so far as to suggest in his separate but concurring opinion that it was the ethical duty of health professionals to assist a competent patient not in a terminal state to end her life by voluntary starvation.[3] That the court affirmed the right of the patient to choose death is not as momentous as one of its justices' assertions that the health professional should assist in bringing about this death.

Present trends suggest that within a few years voluntary direct euthanasia will, under certain circumstances, become legal. When it does, the conflict I have been describing between the health professional's conscience and the law will become much more acute.

Absent a Social Consensus—To Whom Should We Turn?

An important corollary of all of this, and particularly of the current tendency to refer moral and ethical issues to the courts is that, gradually, procedure may be substituted for moral substance whenever consensus cannot be reached. Obviously, there is an ethical dimension to the procedures we employ in making moral choice—respecting the autonomous wishes of the patient, getting morally valid informed consent, refraining from deception or manipulation of the patient's choices, etc. But, adhering to a morally defensible procedure does not ensure that the decision that results will itself be morally correct. As in the Bloomington Baby case, those who lose a court appeal are still left with a moral dilemma.

This is not meant to say that we should not seek consensus. The commendable work and reports of the President's Commission for the Study of Ethical Problems in Medicine and Biomedical and Behavioral Research demonstrated that consensus on important issues can indeed be reached even in a morally diverse society.[4] But consensus does not resolve underlying differences of ethical opinion or the clash of ethical norms. Care must be taken therefore to protect the moral autonomy of those for whom moral substance in a particular decision out-weighs procedure.

Perils and Pitfalls in Withholding Treatment

Let me tum now, as a clinician, to some of the dangers in withholding or withdrawing treatment. To fulfill our obligations to the patient we must take into account the inherent possibilities of error and harm in such decisions. The following are just some of the possibilities:

First, the prognosis may be wrong. Every honest and experienced clinician knows that prognosis since the time of Hippocrates has been the most difficult part

3. *Bouvia vs. Superior Court of Los Angeles*, 225 *Cal. Rptr.* 297.

4. The President's Commission for the Study of Ethical Problems in Medicine and Biomedical and Behavioral Research has issued a number of valuable reports. All are published by the U.S. Government Printing Office, Washington, D.C.

of medicine. Prognoses, like most clinical statements, are probability statements, and like all probability statements they predict the outcome of a given universe of patients, not of this particular patient.

This becomes a matter of utmost importance when trying to determine whether a patient is terminal, preterminal, in a permanent vegetative state, chronically brain damaged, totally or partially brain dead, or "unlikely to recover a sentient conscious state." The criteria for each of these categories may differ even among experienced neurologists as conflicting testimony in some recent court decisions amply demonstrates.[5]

There are many cases in which the prognosis can be stated with a high degree of probability, however. We need not wait for absolute certitude since this is rarely attained in any clinical judgment. But we are morally bound to be sure that the prognosis is as soundly based as possible, that the criteria are clearly stated, and that there is substantial agreement among the appropriate consultants before the possibility of withholding or withdrawing treatment is entertained.

This is particularly the case for definitions of a "terminal" or preterminal state, or decisions as to when the process of dying has begun. Some would define "terminal" as the likelihood of death within a year; others would limit the period to a few hours, days, or weeks. Given the uncertainties of prognosis just mentioned I would limit the designation "terminal" to situations in which death is foreseeable to the best of our scientific knowledge—within a matter of hours, days, or weeks.

None of this is meant to pose absolute prohibitions to withholding or withdrawing life support measures. Rather, I wish to establish the moral obligation to be as clear and as certain about diagnosis and prognosis as is humanly possible. Lacking the time to elaborate my reasons, I feel an obligation to list the following conditions under which I believe medical treatment may licitly be withheld or withdrawn:

(1) The patient's diagnosis and prognosis should indicate that the disease is medically irreversible, or the patient meets criteria for brain death, or permanent vegetative state.
(2) The patient must be in a terminal state, that is, death is foreseeable by qualified clinicians within hours, days, or a few weeks.
(3) Treatment must be futile or excessively burdensome.
(4) Directly intending death is never permissible.
(5) The patient's surrogate, having met the criteria for morally valid surrogateship, or a morally valid living will or power of attorney confirms that withholding or withdrawing treatment would be the patient's likely substituted judgment.
(6) All other measures including ordinary hydration, nutrition, and nursing care are continued as long as they bring comfort, or ease pain and are not excessively burdensome to the patient.

5. *In Re: Jobes* No. C-4971-SSE, Sup. Ct. Ch. Div., NJ (April 23, 1987).

Obviously, these criteria are grounded in philosophical and theological presuppositions that I have not the time to explicate. Nonetheless, I hope these criteria will make my personal positions clear on some of the crucial issues I have raised.

Even if the diagnosis and prognosis are correct, and the criteria I have listed are met, there are additional dangers. Making the decision to withhold or withdraw treatment may deny the patient the possible use of a new treatment for what was previously an incurable disorder. Some years ago this was fanciful, but today the rate at which medical progress occurs makes it a realistic possibility. In certain diseases such as HIV which are under intensive investigation, a few more months could make a significant difference at some point in the future.

Another more subtle, but serious, danger is the cumulative effect on the attitudes and values of physicians and society of deciding to forgo life support measures. Gradual erosion may occur in our respect for the lives of especially vulnerable members of our society: the aged, people with disabilities, the chronically ill, the homeless, the outcast. The "slippery slope" is suspect as an argument in the eyes of many ethicists, but it is a reality in human affairs. Moral sensitivities are easily blunted once a previously forbidden act is given sanction. One need only read Robert Jay Lifton's chilling story of how Nazi physicians slid into rationalizing the most heinous acts once the first steps had been taken. They distorted the Hippocratic Oath so that killing became healing for the victim and for society.[6]

Rationing and Allocation

Professor Gorovitz posed some very difficult dilemmas in which economics and economists confront each other in some tragic choices becoming all too common in healthcare decisions today. Dr. Gorovitz asks bluntly, where is the limit, economically and morally, in sustaining the life of a chronically and irretrievably ill older person? Should economics enter into our decisions? At what level? Who should weigh the economic factors?

Physician's Dilemma

I will focus on the doctor's dilemma first, then on the quality-of-life criterion, and then I will discuss briefly the ethics of rationing and resource allocation. My first point is that the prime ethical commitment of the physician and other health professionals must be to the patient whose care they have agreed to provide. At present, at least, the covenant between physician and patient binds the physician to the patient's good and not to the good of society, the family, or the institution. This is what the patient expects, and unless the patient says otherwise it is an expectation the physician must fulfill. So far as economic considerations go, therefore, they

6. Robert J. Lifton, *The Nazi Doctors: Medical Knowledge and the Psychology of Genocide* (New York: Basic Books, 1986).

should be a factor only with the competent patient. For example, a man may validly decide that he does not wish to expend his life's savings, or his wife's or family's inheritance on futile treatments in the last months of his life. The only limitations on his autonomy in this respect are those I have outlined above.

For the incompetent patient, economics can enter the decision only if the patient, while competent, executed a living will, made some verifiable statement, or appointed a durable power of attorney, through which she expressed a particular wish with respect to cost and expenditures. For the surrogates, economic considerations must not enter—only the other criteria for withholding or withdrawing treatment outlined above may be considered. If this is the case, the economic factors, so far as individual decisions go, are not morally significant.

I believe it is essential to preserve the physician's role as patient advocate. That role must not be jeopardized by making the physician a double or triple advocate, that is, of his own selfish interests (for-profit medicine), of social or fiscal policies (rationing, gatekeeper roles), and of the patient. Because I have developed my detailed reasons for this stance in a paper devoted to the ethical issues in "gatekeeping," I will not develop them further here.[7]

Who Shall Be the Rationer?

The second issue is that of rationing. For reasons I have given, the physician should not be the rationer because it conflicts with his primary moral obligation as the patient's agent. There are, however, certain conditions under which rationing itself might be morally admissible. Let me list them and then outline the physician's role in morally valid rationing.

Rationing of health care could be morally defensible as a national policy if:

(1) There was a true national emergency like war, natural disaster, or pandemic.
(2) It could be shown that healthcare expenditures were dangerously compromising other socially important goals.
(3) All other measures short of rationing have been exhausted. This would imply at least the following: (a) that physicians were practicing truly rational medicine, that is, ordering only those tests and treatments that demonstrably could alter the natural history of the patient's illness. (Unnecessary treatment, operations, tests, and examinations now consume billions of dollars that could be reallocated to provide needed care. Doing only those tests and treatments that are scientifically warranted is a moral obligation which medicine is not fulfilling as well as it might. Yet this is the first obligation of the competent physician. It turns out also to be the physician's best contribution to economic care.) And (b) that all reasonable economics of scale, efficiency in management, and productivity of the healthcare system had been implemented.

7. E.D. Pellegrino, "Rationing Health Care: The Ethics of Medical Gatekeeping," *The Journal of Contemporary Health Law and Policy* 2 (1986): 23–45.

(4) The principles of distribution (that is the definition of distributive justice) to be used in rationing are clearly stated, known to the public, and arrived at by a democratic process.
(5) Whatever system is devised would apply equally to all without respect to the individual's capacity to pay.

These are the criteria I would propose for a decision to ration health care as a national or community policy. Rationing is now becoming an issue at the micro-level involving competition among individual patients for a scarce resource, (an intensive care bed, donated organ, etc.) It happens often enough today that several patients are candidates for the same intensive care beds, or a new patient arrives who may be deemed, for a variety of reasons, to have a claim on a bed already occupied. How should such decisions be made?

Again, I can only suggest a set of criteria I believe would be morally defensible. I believe for example that the first step should be a set of medical screening criteria—aimed at assessing which patients on purely clinical grounds have the best chance of benefiting from the needed bed or procedure. Those who pass this more-or-less objective test should have access on the basis of equity—first come first served, or for those needing access simultaneously, a lottery selection would be in order.

I do not think that, in such a scheme, age—except as part of the medical screening—should itself be a factor. The intrinsic worth of a person bears no relationship to his or her age. Nor should a person's ability to pay, perceived social merit, or productivity in and of themselves confer priority of access.

Allocation Rather Then Rationing

This brings me to the final issue, the allocation question. First, it is my conviction that we are not in the kind of economic crisis that would justify rationing on a national scale. We have yet critically to examine the enormous expenditures we make for so many things other than health care, for example, alcohol, tobacco, cosmetic advertising, handguns, gambling, illicit drugs, recreation, etc. If we did examine the billions we spend on these things against our healthcare expenditures, would we be justified in objecting to $2.5 billion per year to keep 70,000 patients alive on renal dialysis, or deny $4.0 billion needed to support life-saving liver and heart transplants? There is much we could do in resource reallocation (even leaving aside our stupendous defense budget) before we can make a strong moral argument for rationing.

Indeed, the way we spend our money as a nation reveals the kind of people we want to be and the kind of society we want to live in. The central question really is the value question—what do we think is important for a good society? I am surely not going to vex you with my answer to that question this afternoon. Still it is illogical to talk of rationing until we have as a nation faced the larger questions.

Cost Containment as a Factor

None of this is meant as an argument against cost containment, or as a "blank check" approach to health care. Every physician has a moral obligation to be cost-conscious, to practice rational medicine, which is the most economical medicine in the long run, and to participate in policy decisions at the local and national level that will assure quality of care, and protect all the sick from policies that are detrimental to quality care. All health professions share a common obligation to provide the best possible care of the individual patient and to participate in national formulation on behalf of all who are, or will be, ill.

17

Ethical Issues in Palliative Care

Edmund D. Pellegrino, MD

THE CARE OF INCURABLY ILL PERSONS has always presented moral challenges to those who attend them. For families, friends, and society, challenges reside in the burdens of ministering to the decline of a fellow human being. To the physician, the challenges lie in the impotence of the medical art to forestall death and in turning the focus of that art from seeking a cure to providing care and comfort and to smoothing the "avenues of death," as John Gregory so pithily put it. Today, these ethical challenges often devolve from the collective responsibilities of a team of specialists engaged in the practice of intensive comprehensive palliative care (ICPC).[1]

Generically, the ethical issues surrounding the care of patients who need palliation in dying are no different from those encountered in the care of any sick person. What is different is the existential predicament of the incurably ill. Their existence as sick persons is markedly altered when the possibility of cure has been supplanted by the realization of the reality of their impending confrontation with finitude. Then, if the good of the patient is to be served, technical proficiency must yield to compassion, comfort, and consolation.

Accordingly, this discussion of the ethical issues in palliative care proceeds as follows: it begins with a phenomenological description of the existential state of the terminally ill patient; it moves from this to the special needs of such patients, the ethical issues encountered by those who care for them, and the resolution of ethical conflicts that emerge from their special needs.

Historical Prologue

It is difficult to discern historically when the moral obligation to care became as important as the obligation to cure. The evidence from the Hippocratic era is

1. Canadian Palliative Care Association, *Palliative Care: Toward a Consensus on Standardized Principles of Practice* (Ottawa: Canadian Palliative Care Association, 1995).

somewhat equivocal. The primacy of beneficence in the Oath,[2] the exhortation to "do away with sufferings of the sick"[3] and to avoid "doing harm by useless efforts"[4] seem to require the care of incurable cases. Yet, elsewhere, there is the suggestion that a physician ought to refuse to treat hopeless cases lest they reflect negatively on his reputation. Edelstein thinks this latter opinion was not the dominant one.[5]

However, when the nobler moral precepts of the Hippocratic ethic were adopted and given spiritual sanction by the teachings of the three great monotheistic religions, care for the suffering and the dying became both a moral and a religious obligation. This synthesis is epitomized in the parable of the Good Samaritan and actualized in succeeding centuries in the traditional commitment of Christian religious orders and foundations to the care of the sick, the poor, and the dying.[6] Hospitals were founded at a time when their major function could only have been care of the dying, since physicians then offered little in the way of truly effective therapeutics. They provided the only refuge for those who had no other place in which to suffer or to die.

John Gregory's view reflects this fusion of the Christian tradition of solicitude for the sick and the Hippocratic ethics. It is a combination that dominated eighteenth-century medical ethics in England and subsequently shaped the ethical code of the American Medical Association.[7] This integration of care and cure is also now the prevalent note in non-religious, humanistic, holistic theories of therapeutics for both dying and non-dying patients.

Some Definitions

At the outset, it is essential to define the way I shall use the term intensive comprehensive palliative care (ICPC).

Palliative implies the fact that cure—that is, return of the patient to a former or a better state of health is not possible and that the patient is fatally ill and death

2. Hippocrates, "The Oath," in *Hippocrates*, vol. 1, trans. W.H.S. Jones, Loeb Classical Library 147 (Cambridge, Mass.: Harvard University Press, 1972), 299.

3. Hippocrates, "The Art," in *Hippocrates*, vol. 2, trans. W.H.S. Jones, Loeb Classical library 148 (Cambridge, Mass.: Harvard University Press, 1981), 193.

4. Hippocrates, "On Joints," chap. 58, in *Hippocrates*, vol. 3, trans W.H.S. Jones, Loeb Classical Library 149 (Cambridge, Mass.: Harvard University Press, 1968), 339; Hippocrates, "Diseases 1," in *Hippocrates*, vol. 5, trans W.H.S. Jones, Loeb Classical Library 472 (Cambridge, Mass.: Harvard University Press, 1988), 113.

5. L. Edelstein, *Ancient Medicine: Selected Papers of Ludwig Edelstein*, ed. O. Temkin and C.L. Temkin, trans. C.L. Temkin (Baltimore: Johns Hopkins University Press, 1967), 97–98.

6. D.W. Amundsen, *Medicine, Society, and Faith in the Ancient and Medieval Worlds* (Baltimore: Johns Hopkins University Press, 1996); O. Temkin, *Hippocrates in a World of Pagans and Christians* (Baltimore: Johns Hopkins University Press, 1991).

7. E.D. Pellegrino, "Percival's Medical Ethics: The Moral Philosophy of an 18th-Century English Gentleman," *Archives of Internal Medicine* 146 (1986): 2265–69. Also reprinted in *The Persisting Osler II: Selected Transactions of the American Osler Society 1981–1990*, ed. J.A. Barondess and C.G. Roland (Malabar, Fla.: Krieger Publishing Company, 1994), 9–21.

foreseeable as a result of the presenting disease. The patient need not be terminal; death may not be foreseeable in the immediate future, but it is firmly established that whatever disease the patient suffers from is an ultimately fatal disease that will cause death unless some other cause intervenes.

To *palliate* in these circumstances means to ease or lessen the violence of a disease while not yet curing it. It means to recognize the physician's or other caregiver's inability to change the natural course of events. To palliate is to follow the Hippocratic injunction to "lessen the violence" of disease and to "do away with suffering."[8] This requires much more than relief of pain, though this is the first step in palliation.

Providing comfort must be *comprehensive*. It must take into account the highly individualized, personal, and unique response of this patient to the predicament of this sickness—a predicament never the same for any two persons. It must be comprehensive in that it encompasses physical but also psychosocial comfort. Nursing care, attention to spiritual and emotional needs, and elaboration of a suitable support structure all must be included and integrated if comprehensiveness is to be a reality.

In a sense, CPC is *healing even while the patient is dying*. This may seem at first a contradiction since healing means "to make whole, to restore."[9] Manifestly this is not possible in any ordinary sense, since by definition a terminal or fatal illness is one in which a dying patient has progressed beyond the point of no return. But, if we regard healing in a broader sense of establishing a new balance, a new accommodation between the progress of the disease and the patient's adaptation to the course of that disease, then healing may occur even when the patient is dying. Healing here means to recompose the person as a person within the confines imposed by the knowledge, inevitability, and actuality of dying.

Comprehensive palliative care is also *intensive care*—that is, it brings all the capacities of medicine and medical care to focus on the end of palliation. Palliative care must be pursued just as energetically, purposefully, and forcefully as care in the intensive care unit. The difference is that the end of cure is replaced by the end of comfort, of easing and preparing for death. In the palliative hierarchy of patient good, the medical good takes a lesser place, while primacy is given to other dimensions of patient good—his own assessment of the good, his notion of quality of life, his good generically as a human being, and his spiritual good.[10]

This definition of CPC specifies the end or *telos of medicine* as it applies to the care of the dying patient. This telos is what determines what actions and treatments are to be taken, as well as the duties incumbent on the physician, nurse, and other health professionals. It unites them as a team, determines how conflicts among them are to be resolved, and defines the virtues each must exhibit. Conscious commitment to this telos constitutes the act of profession, the promise implicit in it, and the

8. Hippocrates, "The Art," 193.

9. *The Oxford English Dictionary*, vol. 5 (Oxford: Clarendon Press, 1961), 153, s.v. healing.

10. E.D. Pellegrino and D.C. Thomasma, *For the Patient's Good: The Restoration of Beneficence in Health Care* (New York: Oxford University Press, 1987).

covenant that binds the health professional when he voluntarily offers or agrees to care for a dying patient.

Hospice programs are the place in which comprehensive palliative care is usually practiced in its most intensive form. They may be free-standing, hospital-based, hospital-affiliated, or community-based. It is a mistake, however, to confine palliative care to these settings and thereby to relieve others of this responsibility. Any patient with a fatal illness, whether or not in a terminal state, is entitled to the kind and range of ministration included under CPC. Indeed, CPC is, in essence, holistic medicine. It is holistic care of the patient as a human person in whom disease and illness are complex, multifaceted phenomena that generate a wide spectrum of needs. If these needs are unmet, true healing cannot occur. To be sure, CPC, like holistic medicine in general, is an ambitious and even in some ways a presumptuous ideal. But it is an ideal that should guide the caregiver—a standard to work toward even if it is not totally achievable. Palliative care, in fact, in its elements, is the kind of care owed to every patient regardless of whether he or she is in a terminal state or in a hospice. It is simply optimum care fitted to the special requirements of individual incurably ill patients.[11]

The Existential State of Incurability

The Changing Society Context

The ethical issues in palliative care occur within the societal and personal contexts of death as it is perceived in modern society. In recent years, much attention has been given by sociologists and historians to our changed attitudes toward death. We have moved in Western civilization from death as an inevitable, public, social event in which all play a central role one day to death as a technological failure in the march of modern medicine toward immortality. As Aries and others have so well documented, we no longer practice the *artes moriendi*, the arts of dying, which, from the late Middle Ages until the twentieth century portrayed death as the drama that prepares a person and her soul for communion with God.[12] The dying person then played the leading role, summoning all to the bedside, issuing dying statements, giving orders to his family, or forgiveness to his foes. Friend, priest, doctor, family, each had a prescribed but secondary role among the dramatis personae of dying.

Modern society has "deprived man of his death."[13] The patient still plays a role, of course. But now the dying person is an autonomous agent who decides when treatment should be stopped and ordains how she should be treated when she loses the capacity to make decisions. She may insist on treatment even when it is opposed

11. R. Woodruff, *Palliative Medicine: Symptomatic and Supportive Care for Patients with Advanced Cancer* (Melbourne, Australia: Asperula, 1993).

12. P. Aries, "The Reversal of Death: Changes in Attitudes Toward Death in Western Societies," in *Death in America*, ed. D.E. Stannard (Philadelphia: University of Pennsylvania Press, 1974), 137–38; P. Aries, *The Hour of Our Death* (New York: Alfred A. Knopf, 1981).

13. Aries, "The Reversal of Death," 143.

by the physician. The patient remains central in one sense, but in a very different way from previously. He or she no longer accepts death as inevitable, since modern medicine holds promise for defeating so many serious diseases. While surrounded by an ever-expanding healthcare team, the patient may well fear more than anything else the prospect of dying alone, isolated by her autonomy. She must make her own decisions, whereas previously those decisions were made in conformity with a supporting societal matrix of custom, ritual, and belief.

Today's patient dies, too, in a social milieu in which being ill is a defeat, an automatic ejection from the cult of youth and health. The dying person intrudes on everyone else's pursuit of business or pleasure. In the modern world, dying is an obscenity, an experience to be avoided at all costs, (whether via pharmacology or euthanasia).

Within this milieu of isolation, alienation, devaluation, and rejection, the dying person must confront his own mortality. Modern societal attitudes toward dying materially add to and accentuate the experiential modalities of being ill, being a patient, and being on the way to personal extinction.

Becoming a Patient

An ill person becomes a patient when she decides that whatever symptoms or signs she has experienced have reached a state that requires that she seek help from a health professional. Being a patient signals one's entrance into a new form of existence, one that forces a special kind of relationship with another human being. Being a patient is being in a special state of vulnerability. This vulnerability results from several things, as follows.

The patient is in an unequal relationship with the caregiver: She is dependent on the caregiver's knowledge, skill, and character. The patient has little choice but to enter this relationship if she wishes to be healed. She must trust that what is recommended and what is done are in her best interests and that she is not being exploited to the doctor's advantage. The doctor, for his part, in offering or agreeing to help makes a promise that the good of the patient, in all its dimensions, will be his primary concern that is, that the medical good, the good as perceived by the patient, and the good of the patient as a human being and spiritual goodwill all be served.

In his healing, the doctor is obliged to respect the patient's dignity as a human being capable of reasoned choice according to her own values and beliefs. Not to do so is to violate the humanity of the sick person, an act per se maleficent. That is why respect for autonomy is consistent with, and not antagonistic to, beneficence. Strong paternalism violates autonomy, but it violates beneficence, as well.

When, in addition to becoming a patient, a person is confronted with a diagnosis and a prognosis of fatality, a new order of magnitude is added to her vulnerability.[14] The usual phenomena of vulnerability that accompany any illness are exaggerated. This, in turn, accentuates the obligations of the physician to mitigate the impact of

14. M. Cacarian and P. Kelly, *Final Gifts* (New York: Bantam Books, 1993); A.L. Strauss and B.G. Glaser, *Anguish: A Case History of a Dying Trajectory* (Mill Valley, Calif.: The Sociology Press, 1970).

that vulnerability. As a result, the usual ethical duties and obligations of the healing relationship are rendered much more urgent and complex. The already existing inequality of power is tipped far in the direction of the doctor, but so, too, is the weight of obligations to protect the patient's vulnerability.

Dying patients must confront their own finitude, not simply as a future, remote possibility but as an actual, foreseeable, and immediately palpable reality. For most humans, this is the greatest anxiety they will ever confront. They often fear the process of dying, perhaps more than death itself. The fatally ill person faces the anticipation of pain, the loss of her value in the sight of others, and a sense of guilt for being a burden to others. She suffers an assault on her self-image and alienation from her own body, which is now seen as a mortal enemy. Every progress or regress of symptoms is a portent of the rate at which death is advancing.

Whether or not one has been a believer, there is a spiritual challenge in dying. The dying patient must come to some terms with the inevitability of her finitude. She can no longer ignore the challenge to accept or reject the question of God's existence and of life after death. Every ill person, and, most acutely, every fatally ill person, asks Job's questions: Why? Why me? Why now? Some will deny, others will affirm, a religious belief. Some will seek atonement; some, reconciliation. None will be indifferent.

Hope, on which so much of the human capacity to face the future depends, is concomitantly compromised. By definition, there is no chance for cure. Hope must be redirected to living the last days as well as possible, to spiritual and moral growth, and to entering new and enriched relationships with friends, families, and foes.[15] Hope is transformed or extirpated, depending on how the emotional and spiritual crisis is addressed. To go from being a patient, even one with a serious illness, to being a patient whose disease has been pronounced as incurable to being a patient who has been told that her death is imminent is to go through progressive stages of existential change. These substantially alter the degree if not the kind of obligations owed to that person by caregivers, whether doctor, nurse, social worker, family member, friend, or minister.

Concrete Ethical Issues in Intensive Comprehensive Palliative Care

It is against this social and personal existential and experiential backdrop that the specific, practical, concrete, everyday ethical issues of CPC must be examined.

Pain Relief

The first moral obligation of any palliative-care program is to be competent in the optimal relief of pain.[16] This is so obvious as hardly to need mention were it not

15. K. Kraus, "Hoping in the Healing Process: An Integral Condition in the Ethics of Care" (Ph.D. diss.; Washington, D.C.: Georgetown University, 1993).

16. F. Randall and R.S. Downie, *Palliative Care Ethics: A Good Companion* (New York: Oxford University Press, 1996).

for the fact that many physicians still do not treat pain properly. Measures for pain relief are available today such that no patient needs to die in pain.[17] Not to use pain medication well is a moral failure that is inexcusable and tantamount to legal malpractice, as well.

Pain medications should be used in doses sufficient to control pain. They should not be withheld or used sparingly for fear of making the patient an addict. This is surely not a problem of consequence in patients whose death is imminent. Nor is the danger of suppressing respiration as a side effect. So long as the intention is not to cause death but to relieve pain, if no other measure is available to relieve pain, if the patient's relief of pain is not dependent on the death of the patient and there is a proportionate reason for running the risk, then high doses of narcotics can be used.[18] The moral principle of double effect would be operative under these conditions.[19] We knowingly run risks of serious side effects with any potent medication. Moreover, the fact of the matter is that patients given increasing doses of narcotics develop considerable tolerance to their respiration-depressing effects.

Pain relief alone is not sufficient. There is a concomitant moral obligation to relieve suffering, a much more complex phenomenon whose elements have already been outlined. Optimal relief of pain and suffering are the first two morally mandatory obligations of any physician who cares for the incurably ill.

These obligations must be more judiciously approached with patients who have a fatal or incurable disease but whose death is not imminent. Such patients may suffer from chronic pain. They surely deserve optimal treatment. But if there are many years yet to be lived in the natural history of the disease, fear of addiction is a justifiable reason for caution in the administration of addictive drugs. Chronic pain is a subject of its own, a nexus of psychosocial and physiological interacting factors that requires the most careful evaluation and treatment. In these patients, the dangers of addiction cannot lightly be dismissed. The physician's moral obligation is to refer such patients to experts in pain management and not to fall into the trap of escalating doses of opiates.

Freedom to Choose, Refuse, and Request

A major ethical issue is the capacity of fatally ill patients to make choices and to give ethically valid consent. An ethically valid consent must meet several criteria: (1) It must be informed, (2) it must be free of coercion, and (3) it must be made by a mentally competent person. This is not the place to review the requirements for an

17. Agency for Health Care Policy and Research, Public Health Service, Department of Health and Human Services, "Acute Pain Management: Operative or Medical Procedures and Trauma," in *Clinical Practical Guidelines* (Washington, D.C.: U.S. Government Printing Office, 1992); W.T. McGuveny and G.M. Crooks, "The Care of Patients with Severe Chronic Pain in Terminal Illness," *Journal of the American Medical Association* 251 (1984): 1182–88.

18. Pius XII, "Allocutio," *Acta Apostolicae Sedis* 49 (1957): 129–47.

19. O.N. Griese, *Catholic Identity in Health Care* (Braintree, Mass.: Pope John XXIII Center, 1987), 246–99.

ethically valid informed consent.[20] Rather I wish to underscore the ethical issues in informed consent peculiar to the existential state of being fatally ill. In this state, for example, patients are exceedingly vulnerable to the suggestions of those around them. The family and the doctor assume central roles of power, which they must exercise with careful regard for patients' personal wishes and values.

The assumption is too easily made by the comprehensive palliative-care team that the patient implicitly understands or accepts the full program of care offered in a comprehensive care setting or a hospice program. This may not be the case at all. As with other treatments, consent must be explicit before the patient enters the program. Special attention is necessary to such things as the degree of sedation a patient may wish in the relief of his pain or suffering. Some patients accept and want any degree of sedation required to attain relative comfort. Others want to preserve a certain degree of consciousness and are willing to trade off some pain to remain in conscious contact with family and friends or to preserve decision-making ability.

Another example is the depth, if any, to which patients may want their psychological, personal, or sexual lives probed by a team intent on ferreting out the causes of a patient's suffering. This is a particularly sensitive problem in the realm of spiritual conflicts and values. Any fatal illness entails grappling with spiritual issues. Fear of eternal punishment, confrontation with the question of God's existence, and desire for reconciliation or atonement may play a role in the patient's fear of dying. For many, if not most patients the religious as well as the psychological aspects of dying are important to recognize. But others reject and resent probing into their spiritual lives as unwarranted intrusions. It is a mistake to assume that entry into a hospital program implies acceptance of every facet of that program for the duration of the illness.

These are some of the personal matter examinations of which the palliative-care teams must not assume will be welcomed by every patient. Respect for patient autonomy requires the patient's suitably informed consent, particularly before the team begins an examination of psychosocial or religious causes of suffering. For some patients, the religious dimension may be the most important way to find meaning in suffering.[21] Regardless of the caregiver's own views on religion, there is an obligation, therefore, to offer the patient the opportunity of receiving help in confronting these issues or rejecting them as she sees fit. There is an equal obligation not to use the patient's vulnerability as an opportunity to convert him or her to the caregiver's belief system. One may offer such help, but only if the patient wishes it and requests it.

In an era in which there is a tendency to psychologize all of life from birth to death, the same respectful attention is necessary in probing the inner psychic life of the dying patient with the patient or with family members. The dividing line between the licit exploration of the psychological aspects of suffering and the illicit

20. R.R. Faden and T.L. Beauchamp, *A History and Theory of Informed Consent* (New York: Oxford University Press, 1986).

21. P. Kreeft, *Making Sense Out of Suffering* (Ann Arbor, Mich.: Servant Books, 1986).

imposition of such probing is especially difficult to locate. There is no clear, bright line that divides unwelcome and offensive psychological paternalism from beneficent identification of a treatable or ameliorable cause of depression.

Similarly difficult is the question of what treatments to offer to the dying patient. On a strong view of autonomy, all treatments, including those that are of the most marginal benefit, or that are futile, should be offered. This is the consumer view of the autonomy of the patient. On a more moderate view of autonomy, futile treatments, those that are ineffective in altering the natural progress of the disease or are excessively burdensome, ought not to be offered. Presumably, a patient who has entered a hospice program understands that its purpose is to ease the process of suffering and dying, not to effect a miraculous cure.

Cardiopulmonary resuscitation, under ordinary circumstances, frustrates this acceptance of the inevitability of death. Yet, there are patients who may see personal benefit in living a few days or a week longer, for example, to see a relative, a baby born, a child graduate. Under such circumstances, patients must be warned of the dangers of cardiopulmonary resuscitation, such as the vegetative state, rib fractures, spleen rupture, and other conditions that may result. Except in unusual circumstances, cardiopulmonary resuscitation should not be offered.

Telling the Truth

Truth-telling poses a problem in the patient's last days, when consciousness and cognitive capacity may wax and wane or be lost entirely. Presumably, the patient who enters a hospice program knows he has a fatal and terminal illness. Clearly, he knows his diagnosis and his likely prognosis. This should be the case with dying patients in similar circumstances outside the hospice program. There is no way patients can participate authentically in their own dying without knowing the truth. With some exceptions, to be mentioned later, truth-telling is a requisite for comprehensive palliative care in any setting.

The major problems with truth-telling arise from two sources: how the truth is communicated and cultural differences that affect the care of dying patients.[22]

As far as the method of truth-telling goes, no formula can be prescribed. Communication is a highly personal interaction in which the value and the psychological convictions of doctor and patient meet and interact. Each meeting is a unique event. The language and level of detail used in telling the truth should be fitted to the patient's culture, educational background, and age. In general, it is wise to allow patients to direct the rate and depth of information by their questions. Patients should be given room enough to navigate their way through the details in their own way. Poor timing, abruptness, impatience, failure to listen, and insensitivity to the gravity and impact of the truth, are lapses of the ethical obligation of truth-telling.

22. E.D. Pellegrino, "Is Truth-telling a Cultural Artifact," *Journal of the American Medical Association* 268, no. 13 (1992): 1734–35.

In some cultures, it is customary not to give a patient bad news or the full truth on the ground that this destroys hope and only makes the patient more miserable.[23] Some physicians, particularly older ones, stubbornly hold to this view for all patients. Cultural differences should be respected when it is clear that the patient has not requested information, is not desirous of violating his cultural mores, and is not harmed by dying within the formulae prescribed by his own culture. Patients need not die the Anglo-American way, which places much more emphasis on personal autonomy than do other cultures.[24]

Sometimes a patient requests in advance that the truth, or at least all its details, be not revealed. Provided such a choice is genuine and that it is made by a fully competent patient, it ought to be respected so long as it does not so impair the process of care in a way that causes identifiable harm to the patient. Physicians who cannot, in good conscience, take patient autonomy this far should so inform the patient. They should decline to enter the therapeutic relationship or withdraw respectfully. To insist on telling the whole truth over a patient's objections seems a contradictory violation of autonomy to preserve autonomy.

Similarly, however good the doctor's intent may be, it is a breach of trust to shape a vulnerable patient's decision by withholding essential facts, selecting or shaping them to achieve what the doctor thinks best. The vulnerability of the patient and the inequality in knowledge, physical stamina, and power between patient and physician must always be a moral brake on doing harm so that good may come of it.

In any case, most experienced clinicians agree that there are very few instances in which patients are genuinely harmed by knowing the truth. Rather, it is the way truth is communicated and the attitudes the doctor exhibits in telling the truth that cause problems. With respect to truth-telling and respect for autonomy, the key idea is to make decisions with and for the patient. To this end, clinical decisions, especially in the case of fatally ill patients, are best when they arise somewhere between doctor and patient. If decisions veer too strongly to the side of the doctor's estimate of what is "good" for the patient, those decisions can be harmful to the patient; if they are too far from the physician's estimate, they may pose moral conflicts for the physician. In their zeal to do good, or in their enjoyment of the satisfaction of doing so, physicians may unconsciously be tempted to intimidate the patient or to "take over" in the name of relieving the patient of the anguish of decision making.

Surrogate Decision Making

When patients lose the capacity for self-governance, their moral right to make decisions is transferred to surrogates. This surrogate may be designated by the patient

23. A. Surbone, "Information, Truth, and Communication: For an Interpretation of Truth-Telling Practices around the World," in *Communication and the Cancer Patient: Information and Truth*, ed. A. Surbone and M. Zwitter (New York: New York Academy of Sciences, 1997).

24. Royal Dutch Medical Society, *Euthanasia in the Netherlands*, 4th ed. (Utrecht: Royal Dutch Medical Society, 1995).

before he loses his capacity to decide. Surrogates can be designated by word of mouth, by execution of a durable power of attorney for health, by legal definition in some states of the United States, or de facto by virtue of being the available next of kin.

However designated, the surrogate is vested with the moral authority to make decisions in the place of the patient. This authority is not absolute.[25] For one thing, the physician remains morally bound to protect the welfare of the patient. Physicians cannot abrogate this fiduciary responsibility. The physician must write the order that affects the patient and thus cannot escape moral complicity. As a result, the physician has an obligation, within reasonable limits, to ascertain the moral validity of a surrogate, regardless of the way that surrogate is designated. The physician must assess whether the surrogate actually knows the patient's wishes or values, has a financial or emotional conflict of interest in the decision, or is making the decision as if she were the patient and not as the patient himself would have made it.

Conflicts may occur between surrogates and physicians or between both and an advance directive, such as a living will. Few people (including physicians) can so accurately predict the clinical circumstances of their final days that they are able while still competent to write precise orders for others to carry out when the time arrives. Doctor, patient, surrogate, and advance directive may in the actual moment of decision making conflict with one another. If this occurs, the determining factor should be the best interests of the patient, determined among and between those who have a fiduciary obligation to protect the patient's interest, including doctors, nurses, chaplains, family, friends, and surrogates. If discussion and dialogue fail to resolve disagreements, an ethics committee often can be useful. Its purpose is not to substitute for the valid decision makers but to assist them to resolve differences. The use and abuse of ethics committees are separate topics too large to be examined in this chapter.

Physician and Surrogate Autonomy

Like the autonomy of the patient, the autonomy of surrogates is not absolute. Autonomy is limited when it is patently not in the interest of an incompetent patient, when it results in injury to third parties, when it violates the physician's moral integrity or her perception of what is scientifically valid medicine, or when the criteria for a decision are themselves not morally defensible, for example, when quality-of-life or economic criteria are used to decide the fate of a patient in a permanent vegetative state who has not previously expressed preferences on these matters.

Economic and quality-of-life criteria are valid if employed by the patient in her living will or transmitted to a morally valid surrogate. They are not valid if used as a reason to withhold a palliative treatment that would be of benefit to the patient. In our era of concern for rising health expenditures and the dominant cult of youth and health, there is great danger of devaluing the lives of whole segments of society (the elderly, the

25. Surbone, "Information, Truth, and Communication."

handicapped, or the terminally ill) because of a "poor" quality of life. If assisted suicide and euthanasia are legalized or socially tolerated, as in the Netherlands, the danger is accentuated, and involuntary and nonvoluntary euthanasia became realities.[26]

Intensive comprehensive palliative care is the rational alternative to euthanasia and assisted suicide. It is an inescapable moral obligation for anyone who opposes assisted suicide and euthanasia as immoral to provide such care. Hospices in particular must be exceedingly careful not to permit any kind of assisted suicide or euthanasia within their programs. They can too easily become society's designated agents of death, not its haven for fully human, sympathetic, truly compassionate care and dying.

This does not in any way imply that ICPC is inconsistent with the cessation of care that is futile—that is, care that is, ineffective, non-beneficial, and/or excessively burdensome. Nor does it imply that pain relief and comfort are in any way to be compromised, even if the use of narcotics or sedatives intended to relieve pain and suffering results in unintended death. The intentional hastening of death and withdrawal of noneffective life-sustaining care must be distinguished carefully.[27] Preserving the distinction is particularly important when dealing with the vulnerability of patients with fatal and terminal illnesses. The moral philosophy of intention is a neglected subject in ethics, where utilitarianism is the dominant theory, yet the intention is an essential element in evaluating the moral status of any human act.[28]

Alternative and Experimental Treatment

Patients in whom the diagnosis of a fatal or terminal illness has been made are understandably desperate for some alternative to standard or scientific medicine. Families and friends in an effort to be helpful may urge nonstandard therapies on patients. These may range from relatively innocuous high-dosage multivitamin, herbal, dietary, or behavioral therapies to toxic, highly expensive, or burdensome treatments.

The ethical obligation of the physician, mindful of the desperation of family, friends, and patient, is always to present the scientific viewpoint, nonjudgmentally and respectfully. In general, alternate therapies that are not positively harmful can be tolerated. But, alternative therapies that are harmful or interfere with proven beneficial therapy should always be refused. Physicians will vary in where they draw the line. Neither absolute capitulation to patient requests nor absolute rejection seems justifiable in light of the aims of comprehensive palliative care.

Similarly, patients in a clinical state that requires palliative care are often eager to be experimental subjects in hopes that an untried treatment might alter the fatal

26. Royal Dutch Medical Society, *Euthanasia in the Netherlands*; P.J. Van der Maas et al., "Euthanasia and Other Medical Decisions Concerning the End of Life, an Investigation Performed upon Request of the Commission of Inquiry into the Medical Practice Concerning Euthanasia," *Health Policy* special issue (1992): 22.

27. D.P. Sulmasy, "Killing and Letting Die" (Ph.D. diss.; Washington D.C., Georgetown University, 1995).

28. A. Donagan, *Choice: The Essential Element in Human Action* (London and New York: Routledge and Kegan Paul, 1987).

thrust of their illness. Fully competent patients can legitimately participate as experimental subjects if they give informed consent, if the procedure is relatively safe, and if it is potentially of value for the patient or for others similarly afflicted. For reasons already outlined, the complexities of informed consent for experimental care in patients receiving palliative care require particular attention.

With patients who cannot give consent, research is admissible only in the most narrowly defined conditions—if danger and discomfort are nonexistent and the benefits to others significant. For such subjects, consent must be given by morally valid surrogates, as well as by an institutional review board. Anticipatory declarations made by patients while competent that give unspecified permission for research are too easily susceptible to abuse to be morally admissible.

Conflicts between the physician's role as a care provider and his role as a clinical scientist are complicated with any kind of human subject research.[29] These conflicts are yet to be resolved in any definitive way. With terminally ill patients and patients in vegetative states, they are magnified. The caregiver physician's role is primarily to serve as guardian of the patient, given the enormous possibility for abuse in research with terminally ill patients. If any such research is undertaken under rigorous control, the investigator must not be the caregiver physician.

Heroic treatments are sought by some patients and physicians who hope with one throw of the dice to reverse the expected course of the disease. This is a scenario more imagined than actualized. Under rigorously controlled conditions, where preliminary results are promising, where death is inevitable, where the patient is fully competent, and where conflicts of interest on the part of the investigator are minimal, such research may be allowable. On the whole, the danger of turning a fatally ill human being into a test object is sufficiently great to make heroic treatments allowable only, if ever, under the rarest of circumstances.

Ethics of the Palliative-Care Team

Intensive palliative care of necessity requires a ream of caregivers, and this introduces the special ethical issues that arise in collective activity.[30] No one physician or nurse can possibly meet the requirement to be "comprehensive" in the sense the term is used, that is, to designate care aimed at meeting the combined physical, psycho-social, spiritual, and personal needs of the dying person. The advantages of cooperative team care bring with them a variety of ethical issues, some general to team care and some specific to the care of incurably ill persons.[31]

29. O. Guttetag, "The Physician's Point of View," *Science* 117 (1953): 207–208.

30. E.D. Pellegrino, "The Ethics of Collective Judgments in Medicine and Health Care," *Journal of Medicine and Philosophy* 7, no. 1 (1982): 3–10.

31. D. Clark, ed., *The Future of Palliative Care: Issues in Policy and Practice* (Philadelphia: Open University Press, 1993); D.J. Higby, ed., *Issues in Supportive Care of Cancer Patients* (Boston: Martinus Nijhoff, 1986); D. Jeffrey, "There Is Nothing More I Can Do," in *Introduction to the Ethics of Palliative Care* (Cornwall: Petten Press, 1993).

To begin with, there is the question of how accountability is to be divided. Three possibilities exist: Responsibility may reside in one team member; responsibility may be assigned to each member relative to his or her function in the team, and there may be true collective responsibility in which each team member shares responsibility for the whole team's actions. There seems no clear way to disentangle them entirely.

The physician writes the orders and is ultimately responsible for the patient's care since those orders affect the welfare of the patient even if they are recommended and agreed to by other members of the team. On the other hand, nurses, chaplains, social workers, and psychiatrists are professionals in their own right; all have a code of ethics that makes them responsible for that part of the care they provide. They cannot be asked to violate their professional ethics by any other member, even the doctor. Any member who co-operates in, or fails to object to, any harmful act is a moral accomplice. Every team member in that sense shares in the collective responsibility of the whole team for the care provided.

Another complicating factor is the considerable overlap of roles and responsibilities in palliative care. To be sure, each member has special expertise that defines his profession, and each is responsible for applying that expertise to achieving the ends of palliative care. But it would be artificial to confine the physician to the realm of technical expertise and to assign the work of caring and advocacy to the nurse, the psycho-social aspect to the social worker and the psychiatrist, or the spiritual aspect solely to the chaplain. Each team member has a responsibility for palliative care, which, unlike medical or other technical procedures, can never be discontinued or abandoned, or neglected by any team member. All team members must apply their expertise humanely, sensitively, and compassionately.

An additional issue in team ethics is when to "blow the whistle"—to object if harm is being done or incompetence is observed. Everyone is obliged to be on the patient's "side"—to be an advocate. Team members must beware of seeing themselves as the only ones who understand the patient's plight and needs. Each must also feel responsible for the total team effort and refuse to let territorial prerogatives interfere with a patient's access to necessary care no matter who makes it available. Each must remember that intra-team conflict has moral implications because it affects the care the patient receives.

Another set of issues revolves around team dynamics. In the interest of team harmony, members can become too compliant. Members can be enveloped in the satisfaction of their own personal needs, in doing "good." Others may complain of the "anguish of the caregiver" manifested in themselves and their colleagues.[32] Self-pity and self-righteousness are subtle tendencies in those who attend dying and suffering patients. Some may be so infatuated with the superiority of group wisdom that the patient's wisdom about his own body and psyche are ignored.

32. A. Nash, "A Terminal Case? Burnout in Palliative Care," *Professional Nurse* (June 1989): 443–44.

Finally, there is the proper exercise of the role of team leader. This is becoming less the role of a "captain" or leader and more that of facilitator or convener. Patients and family members have roles to play as members of the palliative-care team. As always, the ethical imperative must be the impact of any decision on the comprehensive care of the patient, keeping the goal of palliation always in view.

Too often the palliative-care team develops such a conviction that it is "doing good" that it becomes self-justifying, in a way invading the privacy as well as the autonomy of the patient. Dying patients fear dying alone. They also fear "do-gooders" who appropriate patients' misfortunate to serve their own psychological needs. This is a distorted notion of beneficence that reduces patients to projects of others' need for expressing their own altruism. The result is infantilized patients who are smothered by a "system" that can all too easily become standardized in just the situation where individualized, personalized, and differentiated care is most urgently needed.

Comprehensive Palliative Care Ideologized

As a concept and as a moral obligation, ICPC is an admirable and necessary method for the care of the terminally and incurably ill patient. Its ethical status is firmly grounded in the existential and experiential predicament of patients who are directly confronting their finitude. Yet, as with any powerful concept, ICPC can become an ideology, a self-justifying way to deal with a ubiquitous human problem. This "ideologization" is manifest in several ways.

First of all is the medicalization of care for the dying, which is both a unique and universal human experience. For centuries caring for the moribund was the responsibility of families, friends, and family doctors and nurses. Now it has become a professional specialty. There are obvious advantages to such a transformation but also some dangers. Where there are specialists, we tend to turn the task over to them and relieve ourselves of responsibility. This is unfortunate and may not at all serve the needs of the sick.

All the needs encompassed by comprehensive palliative care cannot be met, since the number of specialists is limited. But even if they were available, the crucial aspects of palliative care rely on humane, compassionate, and sensitive responses to the predicament of suffering. These are not necessarily technical skills. They depend on qualities of personal commitment that belong to those who have a genuine, friendly interest in the patient. A professional relationship cannot substitute for nor even approximate this kind of commitment. While family and friends are usually incorporated into an ICPC program, they are necessarily under the control of the team. This may result in a tendency to manipulate patients, family, and friends so that they conform to a formularized way of dying. When ICPC is ideologized, it can tyrannize and psychologize every aspect of dying, not always beneficently, it then discourages and displaces those other nonprofessionals to whom the patient has turned in other life crises.

A certain amount of denial, depression, and alienation accompanies the knowledge one is dying. Some of this is quite normal. Many patients can deal with this knowledge informally, allowing a certain amount of denial or a certain degree of wanting to "hang on" rather than give in. Some patients want to protect their privacy against invasion and probing by zealous professionals determined to do "their thing," to make the patient die according to formula. Some patients want to die "facing the wall" and not the team. This was once "normal" and may still be for some patients.

None of this is to contravene the enormous value of ICPC. It points, rather, to the frustration of the ends and purposes of palliative care if it is applied mindlessly or more for the satisfaction of those who have made it their specialty than for those whom it is meant to serve. Like every other powerful therapeutic tool, ICPC used within ethical constraints is a great help to the dying person. Used inappropriately, it becomes a travesty, a distortion of the compassion and comfort it so proudly promises and the patient so desperately needs.

Food and Hydration

Withholding or withdrawing food and fluid from dying patients is a much-debated ethical dilemma.[33] That debate can be touched on only sketchily here. Some of the disputed issues are these: Are food and fluids similar in all respects to medical treatments and therefore to be withdrawn under the same conditions as respirators? Are they symbolically and actually such ordinary measures and signs of care as rarely to be discontinued? These questions may be answered differently when the patient is fully competent, when he is in a permanent vegetative state, or when he is in a state of waxing and waning consciousness.

These questions also occur, of course, outside the perimeters of palliative care, although, as with the other ethical issues we have been discussing, they take on special urgency when death is imminent and unavoidable. For the purposes of this chapter, [I] can only summarize the ethical issues that I have discussed in more detail elsewhere.[34]

Anorexic patients who refuse food by mouth or take it in small and insufficient quantities should not be forced to eat more; nor should they be fed artificially against their wills. On the other hand, a lax and negligent attitude that makes no reasonable effort to help the patient take in food and fluids is not defensible, particularly if there is an intention of hastening death.

With patients who are in a permanent vegetative state, food and hydration should be maintained unless the process becomes community-based and very burdensome—for example, if aspiration pneumonia, esophageal or nasal ulcerations,

33. J.F. Tuohey, *Caring for Patients with AIDS and Cancer* (St. Louis, Miss.: Catholic Health Association of America, 1988), 157–69.

34. E.D. Pellegrino, "Withholding and Withdrawing Treatments: Ethics at the Bedside," *Journal of Clinical Neurosurgery* 35 (1989): 164–84.

or pulmonary edema occur. These may distort the balance between putative harm and benefit to such an extent that the normal presumption in favor of sustaining nutrition and hydration becomes injurious.

Conclusion

Intensive comprehensive palliative care is, in essence, the extension of the comprehensive and holistic care all patients are entitled to. It is revised to meet the special needs of the incurably ill. In meeting those needs, palliative care presents ethical issues similar to those encountered in the care of any seriously ill patient. But these ethical issues take on special dimensions and emphases because of the existential realities of modern-day societal attitudes toward dying and the predicament of the dying patient within that social context.

18

Futility in Medical Decisions: The Word and the Concept

Edmund D. Pellegrino, MD

Introduction

The thesis I want to set forth in this commentary is as follows: The term "futility" is in a parlous state, suffering from vagueness in definition, clinically unpleasant connotations, and intense criticisms by credible bioethicists. I agree with the discomfort and confusion the word has produced and would hope for a better term, which would dispel the nimbus of gloomy prospects—of hopelessness, abandonment, and giving up—with which it is associated. Despite these adverse connotations, the concept of clinical futility must not be lost out of eagerness to dispel the unhappy connotations of the word. The clinical concept futility embodies cannot be eradicated from clinical medicine as long as patients are mortal. It is important, therefore, to clarify the concept, rather than hoping to dispose of it by the erasure of futility. This commentary attempts to retrieve and retain the concept and to suggest a functional definition of that time when medical interventions are no longer serving the good of the patient.

The majority of patients dying in hospitals today die as the result of a conscious decision to withhold or withdraw medical interventions. Formally or informally, sooner or later, these decisions turn on an assessment of whether initiating or sustaining these interventions serves the good of the patient. When the good is no longer attainable, the intervention in question is futile—that is, it cannot attain the desired goal.

In the last decade, futility, a commonplace word, has been badly battered, distorted, and abused. Many bioethicists have banished it to the dustbin of outmoded archaisms. Others have circumvented it with euphemisms. Still others have buried it in committees, decision-making procedures, or institutional policies. Unfortunately, in discrediting an admittedly repellant *word,* an ineradicable clinical *concept* is in

---

Reprinted with permission of Springer Nature BV from Edmund D. Pellegrino, "Futility in Medical Decisions: The Word and the Concept," *Healthcare Ethics Committee Forum* 17, no. 4 (2005): 308–18. Permission conveyed through Copyright Clearance Center, Inc.

danger of being obscured as well. This is the *concept* of *futility*. No verbal legerdemain can eradicate the fact that in the course of a serious or fatal disease a point is reached when medicine can no longer serve any recognizable good for the patient. This is when physician, patient, and/or family must confront the universal fact of human finitude. At that point, cure, full restoration of health or function become unattainable by the methods of medical science and technology. Care, comfort, pain relief, and relief of suffering, however, are never futile. Care is a continuing moral obligation, while continuation of medical interventions may not be.

For centuries, this state of affairs has been embedded in the concept of futility. It cannot be erased by changing the name to something more palatable. Camouflaging it in some complicated, multi-step, decision-making procedure will not do. Futility is an inevitable corollary of the fact of human mortality. Anderson-Shaw et al., Brett, Fleming, and Zientek have very carefully set out the difficulties with the way the word and the concept are used in today's medical and societal milieu.[1] To acknowledge the reality of these difficulties does not justify abandonment of the concept or the actuality of the state of clinical affairs that the concept embodies. The understandable attempt to obviate the widespread revulsion for the word, by expunging it from clinical parlance, is itself a futile enterprise.

This essay offers (1) some comment on the logical and historical reasons for such widespread dissatisfaction with the use of the term "futility" in clinical ethics, (2) a brief critique of the proposed substitutes, (3) a proposal for an operating definition of futility suited to the clinical realities of withholding and withdrawing decisions, and (4) some caveats about the abuse of the term "futility," however denoted or connoted.

Sources of Dissatisfaction with the Term "Futility"

A significant part of the problem with futility is with the connotations which surround its usages and obscure its denotation. "Denotation" refers to the specific object to which a word is connected, that is, to that which makes it what it is. "Connotations," on the other hand, are the meanings attached to the usages of a word besides what it primarily denotes. Thus, the denotation of futility is simply inadequacy in producing a result or bringing about a required end.[2] Connotation, on the other hand, is much broader. It includes things in the meaning of a word besides what the word primarily denotes.[3] Connotations are the meanings a word may suggest, that is, the encrustations of meanings acquired in the history of its usage.[4]

1. L. Anderson-Shaw, W. Meadow, H. Leeds, and J.D. Lantos, "The Fiction of Futility: What to Do with Public Policy?" *HEC Forum* 17.4 (2005): 294–307; A.S. Brett, "Futility Revisited," *HEC Forum* 17.4 (2005): 276–93; D.A. Fleming, "Futility: Revisiting a Concept of Shared Moral Judgment," *HEC Forum* 17.4 (2005): 260–75; D.M. Zientek, "The Texas Advance Directive Act of 1999: An Exercise in Futility?" *HEC Forum* 17.4 (2005): 245–59.

2. *The Compact Oxford English Dictionary*, 2nd ed. (Oxford: Clarendon Press, 1994), 644.

3. Ibid., 317.

4. H.W. Fowler *A Dictionary of Modern Usage* (Oxford: Oxford University Press, 1965), 105.

The connotations of clinical futility are many. They are personal, often idiosyncratic, and imaginative. For many, they suggest abandonment, devaluing the patient, considering her "unworthy," or her life "not worth living." Such connotations seem to justify neglect of care or provide an excuse for not visiting the patient, and so forth. In today's climate of distrust for authority, patients may fear that once declared futile, they are at the mercy of the doctor's arbitrary decision about quality of life, or are being denied treatment, or even care, for economic reasons.

These are legitimate concerns. They would persist, however, with any synonym or euphemism for futility and with any construal of the term limited to purely medical or purely subjective criteria. They would persist whether the decision was made by an individual health professional, or by a committee, an ethics consultant, a legal or judicial edict, or a standardized policy or procedure. These concerns can only be addressed adequately if a patient, or surrogate, collaborates with health professionals in the determination of the point of futility.

Devaluation of the Concept of Futility

For more than 3,000 years of recorded medical history, the concept of futility has served as a criterion for the withholding or ceasing of medical therapy. It is clearly described 3,000 years or more ago in the Edwin Smith surgical papyrus.[5] In that account of some forty-five cases of trauma, the Egyptian surgeons distinguished those in whom prognosis was favorable, those in whom it was doubtful, and those in whom it was hopeless.[6] A little later, the Hippocratic physicians recognized that when patients were "overmastered by the disease," physicians should realize that medicine is powerless.[7] Moreover, the Hippocratic physicians held that patients must not seek to be treated in such cases.[8] Doctors, however, should study why treatment fails, try to correct the defect in medicine, and prevent harm by futile treatments.[9]

This was pretty much the meaning of futility in clinical practice until the beginning of the modern era of bioethics in the seventies of the last century. Futility, until then, had been a unilateral decision based solely on the prognostic abilities of the physician. It corresponded to what contemporary bioethicists today call "medical" or "physiological" futility. This changed when emphasis began to be placed on the principle of autonomy—the legal and moral right of patients to refuse even life-saving treatment. That legal right had been established earlier—in 1914 in the case of

5. A. Castiglione, *A History of Medicine,* ed. and trans. E.B. Krumbhaar (New York: Alfred Knof, 1941); Hippocrates, "On the Art," in *Hippocrates,* vol. 2, trans. W.H.S. Jones (Cambridge, Mass.: Harvard University Press, 1981).

6. Castiglione, "*A History of Medicine,*" 55.

7. Hippocrates, "On the Art," 193.

8. H.E. Sigerist, *A History of Medicine,* vol. 1 (New York: Oxford University Press, 1951), 306, 420 similar formulae.

9. Hippocrates, "On the Art," 205; Hippocrates, "Joints," in *Hippocrates,* vol. 58, trans. E.T. Withington (Cambridge, Mass.: Harvard University Press, 1958), 339; Hippocrates, "Diseases I," in *Hippocrates,* vol. 5, trans. P. Potter (Cambridge, Mass.: Harvard University Press, 1988), 113.

Schloendorff vs N.Y. Hospital.[10] In the seventies, autonomy became morally impelling as one of four prima facie principles in Beauchamp and Childress's extremely influential *Principles of Biomedical Ethics.*[11]

Autonomy gained further strength when the right of refusal was extended to family surrogates in the Karen Ann Quinlan Case in 1972,[12] and sustained by a subsequent series of similar court decisions.[13] In 1983, the concept was further reenforced by the President's Commission, which encouraged participation in end-of-life decisions by the use of the living will.[14] Gradually, over the intervening years, the right of autonomy, which began as a negative legal right of refusal of treatment, underwent metamorphosis to a presumptive moral right to request and demand treatment. Some patients and surrogates now carry "participation" to the point of micromanagement at the bedside.

The transition of autonomy from a negative right of refusal to a putative right to demand treatment has been accompanied by a shift in the focus of conflict between physicians and patients or their surrogates. A little over a decade ago, patient autonomy was invoked because physicians were perceived as over-treating patients, keeping them alive unnecessarily, and preventing them from a natural or "dignified" death. Today, the focus has shifted to under-treatment. Families and surrogates now often demand that "everything" be done even when death is inevitable. They see and read about new life saving procedures and treatments in the visual and print media. They want these technological marvels for themselves or those they represent.

This shift occurred almost simultaneously with the movement to discredit or abandon the concept of futility. Physicians are now accused of doing either too much or too little, or both. As a result, there is a move to restrict the physician's role. Many want futility decisions moved from the bedside to committees, institutional policies, and even statutes. Whether the problem is over-treatment or under-treatment, it centers on some form of futility decision. It can only be resolved, ultimately, by clarity on whether or not sustaining life or continuing treatment is in the patients' interests. This is the question any procedure substituted for the clinical concept of futility must answer. That question remains the same—is some good for the patient

10. *Schloendorff v. Society of New York Hospital* 211 N.Y. 125, 105 N.E. 92 (1914).

11. T. Beauchamp and J. Childress, *Principles of Biomedical Ethics,* 5th ed. (New York: Oxford, 2001).

12. *In re Quinlan* 70 N.J. 10, 355 A.2d 647 (1976), cert. denied, 429 U.S. 922 (1976).

13. *In re Quinlan; Superintendent of Belchertown State School v. Saikewicz,* 373 Mass. 728, 370 N.E.2d 417; *In the Matter of Dinnerstein,* Mass. App, 380 N.E. 2d 134 (1978); *In re Conroy,* 190 NJ super 453, 464 A2d 303 (NJ App 1983) appeal docketed, No. 21, 642 (NJ Sup Ct); 98 N.J. 321, 486 A. 2d 1209 (1985); *Patricia Brophy v. New England Sinai Hospital* (Norfolk Probate Ct., No 85 E0009-GI); *Brophy v. New England Hospital* 497 N.E.2d 62 (Mass., 1986); *Cruzan v. Harmon,* 760 S.W.2d 408 (Mo. Sup. Ct. 1988); *Cruzan v. Director, Missouri Dept. of Health,* no. 88-1503, cert. granted, 109 S.Ct. 3240 (1989).

14. President's Commission for the Study of Ethical Problems in Medicine and Biomedical and Behavioral Research, *Deciding to Forego Life-Sustaining Treatment: A Report on the Ethical, Medical, and Legal Issues in Treatment Decisions* (Washington, D.C.: U.S. Government Printing Office, 1983), 554.

being served by continuing treatment? If it is, are the burdens of providing that treatment in any way proportionate, or commensurate, with the perceived benefits?

This brief history of the semantics of futility and the devaluation of the traditional doctor-centered definitions is the shift in the direction of all clinical decisions from the physician to the patient. The answer to the question, "Whose futility?" is that it cannot be one or the other. In an age of patient autonomy, subjective and psychological elements must interdigitate with objective clinical elements. Futility as a criterion applicable to decisions to withhold or withdraw medical interventions must arise between physicians and their patients or the patient's surrogates.

The Assessment of Futility[15]

The first step in any ethically defensible procedure for assessing futility is collaborative definition of the end or purpose which an intervention is presumed to serve. Futility, after all, can only be assessed in terms of an end or purpose. The end questions will be both medical-technical and personal-value specific. Physicians, patients, and surrogates need to agree on ends and the possibility and probability of their being achieved by the proposed intervention.

To ascertain whether or not some defined goal of patient care is attainable in terms of the patient's good requires answering three questions: Will the proposed intervention be effective? Will it be beneficial? And will the burdens imposed by the intervention be justified if the desired end is attained?

Crucial to the collaborative decision making required to answer these questions is the establishment of the boundaries of expertise of each participant. This is not meant to generate competing foci of moral authority but rather, to recognize the different weight of each participant's opinion and the necessity of a synthesis of both subjective and objective components.

Expertise in gauging effectiveness lies within the physician's domain. If he is competent, his diagnosis, prognosis, and selection of treatment should be as objective as possible and within the scope of his professional training. Effectiveness depends on whether or not the proposed intervention is known, on the best scientific evidence available, to be capable of achieving the desired end, and to what degree. Clinical effectiveness is a necessary, but not a sufficient determinant of a futility decision.

The element of patient benefit must be factored into the relationship. This is in the domain of the patient's, or his surrogate's, subjective and affective evaluation. That evaluation will be personalized, often idiosyncratic, reflective of the patient's values, education, and cultural background. The connotations of a term like futility in the clinical setting will vary from patient to patient. Effectiveness and benefit are often conflated, but this is unfortunate and misleading. It confuses the roles of

15. E.D. Pellegrino, "Decisions at the End of Life: The Use and Abuse of the Concept of Futility," in *The Dignity of the Dying Person* (Proceedings of the Fifth Assembly of the Pontifical Academy for Life, February 24–27, 1999), ed. Juan De Dios Vial Correa and Elio Sgreccia (Vatican City: Libreria Editrice Vaticano, 2000), 219–41.

doctor, patient, family, nurse, social worker, etc. Each has his or her own value system about what is "good" in a given situation and, consciously or not, tends to impose it on the patient. Keeping assessments of effectiveness and benefits separated allows a better appraisal of the subjective and objective, the personal and the professional definitions of futility in particular cases.

In the assessment of the burdens imposed by the proposed intervention, both the doctor and the patient have roles to play. The physician describes the known or anticipated physical, emotional, and fiscal burdens peculiar to the intervention in question. It is the decision of the patient, or her surrogate, to assess whether or not the burdens are proportionate to the benefits expected.

This mutual assessment of "futility" permits a proper partitioning of expertise, and moral authority between, and among, the persons most intimately involved in a decision. It makes the process of decision making more transparent, more orderly, and less vague. It meets many of the objections to unilateral definitions of futility, that is, either doctor- or patient-centered.

To be effective, the described dialogue must be used in an anticipatory way—neither too soon nor too late. Changes in the assessment must change as the clinical situation changes and outcome goals are fitted to clinical realities. The three elements being assessed are qualitative, as are most decisions in other aspects of life, so that mathematical certitude is not attainable. Efforts to quantify the definition of futility have not been successful.

The focus must be primarily on the patient, the physician, and the patient's surrogate. Other members of the healthcare team will, of course, be involved, each contributing her own area of expertise on each of the major elements of assessment. Each will have some personal knowledge of the patient's perceptions, especially the nurse, the pastoral counselor, or the social worker. Many times these other team members will know more of what the patient feels or wants than the physician. Nonetheless, the physician signs the final order and is morally accountable for being aware of the findings of other team members. Consultation with these other team members is a moral obligation if the physician is to make an ethically defensible assessment of futility. In the end, however, the doctor is accountable if harm comes to the patient in the form of over- or under-treatment.

Procedural Alternatives

Most recently, bioethicists have tried to circumvent the objections to definitions of futility by elaborate decision-making procedures.[16] These usually involve a set number of articulated steps carried out by healthcare teams, ethics committees, or as

16. Council on Ethical and Judicial Affairs, American Medical Association, "Medical Futility in End of Life Care," *Journal of the American Medical Association* 281 (1999): 937–41; A. Halevy and B. Brody. "A Multi-institutional Collaborative Policy on Medical Futility," *Journal of the American Medical Association* 276 (1996): 571–74; P. Singer, G. Barker, and K. Bowman, et al., "Hospital Policy on Appropriate Use of Life-sustaining Treatment," *Critical Care Medicine* 29 (2001): 187–91.

in Texas, by a legally designated body.[17] The number of steps is variable, ranging from seven to eleven. Common to all is an interdisciplinary and inter-professional approach.

These procedures are designed to remedy defects in bedside decisions, like lack of standardization, the hegemony of professional values, or insufficient transparency. They are commendable in their attempts to compensate for the vagueness of definitions of futility and to protect the vulnerable patient from physician or surrogate bias.

The proposed procedures must however overtly or covertly accept the clinical reality; that is, the concept of futility. To establish a procedure for a decision, after all, is to recognize implicitly that *something* must be decided in an orderly way. The procedure itself is only a means. Unavoidably, it must turn on some set of criteria for action.

Whether a group decision of this kind is morally superior to a decision by those most immediately involved (the patient and/or surrogate with the physician) is a debatable point. There is no evidence that group decisions are more reliable or morally superior to the kind of process described above in this essay. Distance from the bedside could, in some cases, assure better patient protection. In others, it might simply substitute group bias for a clinician's bias.

In general, the closer a decision is to the patient, his surroundings, and life situation, the closer it will be to meeting his personal needs. Empirical studies may, or may not, tell us the answer. What is central to formalized group procedures and the decentralized versions suggested here is the recognition of, and preservation of, the concept of futility as an inescapable clinical reality. While the concept of futility used may be covert, these procedures are more realistic than the effort to reduce futility to mythology.

To refer the decision to a legal process in what amounts to binding arbitration may seem desirable to end disputes. The danger is that the participants might believe that a legal decision has moral authenticity—a dangerous precedent.

Abuse and Misuse of Futility

Any concept of futility, however defined and however used in clinical decisions, is subject to moral misuse and abuse. This is the major reason many want to abandon the concept entirely. Since the clinical state of affairs futility attempts to describe is not eradicable from clinical decisions, it is better to recognize these dangers than to deny the existence of the clinical phenomenon of futility.

First, is the danger of eroding the fundamental moral obligation of those who profess to be healers. The aim of healing is always to relieve pain and suffering and provide care, comfort, and emotional support. Care is never futile. It is a goal always attainable to a degree, usually reflective of the professional's sense of compassion. Indeed, to recognize the dimensions of futility greatly magnifies the obligation to provide care. When technical measures fail, true healers make health care a truly human undertaking.

17. "Texas Advance Directives Act," *Texas Health and Safety Code,* Chapter 166, Sections 166.001–166.166, Vernon.

Second, is the extreme vulnerability of certain patients to personal agendas involving social and economic values. A diagnosis of futility must never justify deprivation of any human being of his or her inherent dignity as a human person. Sadly, today there are those who argue that dignity is a "useless" concept. This attitude, combined with a misuse of the concept of futility, puts all of us at peril of moral abandonment when we are most vulnerable.

A third moral danger is the temptation to use futility, out of a distorted notion of compassion, to justify physician-assisted suicide, or euthanasia, voluntary and involuntary, even in infants.[18] This is already the case in some countries in which these procedures are legal. Some moralists even argue that intentional shortening of the lives of patients when medical interventions are futile is morally mandatory. The slippery slope, decried by many bioethicists as a myth, is another reality that cannot be abolished by wishful or artful thinking.

Patients in permanent vegetative states, those with dementia of various types, Alzheimer's or Parkinson's disease, are especially susceptible to abuses of the concept of futility. A diagnosis of futility per se is no warrant for automatic discontinuance of food, hydration, respiratory assistance, or of pacemaker support of cardiac rhythm. Here is where the proportionality of the treatment in relation to the burdens it imposes becomes crucially important. Futility is never a unidimensional determination. Effectiveness, benefit, and burdens all must be weighed in relation to each other.

A fourth source of abuse lies in the overly strict or overly lax interpretation of the way a diagnosis of futility is to be applied. Some argue that once medical interventions are considered futile, then all supportive measures should be stopped immediately. They would make cessation of all treatment morally mandatory. Such an interpretation is too peremptory. It fails to allow time for surrogates or family to adjust emotionally to the fact that a loved one has no chance of recovery.

A permissive application of the concept is more compassionate and prudent than an overly rigorous one. The parents of a neonate, or the children of an elderly parent, need time to adjust to the loss of a family member. They must be helped by physicians, nurses, pastoral counselors, and others to make this adjustment. It takes time to realize that further medical interventions are burdensome and harmful and that death is inevitable and the natural end of life for all of us.

The permissive approach is best managed by setting timelines for limited goals. The physician should take responsibility for recommending what he thinks best for the patient. Physician and patient or surrogate then share the decision, each with his or her own measure of responsibility. To transfer all responsibility to the surrogate,

18. E. Verhagen and P.J.J. Sauer, "The Groningen Protocol—Euthanasia in Severely Ill Newborns," *The New England Journal of Medicine* 352 (2005): 959.

as some physicians do, in the name of respect for autonomy, is a species of moral abandonment.

Deciding upon, and eliciting, a DNR order must by its nature involve a conscious assessment of futility, or it would not be ordered. The same is true of withholding, or discontinuing, cardio-pulmonary resuscitation. In these instances, death is considered inevitable. No clearer statement of the futility of further medical interventions occurs in clinical medicine.

But here, too, effectiveness, benefits, and burdens must be weighed against each other. For example, CPR might restore heart beat even while it fails to change the inevitability of death. Yet, just restoring the heart beat might be of benefit to a patient who would like to see a child graduate, a grandchild born, or have a last meeting with family or friends. Burdens would be considered, too, since CPR might be prolonged, end in fracturing of ribs, or rupture of spleen. Also, the heart beat might be restored, but not full consciousness, resulting in a permanent vegetative state.

Many other aspects of the use and abuse of the concept of autonomy deserve discussion, but cannot be addressed here. Especially important are the resolution of conflicts between, and among, the participants in the decision—that is, patients, families, surrogates, healthcare team members, and legal advisors. These conflicts must also be resolved in an ethically defensible way. Special issues arise with resuscitation in which the question of cardio-pulmonary resuscitation must be decided, often in emergency conditions.

What I have tried to show is that despite the obvious difficulties with the term, the concept of futility cannot be erased from clinical decisions. With increasing capabilities of medical technology, it will continue to demand careful moral deliberation. To that end, it is better to recognize the reality, categorize the elements that enter into futility decisions, and make those decisions in a collaborative way. The effort to expunge the term "futility" cannot obscure the reality it embraces. Indeed, the effort to expunge futility must itself end in futility.

This paper was written before the author was appointed to the President's Council on Bioethics. The opinions expressed here are his own and do not represent the council in any way.

19

Decisions at the End of Life: The Use and Abuse of the Concept of Futility

Edmund D. Pellegrino, MD

MODERN MEDICAL KNOWLEDGE AMELIORATES, sustains, and postpones the natural course of dying in patients who, in previous years, would have died in a short time. These patients now become subjects of prolonged lives, recurrent episodes of acute complications, and of new superadded diseases. As a consequence, in almost every case, other than sudden, unwitnessed death, some decision must be made about how vigorously to treat and about when it is morally permissible to withhold or withdraw life-sustaining measures. In the U.S. alone, 2.2 million deaths occur in hospitals annually; and in 1.5 million of these, an explicit decision is made to withdraw or to withhold treatment.[1]

With remarkable prescience, His Holiness Pope Pius XII recognized this dilemma before it became the urgent problem it is today. On several occasions, Pius XII clearly set forth the foundations of Catholic teaching about care at the end of life. His teaching was based in the dignity of the human person, the sacredness of all human life, and the duty to use medical knowledge wisely, well, and within certain ethical constraints. He added that under certain circumstances, when treatments were "extraordinary" and excessively burdensome, they might licitly be withdrawn.[2]

Distinguishing between "ordinary" and "extraordinary" treatments, Pope Pius enunciated a carefully nuanced approach to decisions to withhold or withdraw treatment within the context of the particularities of the patient's whole life. When all things are considered, Pius said, treatments can be discontinued if they are deemed to be "extraordinary," that is, excessively costly, dangerous, painful, difficult, or

---

Reprinted with the permission of the Pontifical Academy of Life and Libreria Editrice Vaticana from Edmund D. Pellegrino, "Decisions at the End of Life: The Use and Abuse of the Concept of Futility," in *The Dignity of the Dying Person*, ed. Juan De Dios Vial Correa and Elio Sgreccia (Vatican City: Libreria Editrice Vaticana, 2000), 219–41.

1. Leon R. Kass, "Is There a Right to Die?" *Hastings Center Report* 23, no. 1 (1993): 34–43.

2. Pius XII, "The Prolongation of Life" (November 24, 1957), *The Pope Speaks* 4, no. 4 (1958): 395–96.

unusual when weighed against anticipated benefits.[3] This teaching was reaffirmed notably in the "Declaration on Euthanasia" of 1980[4] and the encyclical *Evangelium Vitae*.[5] In recent years, debates have centered on the precise definitions of the terms "ordinary" and "extraordinary" and the substitution of the terms "proportionate" and "disproportionate" in their place.[6] As medical technology and capabilities expand, what was once extraordinary is thought to have become ordinary.

For Pope Pius XII, "ordinary" and "extraordinary" were ethical categories, not clinical or physical formulae for withholding or removing treatments. John Paul II has specifically rejected that kind of teleologism, which determines the morality of human acts by weighing non-moral or pre-moral goods against harms. As an ethical norm, maximization of good and minimization of harms in order to produce a "better" state of affairs is a denial of the possibility of universal and absolute prohibitions against intrinsically wrong acts.[7]

Much of the difficulty in applying the concepts of ordinary and extraordinary relates to the determination of benefits and burdens in an actual clinical situation: "In the past, moralists replied that one is never obliged to use 'extraordinary' means. This reply, which as principle still holds good, is perhaps less clear today by reason of the imprecision of the term and the rapid progress made in the treatment of sickness."[8]

In this essay I wish to suggest that the *proper* use of the term "futility"—not as a moral principle but as a means for prudential clinical judgment—can be a useful bridge between the ethical formulation of ordinary and extraordinary and the decision in a particular case at the end of life. Futility, taken generically, simply means inability to achieve stated purpose. Futility in the clinical sense simply means that an illness or disease process has progressed to a point such that a proposed medicsal intervention can no longer serve the good of the patient.

Today, in clinical and ethical parlance, *futility*, like ordinary and extraordinary means, has been the subject of very intensive debate.[9] Futility is a very ancient concept clinically which is now being reinterpreted by secular bioethicists and given

3. E. Healey, *Medical Ethics* (Chicago: Loyola University Press, 1956), 67.

4. Congregation for the Doctrine of the Faith (CDF), "Declaration on Euthanasia" (May 5, 1980), in *Vatican Council II: More Postconciliar Documents*, vol. 2, ed. Austin Flannery (Grand Rapids, Mich.: Eerdmans, 1982), 510–16.

5. John Paul II, *Evangelium vitae* (Vatican City: Libreria Vaticana, 1995).

6. D. Cronin, "The Moral Law in Regard to Ordinary and Extraordinary Means of Conserving Life," in *Conserving Life*, ed. Russell E. Smith (Braintree, Mass.: The Pope John Center, 1989), 33–76; A. Moraczewski, "The Moral Option Not to Conserve Life under Certain Conditions," in Smith, *Conserving Life*, 235–75.

7. John Paul II, *Veritatis splendor*, nos. 74–75.

8. CDF, "Declaration on Euthanasia."

9. N.S. Jecker and L.J. Schneiderman, "The Duty Not to Treat," *Cambridge Quarterly of Health Care Ethics* 2 (1993): 151–59; W. Harper, "Judging Who Should Live: Schneiderman and Jecker on the Duty Not to Treat," *Journal of Medicine and Philosophy* 23 (1998): 500–15; J.D. Lantos, P. Singer, R.M.Walker et al., "The Illusion of Futility in Clinical Practice," *New England Journal of Medicine* 326 (1992): 1560–64.

moral weight in decisions to discontinue treatments. Its validity for such decisions, who determines it, and where it fits in moral judgments are in a state of flux. Nonetheless, clarification of the use of futility criteria is important for Catholic Christians, since its use or abuse must be judged ultimately within the framework of Catholic principles of medical morality.

The fulcrum of this essay is the concept of futility for several reasons: (1) Futility is an ineradicable fact in clinical medicine and its language is widely used by health professionals. (2) Futility is subject to morally proper use and abuse and these must be distinguished. (3) The idea of futility is implicit in Pius's teaching about ordinary and extraordinary means, but its explicit relationship to that teaching is yet to be fully examined. Properly interpreted, futility can help to recover some of the full moral import of the terms *ordinary* and *extraordinary*. (4) It can link the clinical with the medical and theological construals of *ordinary* and *extraordinary*. Properly and prudentially used, futility can avoid some of the dangers of automatic stigmatization and devaluation of the lives of certain vulnerable patients.

After a brief theological propaedeutic, this paper is divided into three parts: (1) delineation of the concept of futility, (2) delineation of its abuses in particular clinical decisions, and (3) definition of its proper use within the context of Catholic Christian anthropology and medical morality.

Theological Propaedeutic

Futility involves a prudential judgment about what is right and wrong behavior in deciding how vigorously to treat and when to desist from treatment in a given concrete clinical situation. It does not, and cannot, stand alone as a determining criterion. Although it has empirical dimensions, futility is not solely an empirical determination. Its morally responsible use must be grounded in our deepest perceptions about the nature of human beings and their existence as both material and spiritual creatures.

For the Catholic Christian, that deeper structure must be Christian anthropology and the moral principles that derive from it, that is, the recognition of human beings as creatures made in the image of God, endowed by Him with the gift of life and with a spiritual destiny beyond this world. From these elements flow the inviolable dignity of each human person, of the equal worth of all persons in the eyes of God, and the sanctity of human life in every stage of its development, from first to last. Thus, human life is among the highest of goods, but it is not an absolute good. As Pope Pius's and subsequent teachings indicate, there is a time when the natural history of a disease may be allowed to result in death. Death, too, is the natural end of life and an expression of human finitude. At some point, it must be accepted and surrendered to.

According to Gospel teaching, among human persons the sick and vulnerable have a special claim in charity on our solicitude. Healing, helping, and caring for the sick is an obligation shared by all Christians, for it is Christ as healer (*Christus medicus*)[10] and Christ as patient and sufferer (*Christus patiens*) who is our model. His life gives meaning to our pain, suffering, and death—these things are not to be sought, but when they are inevitable, they have their place in God's ordering of the world and our individual human lives. Through the Incarnation Jesus entered into our suffering in order to give it meaning.

On this view, any intentional hastening of death by physician or patient would be morally inadmissible, even for what might appear to be beneficent reasons of compassion, mercy, and relief of suffering. Under all circumstances, man's stewardship of the gift of life demands that human life be nurtured, cared for, and protected. Other things being equal, there is the expectation that treatable disease will be treated. To violate that stewardship is to challenge God's sovereignty. On the other hand, not to accept the fact of human finitude and to prolong life when death is inevitable is, in its own way, a challenge to God's sovereignty and an act of hubris. Recognition of clinical futility is a crucial element in deciding the moral status of acts of continuance or discontinuance of end-of-life treatments. Catholics and other Christians are obliged therefore to interpret futility within the constraints of a Catholic Christian view of the meaning of human life and health care.[11]

In this regard, it must be re-emphasized that the terms *ordinary* and *extraordinary* were not intended as purely technical judgments, although they have been misused in that way. Rather, they were proposed as moral judgments, that is, as criteria for a morally good or bad decision to withhold or withdraw treatment. Changes in medical technology since then do not change the moral impetus of the traditional language. If these terms are abused, they become morally problematic. This essay wishes to preserve the moral content of *ordinary* and *extraordinary* and to suggest that proper use of the concept of futility can serve as a prudential hermeneutical device for linking traditional moral teaching with its application in a particular case.[12]

Futility: The Evolution and the Definition of the Concept

Evolution of the Concept

Every clinician knows that, at some point in the natural history of any serious disease, further treatment is beyond the powers of medicine and no longer in the patient's interest. Sooner or later, this ineluctable fact becomes apparent to families and patients as well. That is why, from earliest times, the concept of futility has guided clinicians' decisions to treat or to desist. To treat under these circumstances

10. Peter Chrysologus, *Collectio sermonum*, ed. Alexander Olivar, Corpus Christianorum Series Latina 24 (Turnholt: Brepols, 1956), pt. 1, Sermo 50, line 63; Augustine of Hippo, *Enarrationes in Psalmos*, Corpus Christianorum, Series Latina 40 (Turnholt: Brepols, 1975), pt. X,3, Psalm 130, p. 7, line 20.

11. R.E. Smith, ed., *The Gospel of Life and the Vision of Health Care: Proceedings of the Fifteenth Workshop for Bishops* (Braintree, Mass.: The Pope John XXIII Center, 1996).

12. Pontifical Academy of Sciences, "Report of the Pontifical Academy of Sciences on the Artificial Prolongation of Life," *Origins* 15 (December 5, 1985).

violates the first principle of traditional medical ethics, that is, beneficence—acting for the good of the patient.

Futility was recognized as a clinical fact with medical and moral implications as long as 3500 years ago. The Smith papyrus, for example, cites fives cases of high trans-section of the cervical spinal cord in which, given the therapy of the day, treatment would have been futile. In the same papyrus there is mention of additional cases, and still others can be found in the Ebers papyrus.[13] Later, the Hippocratic physicians recognized futility and advised against treatment when patients were "overmastered" by the disease. Patients were admonished not to expect treatment under those circumstances.[14] The Hippocratics also urged physicians to ameliorate these diseases that were untreatable, to learn when they were untreatable, and to avoid harm by "useless" efforts.[15]

Even in these ancient texts, the abuse of futility is evident. In several places, it seems that the physicians of the Hippocratic School were advised not to undertake treatment of incurably ill patients because they would die inevitably and tarnish the physician's image of therapeutic infallibility. This attitude was as condemnable then as it would be today.

Since those ancient times, physicians have used futility as a clinical criterion in unilateral decisions about prolonging life. But such decisions have always been fraught with moral consequences since medicine is, at heart, a moral enterprise. Still, the morality of futility determination did not become a debated issue until a quarter of a century ago when the emergence of patient autonomy challenged the physician's authority. Emphasis on self-determination shifted the locus of all clinical decisions from doctor to patient or surrogate, or at least to some locus between them. The current trend to virtual absolutization of patient or surrogate autonomy in the U.S., and to a lesser extent in other countries, now makes the criteria of futility, and especially the way in which they are determined, a manner of the greatest practical moral significance since patients can demand overtreatment or undertreatment.

Futility, etymologically, means "inadequacy to produce a result or bring about a required end; ineffectiveness."[16] In medical care, the required end, that is, the *telos* of the physician's activity, is the good of the patient.[17] This is the moral center of the healing relationship. It is what patients seek and expect. It is what doctors promise implicitly by offering themselves as healers. When the good of the patient cannot be

13. H.E. Sigerist, *A History of Medicine* (New York: Oxford University Press, 1950), 307.

14. Hippocrates, "On the Art," in *Hippocrates*, vol. 2, trans. W.H.S. Jones, Loeb Classical Library (Cambridge, Mass.: Harvard University Press, [1923] 1981), 193, 205.

15. Hippocrates, "Diseases I," in *Hippocrates*, vol. 5, trans. P. Potter, Loeb Classical Library (Cambridge, Mass.: Harvard University Press, 1988), 113; Hippocrates, "Joints," in *Hippocrates*, vol. 3, trans. E.T. Withington, Loeb Classical Library (Cambridge, Mass.: Harvard University Press, [1923] 1968), 339; and Hippocrates, "On the Art," 203–205.

16. See "futility" in *The Oxford English Dictionary* (Oxford: Oxford University Press, 1989), 626.

17. E. Pellegrino and D.C. Thomasma, *For the Patient's Good* (New York: Oxford University. Press, 1987).

attained, treatment should not be offered, or, if in use, it should be withdrawn. To treat in the presence of futility is to act against the patient's good, and, if such treatment is burdensome, to act maleficently as well.

Futility is, however, not a moral principle. It is an empirical appraisal of probable clinical outcome, benefit, and burden. Thus it instantiates and specifies the principle of beneficence in a particular clinical event. It becomes a decision-making criterion because it offers a definable approximation of the patient's good. It is arrived at by use of our limited human intelligence and is fallible. Properly determined, the idea of futility helps us to attain the good of a particular patient, here and now, at the moment of a withhold/withdraw decision.

Futility derives moral force from its status as a specification of the principle of beneficence—the first principle of clinical ethics. Beneficence is the first precept of the Hippocratic Oath.[18] It recognizes the vulnerability, dependence, and need of the sick person as the source of the doctor's obligation to act always to optimize the welfare of the patient. In addition, when healing is pursued as a Christian vocation, ministry, or apostolate, it can become an explicit manifestation of God's grace. Then, beneficence and benevolence can become acts of loving charity. Futility is then interpreted in the light of the Christian view of human life and its destiny. This construal of charitable beneficence is at considerable variance with the trend of contemporary bioethics, which often puts patient autonomy before beneficence or subverts that autonomy for utilitarian, economic, or social reasons.[19] This divergence can best be delineated by looking at the most intensively debated questions: How is futility defined? Who defines it? How are competing interpretations resolved? How does futility as an instantiation of beneficence relate to competing principles of autonomy in three illustrative clinical situations (do not resuscitate orders, determinations of death, and care of the very young and very old)?

Definitions of Futility

As noted above, for most of the history of medicine, futility was taken to be an objective medical judgment which only physicians were qualified to make. This changed thirty or so years ago with the emergence of autonomy, which granted rights of decision and participation to patients and their valid surrogates. This movement began in the U.S. as legal right to refuse treatment.[20] It was strongly reinforced by a report of the President's Commission in 1983.[21] Since then, autonomy has less

18. Hippocrates, "The Oath," in *Hippocrates*, vol. 1, trans. W.H.S. Jones, Loeb Classical Library (Cambridge, Mass.: Harvard University. Press, [1923] 1972), 299–301.

19. E. Pellegrino, "*Agape* and Ethics: Some Reflections on Medical Morals from a Catholic Christian Perspective," in *Catholic Perspectives on Medical Morals*, ed. J.P. Langan and J.C. Harvey (Netherlands: Kluwer Academic Publishers, 1989), 277–300.

20. *Schloendorf v. Society of N. Y. Hospital 211*, NY 215, 105 N.E., 92, 1914.

21. President's Commission for the Study of Ethical Problems in Medicine and Biomedical and Behavioral Research, *Defining Death* (Washington, D.C.: U.S. Government Printing Office, 1981).

appeal outside the U.S., the American view of autonomy is beginning to be an issue worldwide.

In the 1980s, the major issue in professional ethics in America was the medical profession's adaptation to participation by patients and families in clinical decisions. The degree of authority which should be vested in patients or their surrogates through the instrument of informed consent increases in scope, even threatening the physician's moral integrity. In the '90s, that authority has come to include participation in the definition of futility as well as micro-management of bedside decisions.

As a result, futility is no longer defined solely in medical terms but also in terms of the patient's goals, values, and beliefs, that is, those things by which we determine whether the decision is indeed "worthwhile" from the patient's point of view. A new debate now centers on who should define futility, how it should be defined, and what to do when physician and patient or surrogate disagree on its definition. The range of the debate is wide and, in many cases, the opposing views are mutually incompatible.[22] For example, there are those who argue that the moral weight of the patient's values is such that the idea of futility is no longer sustainable. The physician is thought to be so unable to disentangle his or her own values sufficiently from futility judgments that they should be abandoned.[23] Another view is that the debate and the definition of futility are fatuous exercises in hair-splitting and detrimental to good clinical decisions. Some others hold that the idea should be retained only for obvious situations like total brain death or permanent vegetative state.

Opposing those views is the belief that the traditional idea of futility should be retained but refined by explicit criteria. One proposal suggests that a treatment should be considered futile if it has been ineffective in the last hundred cases, does not restore consciousness, or does not remove the need for intensive care. The increasing availability of studies of effectiveness and ultimate outcome of treatments like cardiopulmonary resuscitation, strengthen these suggestions. However, the potential shortcomings of objective criteria cannot be ignored (that is, errors of diagnosis, prognosis, and medical information, or the problem of applying statistics to individual cases). Newer, empirically based algorithms and models that predict outcomes are helpful in objectifying prognosis, but they too unavoidably include value judgments and thereby lose some of their objectivity.

To mitigate the influence of the provider's or third party's value judgments, to protect patient autonomy, and to avoid the potential abuses of unilateral judgments, some have proposed that the criteria for futility be institutionalized in hospital policy or ethics committees.[24] There are, however, objections to institutionalizations, such

22. S. Younger, "Who Defines Futility?" *Journal of the American Medical Association* 260 (1988): 2094–95; The American Medical Association Council on Ethical and Judicial Affairs, "Medical Futility in End-of-Life Care, Report of the Council on Ethical and Judicial Affairs," *Journal of the American Medical Association* 281 (1999): 937–41.

23. L.K. Stell, "Stopping Treatment on Grounds of Futility: A Role for Institutional Policy," *St. Louis Univ. Public Law Review* 11 (1992): 481–97.

24. Ibid.

as that institutions depersonalize the decision, that standards vary between institutions or committees, and that the very fact that a policy is needed implies a patient's right not only to reject but also to demand treatment.

Futility as a Prudential Guide

To obviate some of these difficulties, a combination of subjective and objective criteria and a joint determination of futility by physicians and patients or surrogates seems most reasonable. This approach strikes a balance between three criteria: *effectiveness, benefit,* and *burden.* This balance is not a mathematical but a moral calculation, based on clinical assessment, which gives a weight to each of these three dimensions in relationship to the other and, ultimately, to the patient's good.

Effectiveness: for each treatment intervention, an estimate of its capacity to alter the natural history of the disease or symptom in a positive way. Does the treatment make a difference in morbidity, mortality, or function? This is an objective determination, dependent upon outcome studies and within the physician's domain of expertise. Effectiveness centers on medical good and on measurable clinical data about prognosis and therapeutics.

Benefit refers to what is valuable to the patient as perceived by himself or his valid surrogate. This is a subjective determination and not within the doctor's domain but in that of the patient or his surrogate. Benefit centers on the patient's assessment of his own good—his goals and values in undergoing treatment. It asks the question, Is this treatment worthwhile for me, the patient? It is not quantifiable.

Burden refers to the physical, emotional, fiscal, or social costs imposed on the patient by the treatment. Burdens are both subjective and objective and within the domain of both the doctor, when factual, and patient, when subjective and personal. Burdens imposed on the medical team or society would, in certain rare circumstances, be considered as well as burdens on the patient. The question here is, What will effectiveness and benefits cost, not just in dollars but in their totality? Like benefit, burden is not readily quantifiable.

When the assessment of these three phenomena is favorable to the patient's good, other things being equal (*ceteris paribus*), treatment is morally justifiable; when it is unfavorable to the patient's good, the treatment in question is not morally justifiable.

This approach combines subjective and objective components and integrates the expertise and authority proper to each of the major participants—physicians, nurses, patient, and surrogates. It cannot be a unilateral decision. It requires a joint determination and agreement if the total good of the patient is to receive the consideration it deserves. This approach also gives some concrete clinical expression to the terms *ordinary* and *extraordinary*. *Ordinary* treatment is effective, serves some beneficial goal of the patient, and/or carries burdens which can be outweighed by the effectiveness and benefit. *Extraordinary* treatment would be futile treatment as determined by the above criteria, that is, ineffective, not consistent with the patient's goals and values, and/or so costly, dangerous, painful, or otherwise so burdensome as to outweigh effectiveness and benefit.

On this view, *ordinary* treatment becomes beneficial treatment that can vary with current technological capability; in extraordinary treatment there would be little or no probability of a beneficial outcome for the patient. No matter how *high* the technology, the availability and non-availability of technology per se is not the determinant of what is ordinary or extraordinary. The meaning of ordinary and extraordinary thus is not tied to the state of technological progress. Technology is a means which is itself judged by its effectiveness, benefits, and burdens. This approach should be helpful in actualizing the notions of *proportionate* and *disproportionate* as they are used in the Vatican "Declaration on Euthanasia": "Thus, some people prefer to speak of *proportionate* and *disproportionate* means. In any case, it will be possible to make a correct judgment as to means by studying the type of treatment to be used, its degree of complexity, and comparing these elements with the result that can be expected by taking into account the state of the sick person and his or her physical and moral resources."[25] Thus, on the definition of futility that I have suggested, a *disproportionate* means would be a futile means, remembering always the misuse of these terms by proportionalists as already noted above.

This approach to futility avoids the stigmatization and devaluation inherent in equating futility with any particular diagnosis, clinical condition, or category of patient. Too often, patients in a "permanent vegetative state," those with lethal genetic disabilities or mental "retardation" are relegated automatically to dangerous under-treatment or neglect of remediable conditions. This schema also avoids automatic negative quality-of-life determinations or denials of personhood to brain-damaged infants, demented adults, or handicapped people generally. Instead, the focus is on the prudential interplay of effectiveness, benefit, and burden as they relate to the good of the patient with the spiritual good of the patient as the highest priority.

This schema recognizes that the *good* of the patient is a complex notion. As Thomasma and I have suggested,[26] it includes at least four components, hierarchically arranged. The lowest good is the medical good, that is, the well-functioning of the human organism as organism. This includes psycho-social as well as physical functioning. This is the realm in which the physician has major expertise. The next level of good is the patient's own assessment of his or her personal good, a definition of the patient's preferences, goals, the kind of life he or she wishes to live. In this realm, the patient or his or her designated surrogate is the point of reference. Next is the good of the patient as a human person, an assessment in terms of the natural law's grasp of what is proper to the life of humans as humans—this level of patient good is not defined by the doctor or patient. It is built into what it is to be human. Its point of reference is the natural law. Finally, the highest good is the spiritual good, that which derives from the fact that humans are created and destined by a

25. CDF, "Declaration on Euthanasia," 510–16.

26. Pellegrino and Thomasma, *For the Patient's Good.*

personal God to a life beyond this world in union with Him. The point of reference here is Scripture, Church teaching, and tradition. These are not definable by patient or physician. This is the level entirely negated or ignored in secular bioethics despite the fact that every patient, physician, or surrogate will have some faith commitment or faith rejection.

Moral Dangers in the Application of the Concept of Futility

Abuses of the concept of futility can be the result of a wrong or bad assessment of the balance between effectiveness, benefit, and burden. A *wrong* assessment could result from objective errors of observation, prognosis, probability estimates, incompleteness of knowledge of the patient's life context, or illogical reasoning. A *bad* assessment, on the other hand, would be the result of the erroneous moral, metaphysical, or theological presuppositions about human nature, moral philosophy, or the nature of human life or its meaning. In the examples that follow, most emphasis will be put on the importance of the proper theological and metaphysical starting points. In the two examples, namely, cardiac resuscitation and brain death, the emphasis is on objectively wrong judgments. Usually the two forms of error will be intermingled and will have to be carefully disentangled in order for us to discern the sources and possible remedies for any specific instance of abuse of the concept. Futility has been defined here as a prudential guide to moral assessment of the good of the patient and to the moral permissiveness of withholding or withdrawing particular treatments in seriously ill or dying patients. On this view, if a treatment is judged to be futile after weighing its benefits, burdens, and effectiveness, it need not, and ought not, be offered or used. However, like any prudential guide, there is a danger of misuse if the metaphysical, theological, and ethical presuppositions upon which the judgments are based are faulty.

This is often the case when secular bioethicists employ the concept in ways totally opposed to Catholic Christian medical morals. Thus, futility has been used to justify euthanasia; assisted suicide; refusing treatment to seriously handicapped infants, to the aged, and to the infirm in order to spare parents, families, or society the burdens of caring for such patients. Without a foundation in Christian anthropology and the Gospel vision of health care, such things as quality-of-life, economics, sacrifice for others, and spiritual belief are translated into terms of mere utility, economics, pleasure, or absence of all suffering.

Each of the three elements of the relationship, that is, effectiveness, benefit, and burden, must be judged within a moral context, that is to say, within the context of treatments that are not intrinsically wrong. Thus, abortion, tubal ligation, or assisted suicide could, on purely secular grounds, be classified as *effective*. A patient might see euthanasia or assisted suicide as highly *beneficial*. A marginally painful but highly effective treatment might be judged as *burdensome*. These three variables are to be judged clinically but always within the boundaries of the morally permissible. Futility assessments, thus, are prudential judgments about specific clinical situations to assure that the moral judgments of *ordinary/extraordinary* and *proportionate/disproportionate* are grounded empirically.

Let us look briefly as some examples of the misuse of futility, recognizing that *abusus non tollit usum* and that Catholics need not abandon the idea of futility because it is misused so often in secular bioethics. Indeed, secular misuse of the futility criterion imposes an obligation on Catholic moralists to "rescue" the term from its misuse. To abandon the term is to surrender it to secular definition solely.

Quality of Life

One frequent misuse of futility is to interject the observer's quality-of-life assessments into the judgment, especially with infants, children, and those who cannot express their own views. Usually, the argument is made that the projected disabilities or discomforts are so severe that no one would want to endure them or ought to have to endure them. Out of compassion and mercy, it is insisted that treatment should not be considered since the possibility of a satisfactory life is futile. This is a particular danger with neonates, the retarded, or the comatose patient.

Quality of life, however, is an infinitely malleable term. No two persons have the same definition of a satisfactory life. No one is qualified to make a quality-of-life decision for another, especially for an infant or a child who has had no opportunity to experience life. In Christian charity and morality, there is no such thing as *Lebens unwertes lebens* nor metaphysically wrongful existence.

Quality of life for those who cannot assess that quality for themselves is not a consideration in a Christian view of life, which bestows dignity on every human being, regardless of physical or intellectual limitations. Who among us can discern God's intent or provide initial purposes for any human life or for those in whose midst that life may be placed? Whether a person is a "useful" or "contributing" member of society does not affect his or her dignity or the sanctity of that person's life. Human dignity is intrinsic, conferred by God, and therefore not "lost" or "gained" by human judgments. Indeed, the very disabilities so many fear may be the occasion of spiritual growth among the family or friends called to raise a disabled child. None of this denies the difficulties or suffering that may accompany years of caring for a mentally or physically disabled person. It is to insist, however, that in a Christian view of human life, no life is so disvalued as to be per se futile or "worthless." Confusing the futility of treatment with the futility of a life itself is a serious offense against human dignity and God's providence in our daily lives.

Even a mentally competent person must make a decision to refuse life-saving or effective treatment on grounds of the quality of that life with the utmost care. There are clinical situations in which the burdens of treatment are so heavily fraught with physical, emotional, and fiscal burdens, and the benefits are so remote that conscious refusal can be justified.[27] However, this is not the case with persons in the early stages

27. CDF, "Declaration on Euthanasia," 510–16.

of chronic illnesses, which may be disabling, painful, or fatal in the future. Some patients refuse treatment or seek euthanasia and assisted suicide in anticipation of future changes in the quality of their lives before it is clear what those changes will be or how they will respond to them. There are reasonable limits to how much additional burden or suffering one must assume. But the extent and weight of those burdens must first be known before a decision can be made.

The most opprobrious abuses of the quality-of-life argument are being advanced to justify experimentation with humans in permanent vegetative states.[28] Here, futility, the impossibility of returning to "meaningful" social relationships, is taken to devalue this class of humans as non-persons. Some moralists give such persons less claim on life and respect than anthropoid apes.[29] The same devaluation of persons because of the futility of attaining a *quality life* leads some ethicists to speak of disabled infants as "biological remnants" to be mercifully euthanized. The same fallacious reasoning lies behind the distinction between *having a life* (biological life) and *being alive* (having a biography), which is used to justify euthanasia and assisted suicide.[30] Similar reasoning leads to devaluation of the lives of the frail and the aged. Young people may fail to see any *quality* in a life restricted by the infirmities of age. Medical treatment cannot return a young person's estimate of quality-of-life to the aged. But all treatment is not, on that account, *futile*. As with infants, underlying this abuse of the futility concept is a contradiction of the Christian teaching about the value and dignity of all lives. It also reflects an unwillingness to accept any sacrifice of one's own pleasures, pursuits, or resources in corporal works of mercy.

Economics and Futility

Ours is an age obsessed with economics and with the *drain* on society of caring for those whose lives are socially devalued. To accept that devaluation is to risk the next step, that is, to limit care, undertreat, and accelerate the deaths of those who are an economic burden on society. This is a special danger in countries like the U.S., in which managed care in its commercialized form is taken to be an economic necessity.[31] To be sure, no plan at present openly advocates withholding necessary life-sustaining care, but the temptation to do so is not negligible. Moreover, the definition of *necessary care* is manipulable when money is the issue. *Economic futility* is justified usually by the presumed deprivation of resources for the young and those with better prognoses. Again, to reach such a conclusion is to devalue the lives of a

28. R.G. Frey, "Moral Standing: The Value of Lives and Speciesism," *Between the Species* 4, no. 3 (1988): 191–201. See also R.G. Frey, *Rights, Killing, and Suffering* (Oxford: Basil Blackwell, 1983).

29. P. Singer, *Rethinking Life and Death: The Collapse of Our Traditional Ethics* (New York: St. Martin's Press, 1994).

30. J. Rachels, *The End of Life: Euthanasia and Morality* (New York: Oxford University Press, 1986); J. Lachs, "Active Euthanasia: Theoretical Aspects," *Journal of Clinical Ethics* 1, no. 2 (1990): 113–15.

31. D.P. Sulmasy, "Managed Care and Managed Death," *Archives of Internal Medicine* 155, no. 2 (1995): 133–36.

whole segment of society. A stronger denial of the Christian respect for the dignity of each person regardless of disability cannot be imagined.

On the Christian view of economics and healing, there must be a concern for solidarity, for the mutuality of our responsibilities to each other as members of the human community. This is properly expressed in a positive, rather than a negative, way as the responsibility to assist each other in time of need. We are all expected to use common resources wisely, but also to make sacrifices for the most vulnerable among us. In times of war, famine, or pestilence, rationing of resources might be unavoidable. But, even then, allocation must be on the basis of respect for all humans and a personal willingness to sacrifice some part of one's own resources for others more needy.

Obviously, the same abuse of futility can be directed against life-sustaining care among the poor. Here clinical futility is equated with non-clinical futility. The unlikelihood that a poor person will become self-sufficient or a contributing member of society is used to justify the withholding of all or of expensive treatments. This abuse is in direct contradiction to Christian charity, which gives preferential option to the poor.

The disturbing aspect of economic influences on futility determinations is that they are so evident in affluent countries in the absence of any demonstrated economic emergency. The fear seems to be that expenditures for the aged, the poor, or the handicapped infant will compromise or limit discretionary spending for luxuries, recreation, or personal pleasure. Such an attitude contradicts the idea of a society founded on Gospel teachings, or the social encyclicals of the modern pontiffs.

Abuse of the futility concept does not preclude morally proper considerations of economics in healthcare decisions. Competent patients can, out of consideration of charity, refuse treatment for themselves to spare others the expense of their care or to protect an estate for children, for example. Patients anticipating the loss of competence to make their own decisions can instruct their proxies or surrogates or prepare a living will to impose economic restraints on their terminal illness. The treatment in question, however, must be of marginal effectiveness or benefit. Refusal could then be an act of charity in the interest of one's family or to society at large.

Futility and Autonomy

In American bioethics, autonomy has become the dominant ethical principle. In a short period, it has evolved from a negative right to refuse treatment into a positive right to participate in treatment choices. In the last decade, many have come to demand treatment in the name of autonomy or even to "micro-manage" clinical decisions at the bedside. Some would argue that, in the name of autonomy, patients have a right to *demand* that "everything be done," even when treatment is judged futile by the definition we have suggested.

On philosophical grounds alone, one can argue against such a demand since it would force physicians to practice irrational medicine. This violates even the ancient notion of futility as simply medical futility. It also imposes economic burdens unjustly on others without a proportionate reason and without their consent. A Christian patient should not make such a request because it would be selfish and uncharitable. It would offend against the acceptance of finitude and the sovereignty of God who calls us to Him when He wills. To ask for repeated resuscitation and for futile employment of the full panoply of medical technology when death is inevitable is an act of pride.

There are times, however, when a treatment may be futile in the long term but of benefit to the patient in the short term. A patient dying of disseminated carcinomatosis might desire to live to see a grandchild born or graduate from college or to say a final farewell to his family. He might ask that dying be prolonged by antibiotic treatment for a pneumonia or dialysis for renal failure. Treating pneumonia or using dialysis would be futile in curing the cancer, but not in attaining a benefit for the patient, like having time to complete unfulfilled religious or personal obligations. For similar reasons, a patient might ask to be resuscitated or given transfusions.

Continuing treatment could also be justified in order to allow young parents to adjust to and to accept the inevitable death of a newborn baby. Or, it may be justifiable to some extent when patients or families genuinely believe in and pray for a miracle. Pastoral counselors should be given time to help patients who hope for miracles to comprehend the burdens they may be imposing on a terminally ill, comatose infant or adult.

In all these instances, treatment does not satisfy the full notion of futility since there is some benefit, at least as seen by the patient or his family. Moreover, futility, like any other consideration in decision making, must always be applied humanely, sensitively, and with discretion. Some argue that a treatment judged futile should never be initiated or, if initiated, should be stopped immediately. Applied too rigorously, the futility concept could ignore the obligation to help the patient live the last days of his or her life as serenely and satisfactorily as possible.

In these cases, patient autonomy cannot override a physician's conscientious moral objection. This would be absolutizing the patient's right of self-determination. Patients cannot expect physicians to provide treatments that they take to be medically futile. Physicians are persons, too, and are entitled to respect as such. Like patients, physicians are moral agents bound to follow their consciences and are accountable for failure to do so. Neither doctor nor patient is empowered to impose his will on the other. A civil and courteous discontinuance of the relationship may sadly be the only answer when moral and religious commitments are mutually incompatible.

In any case, a Christian view of autonomy would be based in respect for others as brothers and sisters in Christ. Physicians and patients would recognize their mutual obligations in charity and work together to negotiate the establishment of treatment goals, the conditions of futility, and the timelines for re-evaluation periodically of those definitions. Autonomy modulated by charity is an obligation of Catholic Christian patients, families, and health professionals.

Resuscitation and Brain Death

Do not resuscitate decisions involve the concept of futility very intimately. Cardiopulmonary resuscitation is a treatment like any other. It was developed for sudden cessation of cardiac function usually as a result of failure or disorder of electric activity of an intact myocardium. Under these conditions, CPR is very effective if done promptly—within 2 to 5 minutes, that is, before cessation of cerebral bloodflow has irretrievably damaged the brain.

Cardiopulmonary resuscitation is, however, of dubious, marginal, or no value when used in patients dying of some underlying fatal disease or when the heart muscle itself is seriously damaged. It is now known, for example, that patients with massive intracerebral bleeding, disseminated carcinomatosis, or chronically ill and aged patients with simultaneous failure of three organ systems (cardiopulmonary, renal, pulmonary, or hepatic) do not survive to leave the hospitals even if cardiac activity can be restored. Indeed, they are apt to end up in a permanent vegetative state even if cardiac function is restored. Cardiopulmonary resuscitation is, therefore, not intended for every patient who dies. It must be regarded as a treatment with proper and improper clinical use.[32]

For a Christian Catholic patient or his or her surrogate to demand repeated cardiopulmonary resuscitation in the face of its futility as a treatment would be morally wrong. It would, again, be to deny the fact of human finitude and impose unnecessary effort, expense, and emotional trauma on the patient and on others. One may believe in the power of prayer and miracles without resorting to repeated futile resuscitations. If God wills a miracle, He will intervene in His way and on His time, so long as we do not cease treatment when it is still effective or beneficial. In passing, it must be said here that the so-called slow or chemical code is not justified morally. Here, physicians go through the motions of resuscitation with no intent to succeed. Physicians may wish to please or comfort the family by these incomplete faux resuscitations. But they are acts of deception and, in the end, betrayals of trust. For the Catholic Christian, there is either a *full* code with the intention to resuscitate if possible, or code at all. Cardiopulmonary resuscitation should be withheld when it is not indicated, that is, when it is futile.

Another clinical situation in which the notion of futility is crucial is total brain death.[33] Here the whole brain (cortex and entire brainstem) is irreparably destroyed and recovery is not physiologically possible. In such a situation, treatment could not be effective; no benefit could accrue; resources in personnel time and effort would be used to no discernible purpose. To continue to treat or repeatedly to resuscitate such a patient would be to no spiritual or material purpose and a wrongful intrusion on the natural process of dying.

32. L.J. Blackhall, "Must We Always Use C.P.R.?" *New England Journal of Medicine* 317 (1987): 1281–87.

33. H. Beecher, R. Adams, A. Barger, et al., "A Definition of Irreversible Coma: Report of the Ad Hoc Committee of the Harvard Medical School to Examine the Definition of Brain Death," *Journal of the American Medical Association* 205 (1968): 337–40.

In this essay, we need not confront the debated question of whether a person is dead when the brain is dead. This question deserves re-examination since its metaphysical implications are highly significant.[34] Indeed, it is critical when it comes to organ transplantation. The temptation here is to declare *death*—or futility—hastily in order to procure a needed organ. This abuse is a constant danger in secular bioethics where utility, and not the dignity of the human person, is the dominant criterion. Catholic and secular clinicians and moralists differ on the moment at which death of the person occurs.[35] For the purposes of clinical decision making, as long as it is clear objectively that recovery is not possible, a prudential judgment of futility is defensible and the patient may be allowed to die as a consequence of the natural history of his or her disease.

The situation is different, however, with partial brain death, for example, death of the cortical function but retention of mid-brain function—the so-called permanent vegetative state. Here the patient is unquestionably alive.[36] Some ethicists would equate cortical with whole-brain death because the patient can no longer enter into *meaningful* relationships with other humans or achieve any of his spiritual or physical goals.[37] Some ethicists even argue that the patient is no longer a person, erroneously making a metaphysical judgment not determinable by the clinical state of the patient. This is an especially dangerous conclusion if it is factored into a futility judgment since it puts many vulnerable persons—infants, the demented, the brain-damaged—at risk.

These conclusions are objectively wrong, and they lead to morally bad decisions by imposing metaphysical categories beyond the scope of medicine to determine. Patients in a permanent vegetative state should be approached as seriously ill human persons. Decisions to withhold or withdraw life-sustaining treatments should be made on the basis of whether such treatments will be effective and beneficial or whether the burdens are so great as to be disproportionate. The presumption, as the American bishops have stated, is to provide "nutrition and hydration to all patients, including patients who require medically assisted nutrition and hydration as long as this is of sufficient benefit to outweigh the burdens involved to the patient."[38]

34. J. Seifert, "Is Brain Death Actually Death?" *The Monist* 72 (1993): 157–202.

35. P.A. Byrne and G. Rinkowski, "Brain Death Is False," *Linacre Quarterly* 66 (1999): 42–48; J.L. Bernat, "A Defense of the Whole Brain Concept of Death," *Hastings Center Report* 28 (1998): 14–23; A.A. Howsepian, "In Defense of Whole-Brain Definitions of Death," *Linacre Quarterly* 39 (1998): 39–61; Multi-Society Task Force on the Persistent Vegetative State, "Medical Aspects of the Persistent Vegetative State," *New England Journal of Medicine* 330 (1994): 1499–1508, 1572–79; R. Veatch, "The Impending Collapse of the Whole Brain Definition of Death," *Hastings Center Report* 23 (1993): 18–24; D.A. Shewmon, "Recovery from Brain Death: A Neurologist's Apologia," *Hastings Center Report* 27 (1997): 29–37; P.A. Byrne, S. O'Reilly, P.M. Quay, et al., "Brain Death—The Patient, the Physician, and Society," *Gonzaga Law Review* 18 (1982–1983): 429–516.

36. Multi-Society Task Force, "Medical Aspects of the Persistent Vegetative State."

37. R. Veatch, "The Impending Collapse of the Whole Brain Definition of Death"; R. Troug, "Is It Time to Abandon Brain Death?" *Hastings Center Report* 27 (1997): 29–37.

38. National Conference of Catholic Bishops, *Ethical and Religious Directives for Catholic Health Care Services* (Washington, D.C.: U.S. Catholic Bishops, 1995), 23.

All of these problems are compounded in the case of infants and children. Here, determinations of cortical function, future prognosis, and intervals beyond which recovery is impossible are much more difficult to assess empirically. Prudence in the evaluation of empirical data is an essential moral precaution.[39] This is especially true in cases of trauma or in prognostication of future defects in intelligence.

With infants, we must be especially cautious in applying the criterion of futility. We cannot know what the infant now, or in later life, would take to be the "benefits" of treatment. We can ascertain what the effectiveness of treatment may be in terms of mortality and morbidity. We can also, to some degree, assess the burdens to the infant. In an age driven by utility, economics, and the unwillingness of some parents to "accept" anything less than a perfect child, "futility" can be grievously misapplied. Lacking the infant's participation, the physician is under special moral obligation to protect the welfare of the infant, even in the face of the parents' wishes. Parent "autonomy" is often misinterpreted erroneously as giving absolute dominion over the life of the infant. "Benefit" in these situations cannot mean intentionally accelerating death, involuntary or non-voluntary euthanasia, or neglecting effective treatment which carries little burden. Nor can benefits or lack of benefit to society—economic or relief of social burdens—be considered. If they involve intrinsically immoral acts, they can never be factored into futility determinations.

Both in brain death and resuscitation, there is an integration of objective clinical data (making the technically right decision) and the metaphysical and moral principles (making a good decision). These distinctions can, perhaps, help in clarifying what is "ordinary" and what is "extraordinary" treatment.

Judicial use of the futility criteria avoids stigmatization and the resulting devaluation of brain-damaged patients by confining the decision to a deliberative balancing of effectiveness, benefit, and burdens. Proper use (empirical and moral) of futility criteria would forbid automatic cessation of life-support in patients simply because they are classified as being in a permanent vegetative state, frail and aged, or, as infants, faced with what some would judge as lives of poor or reduced "quality."

The Morally Appropriate Use of Futility

The criterion of futility is today working its way into secular clinical parlance as a component of the ethical decisions to withhold or withdraw treatments. It will undoubtedly attract the interest of Catholic physicians and health professionals since it has strong roots in clinical tradition and empirical observation. Properly interpreted as a prudential guide within specific moral constraints, it can help to recover and explicate the continuing importance of the traditional terms *ordinary* and *extraordinary*, *proportionate* and *disproportionate*. These terms are central to the teaching of Pius XII on end-of-life care, and they have strongly influenced subsequent Catholic approaches to end-of-life decisions.

39. I. Torres and J. Fugate, eds., *Critical Care of Infants and Children* (Boston: Little Brown and Co., 1996), 403–405.

Some of the requirements for the morally proper use of a notion of futility can be formulated as follows:

(1) Every determination of futility must be made within the set of beliefs and commitments that inspire all Catholic health care with the dignity of the human person, the sanctity of life, and the ministry of Jesus as healer and suffering servant. Life can never be willfully ended simply because treatment may be *futile*.

(2) Each judgment of futility must take all aspects of the patient's total life into account—physical, mental, spiritual, preferences, and life-goals included. Futility is not an isolated, empirical, yes-no test. It demands prudential assessments for a particular person in a particular experience of illness and within a particular metaphysical and theological content.

(3) Care, comfort, pain relief, amelioration of suffering must always be provided. Futility does not mean abandonment of care.

(4) Efforts must continue to discover genuine cures or treatments for diseases now considered incurable. Futility is not a justification to limit the progress of medicine for certain vulnerable groups, including the very young, the very old, the disabled, those in permanent vegetative states. Indeed, properly used, the criterion of futility avoids the stigmatization of this group of patients whose lives, rather than their treatments, are too easily regarded as *futile* by others.

(5) Futility determinations cannot be made unilaterally. They are always a cooperative enterprise in which each participant has a defined area of authority; the doctor is best equipped to determine effectiveness, the patient is the authority on benefits, and the doctor and the patient together share the assessment of burdens. Anticipations and working together will prevent the conflicts that arise when decisions are urgent and communication has been lacking.

(6) The concrete judgment of futility must not be so rigorously applied that it precludes prolongation of life in order to meet religious obligations, to see family and friends, and so on. As always, futility must be interpreted within a Christian context of life, death, illness, suffering, and the spiritual destiny of all humans. Charity, not utility, is the final principle and ultimate virtue of care for the dying.

B. Physician-Assisted Suicide

20

Introduction: Physician-Assisted Suicide

I. John Keown, DPhil, PhD, DCL

DR. PELLEGRINO'S 1992 PAPER "DOCTORS MUST NOT KILL" richly repays re-reading (see chap. 21 below). One of around thirty articles he wrote addressing voluntary euthanasia and physician-assisted suicide, it remains an exemplary response to the case for legalization. He listed three concerns: that legalization was not supported by adequate ethical arguments, was incompatible with the doctor-patient relationship, and would produce grave social consequences. The intervening years have witnessed at least five significant developments in the debate.

First, the terminology. Although he rightly maintained that the debate was too important to be obscured by euphemism, discourse is now dominated by the tendentious and muddying "assisted dying."

Second, he discussed the Dutch legalization of euthanasia, and now several other jurisdictions have since followed suit, including Canada, Spain, Portugal, Belgium, and New Zealand. In the US, Oregon permitted physician-assisted suicide in 1997, as do several other US jurisdictions. In 2020, the German Constitutional Court fashioned a sweeping right to assistance in suicide. In several other developed jurisdictions like the UK, the debate has intensified.

Third, campaigners in some jurisdictions, like the US, have switched tactics, successfully calculating that legislators and the public are more amenable to proposals that limit physician-assisted suicide to the "terminally ill."

Fourth, as more jurisdictions have relaxed their laws, the focus of debate has moved the question of the social consequences of legalization to center stage, in particular the ability of legal guidelines effectively to control voluntary euthanasia and physician-assisted suicide. Do permissive laws confirm or confute the claim of their opponents that lead society down an empirical and/or a logical "slippery slope"? While the Dutch experience remains important, the Oregon experience is no less so.

A fifth development has worked against legalization. The rise of the disability movement, which has closed ranks against legalization, has reinforced the argument that voluntary euthanasia and physician-assisted suicide rest on a belief that patients would be "better off dead," a judgment that particularly endangers those with disabilities.

The ethical debate about whether voluntary euthanasia and physician-assisted suicide are justified by arguments based in autonomy or beneficence, or by equating foresight with intention, is still important, but it has lost some ground to the debate over the effects of legalization. Pellegrino's second concern has also lost purchase as more healthcare professionals, and occasionally their professional bodies, drop their opposition to legalization. It is his third concern, the likely social consequences, that dominates the debate.

Regrettably, it is not as if his three concerns have been tested in the ongoing debate and found wanting. They have been more evaded than tested, often swamped by emotional and sympathetic media coverage of patients campaigning for reform, such as Brittany Maynard in the United States. Public opinion, which Pellegrino rightly criticized as one of the feeblest arguments for reform ("plebiscite ethics"), has proved increasingly powerful, especially in jurisdictions like Oregon where voter referenda can change the law.

The literature is even vaster than it was in 1992, though much is either unoriginal or blinkered. Dr. Pellegrino's article set the standard for informed reflection and analysis. What he so cogently argued against, however, is increasingly coming to pass, as misguided compassion trumps rational reflection.

21

Doctors Must Not Kill

Edmund D. Pellegrino, MD

He knew my objective to do good, though he did not believe in the good to be done.

—Anthony Trollope, *The Fixed Period*

Introduction

Is it ever morally licit for physicians to intentionally kill a patient to relieve suffering even with the patient's request? Should the long-held proscription against active euthanasia be relaxed legally and ethically, as it has been in the Netherlands? Are physicians who oppose active euthanasia being unfaithful to their traditional duty of compassion? These questions are being debated in the United States and in several Western European countries. They are highlighted by the legal, ethical, and social tolerance for euthanasia in Holland and the growing public and professional sentiment elsewhere that favors the termination of life by physicians. How the medical profession, society, and the public answer these questions promises to have a profound effect on the moral quality or the physician-patient relationship and of the whole of society.

In this article, I shall argue that physicians should not kill, directly or indirectly, even out of compassion, for three reasons. First, the moral arguments favoring euthanasia are logically inadequate. Second, killing by physicians seriously distorts the healing relationship. And third, the grave social consequences of such killing are morally prohibitive.

Euthanasia literally means "good" or "gentle" death. No one can reasonably oppose a "good" death. But some construals of euthanasia are morally defensible; others are not.[1] It is crucial, therefore, to define the sense in which I shall use the term. I shall argue against what is called "active" euthanasia—the intentional killing of a patient by a physician, with the patient's consent (voluntary euthanasia), or without

---

Reprinted with permission of the Pellegrino family from Edmund D. Pellegrino, "Doctors Must Not Kill," *The Journal of Clinical Ethics* 3, no. 2 (1992): 95–102.

1. D.C. Thomasma and G.C. Graber, *Euthanasia: Toward an Ethical Social Policy* (New York: Continuum, 1991), 1–11.

consent when consent is impossible (nonvoluntary euthanasia), or when consent is possible but not sought (involuntary euthanasia). To include in this definition such qualifiers as "good" or "gentle," and to evade the use of the term "killing," is to beg the central moral question—as Van der Meer does in his definition.[2] Van der Meer also manipulates the definition by including "in the patient's interests," so that involuntary euthanasia also can conveniently be accommodated. The ethical issue is too serious to permit its obfuscation by either euphemism or euphuism.

What is called "passive" euthanasia—allowing a patient with an incurable disease to die either by withholding or withdrawing life-sustaining support—will not be my concern in this essay. Nor will "mercy killing," which includes killing by family, friends, or someone designated by society to play this role. My focus is specifically on killing by physicians, for this is what many seek to legitimate today.

The line of argument I oppose can be summarized as follows. Active euthanasia is a beneficent and compassionate act because it relieves human suffering. Moreover, human beings should, on the principle of autonomy, have the right to end their lives when they wish to terminate their sufferings. Physicians, as agents of the patient's welfare, should assist either by directly killing the patient or by assisting the patient in suicide. Since physicians are best qualified to assist in dying, they should participate if the patient's "right to die" is to be actualized. Physicians already participate in passive euthanasia. Since there is no moral difference between killing and letting die, the opposition to active euthanasia is logically untenable.[3] In addition, the abuses envisioned in slippery-slope arguments can be prevented by legal guidelines and criteria, like those currently used in Holland.[4] No physician who is morally opposed need participate. Medical ethics should be updated to take account of the changed religious, moral, and political milieu. Because of all of these arguments, the proscriptions in law and ethics against physician killing must be relaxed.

I recognize the appeal of this line of argument in a pluralistic, democratic, and secular society. Nothing that follows is meant to impugn the sincerity or motives of the proponents of euthanasia. Their desire to spare human suffering is commendable. It is the means they use to this end that are morally unacceptable.

I will make my case without invoking what, for many, are still the two most powerful arguments against euthanasia: (1) the Jewish and Christian belief that humans are stewards and not the absolute masters of the gift of life, and (2) the Christian belief that even human suffering may have meaning. David Thomasma and I have enlarged on this perspective in a forthcoming volume.[5] The religious

2. C. Van der Meer, "Euthanasia: A Definition and Ethical Conditions," *Journal of Palliative Care* 4, nos. 1–2 (1988): 103–106.

3. J. Rachels, *End of Life* (New York: Oxford, 1986), 106–28.

4. M.A.M. de Wachter, "Active Euthanasia in the Netherlands," special communication, *Journal of the American Medical Association* 23 (1989): 3316–19.

5. E.D. Pellegrino and D.C. Thomasma, *Helping and Healing: Religious Commitment and Health Care* (New York: Continuum, in press). [This book was published in 1997.—Ed.]

viewpoint is unpalatable to contemporary mores, but it still hovers in attenuated form over the debate and is not lightly dismissed. However, since these arguments involve faith commitments, they are rejected by those without a faith commitment. For this reason, I shall confine my argument solely to the philosophical objections. Even without recourse to religious beliefs, active euthanasia is morally untenable.

Adequacy of the Arguments

The burden of proof weighs heavily on those who would abolish a moral proscription so deeply rooted in public and professional morals and in so many cultural traditions. However, a review of the major arguments advanced in favor of euthanasia does not sustain the necessary burden of proof.

The Distinction Between Killing and Letting Die

Advocates of euthanasia reason that there is no moral distinction between killing the patient and letting the patient die by deliberate withdrawal or withholding of life support or lifesaving treatment. This argument ignores the fact that in euthanasia, the physician is the immediate cause of a death that she fully intends. This act differs from withholding or withdrawing treatment. When treatment is discontinued, it is the disease that kills the patient. In so-called passive euthanasia, we remove our intervention for good reasons (for example, it is no longer effective or beneficial and its burdens are disproportionate). To continue treatment would be unethical, since futile, burdensome, and expensive treatments would be forced on the patient in violation of the canons of good medicine and of the patient's best interests. Adherence to these canons is a primary professional obligation. The physician also violates this moral canon if she continues with unnecessary, burdensome, or futile treatment.

James Rachels, who makes the most extended case against the distinction between killing and letting die, admits that the general sentiment that favors such distinction has some weight. He argues that such intuitions should be distrusted, however, because the well-off members of society have more to fear from relaxing the prohibition against active euthanasia than letting people die.[6] This is an unconvincing argument that begs the question of interest.

When the patient is "overmastered by the disease," to use the Hippocratic phraseology, we have a moral obligation to stop treatment, since our interventions serve no beneficial purpose.[7] When we do so, the natural course of the disease supervenes. The disease, not the physician, kills the patient. As Daniel E. Callahan suggests, medical hubris has grown to such proportions that we think nature is totally subservient to us. We take responsibility for its operations as well as our own. However, the AIDS epidemic, the ubiquity of bacterial resistance to

6. J. Rachels, "Killing and Starving to Death," *Philosophy* 54 (April 1979): 159–71.

7. Hippocrates, vol. 2, trans. W.H.S. Jones, Loeb Classical Library Series (Cambridge, Mass.: Harvard, 1981): 193–203.

antibiotics, and Alzheimer's disease, Callahan reminds us, are physical causes of death we do not control.[8]

Proponents of euthanasia also argue that there is no distinction between euthanasia and giving a patient morphine to relieve pain that could end in death from respiratory depression. This is a specious argument, because under these circumstances the primary intention is not to kill the patient as in active euthanasia, but to relieve pain; if death occurs, it is the result of a side effect of a powerful drug. Physicians use hazardous drugs and perform many operations in precarious cases where the intent is to help. The risk of dangerous side effects is tolerable, if benefits can be achieved in no other way. The use of morphine would be euthanasia only if the dose were deliberately calculated to cause respiratory depression. Rather, the effort is to titrate the dose, so as to achieve the beneficial and avoid the harmful effect. This titration is not controllable by formula. It may be in error, but the risks of side effects are outweighed by the proportionate potential for pain relief.

Euthanasia and Patient Autonomy

A second argument for euthanasia justifies it on grounds of respect for the moral right of autonomy and the dignity of the person. The logic of this assertion is suspect. When a patient opts for euthanasia, he uses his freedom to give up his freedom. In the name of autonomy, the patient chooses to eradicate life and consciousness, the indispensable conditions for the operation of autonomy. He loses control over a whole set of options, all of which cannot be foreseen and many of which would be of importance if life—the basis of freedom—had not been forgone. Moreover, if suffering is so intense that it limits all other options, and euthanasia is the only choice, then that choice is really not free. Seriously ill persons suffer commonly from alienation, guilt, and feelings of unworthiness. They often perceive themselves, and are perceived by others, as economic, social, and emotional burdens. They are exquisitely susceptible to even the most subtle suggestion by a physician, nurse, or family member that reinforces their guilt, shame, or sense of unworthiness. It takes as much courage to resist these subliminal conformations of alienation as to withstand the physical ravages of the dying patients coming from being subtly treated as nonpersons. The decision to seek euthanasia is often an indictment against those who treat or care for the patient. If the emotional impediments to freedom and autonomy are removed, and pain is properly relieved, there is evidence that many would not choose euthanasia.[9]

Euthanasia is too often an act of desperation that physician, family, and friends can forestall. To do so, they must provide the understanding, support, and sharing that will assure the sick person that she is still very much a member of the human community going through an experience everyone will eventually share.

8. D.E. Callahan, "Euthanasia Question Complicated by Need to Ration," *Commonweal* 115 (July 1988): 397–404.

9. N. Coyle, "The Last Four Weeks of Life," *American Journal of Nursing* 90 (December 1990): 75–78.

Beneficence

The most powerful argument for euthanasia is based upon compassion, mercy, and beneficence. These are obligations intrinsic to medicine as a healing art. Even those who see meaning in suffering urge comfort and relief of pain for the dying. Compassion and mercy would seem to override all other considerations.

The duty of beneficence indeed obliges the physician to help the patient to a good and gentle death. It is also compassionate and morally commendable to help a person to die well. A good death completes life with a finale befitting our true dignity as thinking, conscious, and social beings. The aim of medicine should be to facilitate a death that is as pain-free as possible but that is also a human experience. We can justify euthanasia for our pets precisely because they cannot possibly understand suffering or dying. They cannot die in a "human" way. But humans can grow morally even with negative experiences. A good death contributes something valuable to the whole human community. It enables us to assess our human relationships and to come to grips with the important and ultimate questions of human destiny, which believers and nonbelievers alike must confront. A good death is the last act of a drama, which euthanasia artificially terminates before the drama is really completed.

Some might object that all of this can be achieved in preparation for euthanasia and, even more effectively, before suffering makes rational thought difficult. But is this really so? How much of the quality of our deaths is dependent on living through the experience as it presents itself? To cut off this experience abruptly is to abort what might be the most important experience of our lives and a contribution to the lives of family, friends, and those who attend us. Often we see ourselves most clearly at the hour of our death. We see what we have made of ourselves. Critical times force us to reveal and confront the inner self. Artifices and circumlocution are no longer acceptable. Euthanasia deprives us of these last insights into who and what we have been, and it closes the door on the final accounting we may want to make.

Finally, there is the illusion that a life that is free of all anxiety, suffering, or misfortune is the "good" life. But, such a life is devoid of opportunity for expression of some of our most characteristically human feelings—mercy, compassion, understanding, empathy, love, and giving of ourselves to others. To be sure, dying persons can teach us good and bad lessons. A "good" death helps all of us to become more humane—something we lose easily in the sanitized, atomistic, hedonistic world in which we live. Miguel de Unamuno may not have been entirely wrong when he wrote, "Suffering is the substance of life and the root of personality. Only suffering makes us persons."[10] This is not an argument for letting patients suffer, but it does suggest that inevitable suffering may serve some purposes that are not immediately apparent. The lives of many of the handicapped, the retarded, and the

10. M. de Unamuno, "The Tragic Sense or Life," in *Men and Notions*, trans. A, Kerrigan, Sollingen Series BS, no. 4 (Princeton, N.J.: Princeton University Press, 1972), 224.

aged teach us much about courage and personal growth and give some substance to Unamuno's observation.

Beneficence and compassion remain the first principles of medical ethics. No one would seriously urge sadism or masochism. Especially to be avoided is the unnecessary suffering resulting from the inappropriate overuse of technology. But, a case can be made that beneficence requires not killing, but alleviation of physical and emotional suffering by optimum palliative care—the comprehensive physical, emotional, and community support that leaves more options to the patient than instant oblivion.

The Necessity of Euthanasia

From the medical point of view, even if euthanasia were morally licit, does it follow that it is necessary? The motivation for euthanasia arises principally from two worries: fear of intolerable pain, suffering, and anguish, and fear of becoming a victim of overzealous physicians and dehumanizing medical technologies. These are legitimate worries, but both are well within the power of medicine to remedy without resort to euthanasia. Improved measures for relief of pain and anxiety are already available. It still remains for physicians to use them effectively.[11] We must be unequivocal about physicians' duty to do so. The current trend to active euthanasia only underscores the moral obligations of physicians to practice competent analgesia, to understand why the patient requests death, and to deal with and remove those reasons in a program of palliative care.[12]

Hospice programs or palliative care offer comprehensive alternatives to euthanasia that are more respectful of beneficence and autonomy than killing. They relieve pain and anxiety, prepare the patient for the experience of dying, anticipate the need and value of advance directives, and establish understanding between patient and physician about which life-support measures are acceptable to the patient and which are not. These programs enlist the help of family and friends to make dying a communal experience, in which the dying person contributes something positive to those around her as well as to her own growth as a person. The physician's obligation to act beneficently and to show respect for the patient's dignity is better served by these measures than by killing the patient.

Plebiscite Ethics

Another line of argument relies on the fact that "polls" show that a majority of people favor legalizing euthanasia. Little is said about the notorious difficulties of conducting polls without contaminating the responses by the way the questions are

11. Agency for Health Care Policy and Research, *Clinical Practical Guideline: Acute Pain Management: Operative or Medical Procedures and Trauma* (Washington, D.C.: US Department of Health and Human Services, 1992).

12. J. Lynn, "The Health Care Professional's Role When Active Euthanasia Is Sought," *Journal of Palliative Care* 4, nos. 1–2 (1988): 100–102.

posed. It is not clear how respondents interpret the question or what question they think they are answering. Many, for example, still confuse killing with letting a patient die when death is imminent and inevitable. Many may favor euthanasia because they think it is the only way to retain control over their dying or to prevent un-necessary overtreatment of terminal illnesses. Others are not aware of the alternatives to intolerable suffering—competent use of analgesics, hospice and palliative care, and changed social attitudes toward the terminally ill. Some among them may be personally opposed, but believe others should be free to choose euthanasia, and so favor legalization. It cannot be denied that public interest in euthanasia is considerable, judging by the number of people who have bought Derek Humphry's book, *Final Exit*.[13] A few have already used it as a death manual to guide their own suicides.

From the ethical point of view, arguments based on the results of polls assume that a plebiscite, a majority opinion, or even true consensus can establish what is right and good. Equally specious is the argument that if a practice exists widely, it should become law so it can be regulated. Or, that what becomes legal is therefore moral. And what is no longer secret gains moral credibility by being exposed to the "light of day."

These arguments are seriously flawed. There have been too many instances in human history—past and present—of immoral laws (segregation, slavery, suppression of the rights of women and children, and so forth). There have also been morally distorted societies in which popular approval and law condoned serious violations of human rights (Nazi Germany, Stalinist Russia, and Fascist Spain and Italy, for example). Something more philosophically cogent than public opinion, national sentiment, or law is needed to make a cogent case for the moral acceptability of euthanasia.

Distortion of the Healing Relationship

Even if euthanasia did not have the deficiencies in moral reasoning I have outlined, it would be morally dubious in the light of the "internal morality" of medicine itself. By internal morality, I mean the moral obligations that devolve upon physicians by virtue of the nature of medical activity.[14] On this score, I am in agreement with Leon Kass[15] that euthanasia is a serious violation of the moral nature and purposes of medicine.

I have discussed my own philosophy of medicine elsewhere and will only summarize a few points here. Medicine is a healing relationship. Its long-term goal

13. D. Humphry, *Final Exit* (Secaucus, N.J.: Hemlock Society, 1991).

14. E.D. Pellegrino, "The Healing Relationship: The Architectonics of Clinical Medicine," in *The Clinical Encounter: The Moral Fabric of the Patient-Physician Relationship*, ed. Earl Shelp, Philosophy and Medicine Series 4 (Dordrecht, Holland: D. Reidel Publishing, 1983), 153–72.

15. L. Kass, "Neither for Love nor Money: Why Doctors Must Not Kill," *Public Interest* 94 (Winter 1989): 25–46.

is restoration or cultivation of health; its more proximate goal is healing and helping a particular patient in a particular clinical situation. Medicine restores health when this is possible, and enables the patient to cope with disability and death when cure is not possible. The aims of medicine are positive, even when death is inevitable. Healing can occur even when cure is impossible. The patient can become "whole" again if the healthcare professional helps him to live with a disability, to face dying, and to live as human a life as circumstances will allow.

Medicine is also ineradicably grounded in trust.[16] The physician invokes trust when she offers to help. The patient is forced to trust, because he is vulnerable and lacks the power to cure himself without help. The patient is dependent upon the physician's good will and character. The physician—to be faithful to the trust built into the relationship with the patient—must seek to heal, and not to remove the need for healing by killing the patient. When euthanasia is a possible option, this trust relationship is seriously distorted. Healing now includes killing. The already awesome powers of the doctor are expanded enormously. When cure is impossible, healing is displaced by killing. How can patients trust that the doctor will pursue every effective and beneficent measure when she can relieve herself of a difficult challenge by influencing the patient to choose death?

Uncertainty and mistrust are already too much a part of the healing relationship. Euthanasia magnifies these ordinary and natural anxieties. How will the patient know whether the physician is trying to heal or relinquishing the effort to cure or contain illness because she favors euthanasia, devalues the quality of the patient's life, or wants to conserve society's resources? The physician can easily divert attention from a good death by subtly leading the patient to believe that euthanasia is the only good or gentle death.

This is not a blanket indictment of the character of physicians. But the doctor is an ordinary human being called to perform extraordinary tasks. Her character is rarely faultless. We cannot simply say the "good" doctor would never abuse the privilege of euthanasia. Whose agent is the doctor when treatment becomes marginal and costs escalate? Will the physician's notion of benevolence to society become malevolent for the older patient? How can the aged be secure in the hands of younger physicians whose notion of a "quality" life may not include the gentler pleasures of aging that Cicero praised in the *De Senectute*? Can patients trust physicians when physicians arrogate to themselves the role of rationers of society's resources or are made to assume this role by societal policy? We already hear much talk of the social burdens imposed by chronically ill, handicapped, terminal adults and children and the necessity of rationing with the physician as gatekeeper.[17]

16. E.D. Pellegrino, "Trust and Distrust in Professional Ethics," in *Ethics, Trust, and the Professions: Philosophical and Cultural Aspects*, ed. E.D. Pellegrino, R.M. Veatch, and J.P. Langan (Washington, D.C.: Georgetown University Press, 1991), 68–89.

17. E.D. Pellegrino, "Rationing Health Care: The Ethics or Medical Gatekeeping," *Journal of Contemporary Health Law and Policy* 2 (Spring 1986): 23–45.

Moreover, there are clinical imponderables that can undermine the physician's judgment. To define a pain as intolerable, to distinguish gradations of suffering, and to prognosticate accurately are difficult enough in themselves. These difficulties are easily compounded when the patient pleads for release or the physician is frustrated, emotionally spent, or inclined to impose his or her values on the patient. When the proscription against killing is eroded, trust in the doctor cannot survive. This is already apparent in Holland, that great social laboratory for euthanasia. According to some observers, older and handicapped people are fearful of entering Dutch hospitals and nursing homes.[18] Older Dutch physicians have confided to some of us their personal fears of being admitted to their own hospitals. There is anecdotal evidence of physicians falsifying data to justify euthanasia, making egregious mistakes in diagnosis and prognosis, entering into collusion with families, and ordering involuntary euthanasia.[19] These anecdotal impressions must be better documented. Further study of the Dutch experience is therefore crucial for any society contemplating euthanasia as public policy. Present evidence indicates that the slippery slope—conceptual and actual—is no ethical myth but a reality in Holland.[20] When the physician who traditionally had only the power to heal and to help can now also kill, the medical fiduciary relationship—one of the oldest in human history—cannot survive.

There is also the serious effect on the physician's own psyche of premeditated, socially sanctioned killing. To some degree, physicians are desensitized to loss of life by their experiences with the anatomy lab, autopsy operating rooms, and the performance of painful procedures. To carry out their duties, doctors must steel themselves against suffering and death to avoid being emotionally paralyzed in the actions they must take daily. Euthanasia reinforces this objectification of death and dying, and further desensitizes to killing. A "gentle" death, as Van der Veer wants to call it, is still a premeditated, efficiently executed death of a living human being.[21]

At this point, proponents of euthanasia may correctly point out that, historically, the prohibition against euthanasia was respected only by one group of physicians, the Hippocratics Other physicians approved the practice of euthanasia as some physicians overtly and covertly do today. Paul Carrick makes clear that in antiquity, doctors were inconstant in their respect for the Hippocratic Oath.[22] This fact does not in any way

18. R. Fenigson, "A Case Against Dutch Euthanasia," *Hasting Center Report* 19, no. 1 (1989): 522–30; R. Fenigson, "Euthanasia in the Netherlands," *Issues in Law and Medicine* 6, no. 3 (1990): 229–45.

19. H. Ten Have, "Euthanasia in the Netherlands: The Legal Context and the Cases," *HEC Forum* 1, no. 1 (1989): 41–45.

20. I. Van der Sluis, "The Practice of Euthanasia in the Netherlands," *Issues in Law and Medicine* 4, no. 4 (1989): 455–65; B. Bostrom, "Euthanasia in the Netherlands, a Model for the United States?" *Issues in Law and Medicine* 4, no. 4 (1989): 467–86; C.F. Gomez, *Regulating Death: Euthanasia and the Case of the Netherlands* (New York: Free Press/Macmillan, 1991).

21. Van der Meer, "Euthanasia."

22. P. Carrick, *Medical Ethics in Antiquity* (Boston: Dordrecht, 1985), 127–50.

weaken the argument I have been making based in the nature of the healing relationship. Medical codes are not self-justifying. They have only recently become the subject of critical examination. I would contend, for reasons stated briefly above, that a critical examination of the moral basis for medical practice must proscribe euthanasia because it contravenes the primary healing purposes of medical activity. Moreover it is capricious to argue that since some physicians in the past and some in the present practice euthanasia, it should be legalized. An immoral act does not become moral because it is common practice. Morality must have deeper roots than mere medical custom or even the code of medical ethics.

The medical profession is a moral community.[23] Its members have a collective moral responsibility to patients and society. For this reason, the whole profession must oppose the legalization of euthanasia as detrimental to the welfare of patients and the integrity of society. Individual physicians cannot abstain on grounds that they oppose euthanasia but believe in free choice. The social nature of the acts of dying and killing do not permit anyone to choose such a socially destructive option. This is why the American Medical Association and British Medical Association have recently reaffirmed the proscription against doctors killing patients under any circumstance.[24]

The Social Impact of Euthanasia

Much as libertarians would like to see euthanasia as an individual decision protected by an absolute privacy right, it is an event inescapably fraught with social significance. A society that sanctions killing must abandon the long-standing tradition of "state's interest" in human life.[25] This devalues all life but especially the lives of certain citizens—the chronically ill, the aged, and the handicapped. Those who do not take the easy exit that legalized euthanasia offers become selfish over-consumers of their neighbors' resources. The vaunted autonomy of the choice for euthanasia withers in the face of the subtle coercion of a social policy that suggests the incurably ill are a social, economic, and emotional burden. Few of us would not feel the pressure to do the "noble" thing and ask for euthanasia, especially if the physician is gently urging us to do so.

The social sanction of euthanasia presumes a responsibility to monitor the killing process to keep it within agreed upon constraints. Killing then becomes bureaucratized and standardized. But even with standardization of criteria, it is impossible to contain euthanasia within specified boundaries. Laws will not prevent

23. E.D. Pellegrino, "The Medical Profession as a Moral Community," *Bulletin of the New York Academy of Medicine* 66A, no. 3 (1990): 221–32.

24. American Medical Association, Council on Ethical and Judicial Affairs, *Euthanasia*, report (Chicago: AMA, 1989); British Medical Association, *Euthanasia: A Report of the Working Party to Review the British Medical Association Guidance on Euthanasia* (London: BMA, May 1988), 69.

25. H. Tristram Engelhardt Jr., "Fashioning an Ethic of Life and Death in a Post Modem Society," *Hastings Center Report*, Special Supplement: "Mercy, Murder and Morality: Perspectives in Euthanasia," 19, no. 1 (1989): 13–15.

abuses, despite the hopes of those who favor legalization in the United States. The Dutch experience shows that even when euthanasia is not legal but is tolerated, expansion of its boundaries—from voluntary to involuntary, from adults to children, from terminally ill to chronically ill, from intolerable suffering to dissatisfaction with the quality of life, from consent to contrived consent—is inevitable.[26] The ethical proscription against killing by doctors is a social sea wall. Once it is breached, it is impossible to avoid inundation. Literature often teaches ethics more effectively than moral philosophy. One needs only to read Trollope's *Fixed Period* to appreciate how the most benevolently generated and seemingly rational policy of euthanasia can corrode human relationships.[27]

Another imminent social danger is the real possibility that euthanasia will converge with the current trend toward rationing of healthcare resources. It is a short way from the need to contain costs to covertly- or overtly-planned euthanasia for those members of our society who present the greatest economic burdens. At the beginning, some might suggest rationing needed care to retarded or handicapped infants, very old people, or those with fatal, incurable diseases like Alzheimer's. Once euthanasia, in any of its forms, is legalized, the temptation to encourage its use, tacitly or overtly, to alleviate one of our most socially vexing problems—the increasing scarcity of healthcare dollars—will be strong. This could be the first slip on the slippery slope, which leads inexorably from voluntary to non-voluntary and involuntary euthanasia. This is evident already in the recent Dutch government report on euthanasia in the Netherlands.[28]

Some of euthanasia's protagonists are so convinced of the individual and social benefits of killing that they believe it is unjust not to make them available to patients who cannot give consent—infants, anencephalics, the retarded, persons in a permanent vegetative state, and so forth. On this view, covert euthanasia—that is, killing patients who could consent but whom it is not deemed necessary to consult—would be permissible. These proponents of euthanasia argue that it is unjust to deny such patients the benefits of relief of suffering just because they cannot, or will not, give consent. Some argue that such patients impose unjust economic, social, emotional, and physical burdens on family, health professionals, and society and, thus, should be killed as a matter of medical duty.[29] Others in the name of "compassion" urge the involuntary killing of the handicapped and retarded, since there is no place on this earth where they can be "happy."[30] Here we have a conceptual slippery slope that prepares the way for the actual slippery slope and

26. Fenigson, "A Case Against Dutch Euthanasia."

27. A. Trollope, *The Fixed Period*, ed. R.H. Super (Ann Arbor: University of Michigan Press, 1990).

28. P.J. Van Der Maas, J.J.M. van Delden, L Pijnenborg, and C.W.N. Looman, "Euthanasia and Other Medical Decisions Concerning the End of Life," *Lancet* 338 (14 September 1991): 669–74.

29. J. Lachs, "Active Euthanasia: Theoretical Aspects," *The Journal of Clinical Ethics* 1, no. 2 (1990): 113–15; J.H. Van Der Berg, *Medical Power and Medical Ethics* (New York: Norton, 1978).

30. K.L. Lyle, "A Gentle Way to Die," *Newsweek* (March 2, 1992), 14.

provides its rational justification. The horrendous nature of these conclusions requires more rebuttal than I can give in this paper. Suffice it to say that the moral aberrancy of these conclusions is ample proof of the immorality of the premises from which they derive. This is precisely the kind of reasoning that led to German physicians' killing of the "unfit" and to the Holocaust.

In the United States, a model state law has been proposed in Iowa that would legitimate all the "abuses" of the slippery slope. The model sanctions euthanasia for children (with or without parental approval), includes an appeal mechanism enabling children to override parental objections, sanctions proxy consent on behalf of incompetent patients, and establishes a registry of people qualified to administer aid in dying. The aim of this law is to provide "quality control" in the termination of life and "a principled means of managing healthcare resources," thus dangerously conflating allocation decisions and euthanasia.[31]

Rachels chooses to rely on the "good sense of judges and juries" and the medical profession to prevent abuses. In any case, he argues, the good results will outweigh whatever abuses might occur.[32] But experience in Holland and in the current debate in the United States suggests that this may not be the case at all. Rachels is right when he says that the slippery slope is the "outstanding argument" against legalized euthanasia, and he is seriously wrong when he disposes of its reality so cavalierly.

The arguments for making assisted suicide a moral option, moral duty, or legal choice are equally indefensible.[33] The arguments adduced here against euthanasia apply as well to assisted suicide. Physicians are de facto moral accomplices in what happens to their patients. Even when they are moved by compassion, as was Dr. Timothy Quill (the physician who recently revealed his role in a euthanasia case in the New England Journal of Medicine), physicians cannot morally justify cooperation in terminating the patient's life.[34] They cannot excuse themselves by a professional "Pontius Pilate Act" if they provide the lethal drugs and the directions for their effective use. This is indefensible moral cooperation, in that it shares the patient's intent to commit suicide. Further, the doctor's cooperation is essential for the patient to achieve his or her purpose. For the same reasons, physicians cannot cooperate with socially "designated" killers, "obitiatrists" or "teliastrists."[35] Even though such cooperation is more remote than assisted suicide, the physician cannot

31. C.A. Brandt, P.J. Cone, A.L Fontana, et al., "Model Aid in Dying Act," *Iowa Law Review* 15, no. 1 (1989): 125–215.

32. Rachels, *The End of Life*, 187.

33. S.H. Wanzer, D.D. Federman, S.J. Adelstein, el al., "The Physician's Responsibility Toward Hopelessly Ill Patients, A Second Look," *New England Journal of Medicine* 320 (1989): 884–89; C.K. Cassel and D.E. Meier, "Morals and Moralism in the Debate Over Euthanasia and Assisted Suicide," *New England Journal of Medicine* 323 (1990): 750–52.

34. T.E. Quill, "Death and Dignity: A Case of Individualized Decision-Making," *New England Journal of Medicine* 324 (1991): 691–94.

35. These are etymologically dubious euphemisms for socially designated functionaries who would administer lethal drugs if doctors refuse to cooperate.

remain at a distance. Someone will have to say this disease is "incurable"; this pain, "intolerable"; this patient is "competent" or "incompetent" to give consent. The proponents of euthanasia are right—legalization is impossible without cooperation of the medical profession. And for all the reasons I have given, physicians cannot cooperate in the killing of their patients directly or indirectly.

Obligations of Physicians Who Reject Euthanasia

What are the moral obligations of physicians who reject all forms of euthanasia? To begin with, we must accept responsibility for confronting the reality of pain and suffering—the fear and emotional traumata of the fatally ill and dying person and the legitimate desire for a good death. We must counter the destructive force of euthanasia with a constructive effort. If, as I have argued, killing is not a good death, what can we—indeed, what must we—as physicians do to help the patient achieve as good a death as possible without killing him? First of all, physicians must recognize that the request for euthanasia is a plea for help and an attempt to regain some measure of control over one's life that fatal illness seems to take away so forcibly. Why does this particular patient want to be killed? Is it pain, suffering, loss of dignity, depression? Is it a challenge to see whether the physician, family, and friends really do care? Is it a test to see if the family really regards the patient as a burden? Is it fear of bankrupting the family, fear of being kept alive artificially to no purpose, or a response to the doctor's attitude of futility or disinterest? There are many reasons for the request to be killed and many remedies once we know the reason.

Clearly, the physician has a moral obligation to ascertain these reasons and to spend the time necessary with the patient to learn about the factors specific to the situation. Too many physicians are still fearful of talking about death. Physicians must engage the help of nurses, social workers, pastoral counselors, family, and friends in discerning the reasons for a patient's request. They must mobilize the forces necessary to remove or ameliorate these causes. Physicians must work out with each patient what the patient's definition of a good death is, and determine—before a crisis comes on—what life-support measures will and will not be acceptable to the patient.

We must assure patients that they can control the starting and stopping of life-sustaining measures by advance directives when they lose their competence to do so as the disease progresses. Physicians should dispose of their own anxiety about making patients into addicts. Pain relief competently applied is a moral obligation, as are all other supportive measures. Addiction in the last weeks of life and even death as an unintended side effect of analgesia are morally defensible. Physicians unable or unwilling to make the investment of time and emotion required for comprehensive palliative care should not care for patients who need such care. They would be better placed in specialties that do not often confront terminal illness. Physicians also have a moral responsibility to be honest about their reasons for accepting or refusing a role in euthanasia. Euthanasia is not the answer to the physician's inadequacy, frustration, or emotional exhaustion as a healer. Refusing to accept one's own finitude or that of one's patients is not justification for ineffective, burdensome, or futile treatment.

All members of the healthcare team should play a part in comprehensive palliative care. Still, the physician remains the focal point of the effort. Harm results from dividing the tasks of terminal care, unless there is someone to coordinate that care, interpret it, and to make changes when needed. Again, physicians unwilling to assume this focal role should not undertake the care of fatally ill patients. They must be honest and morally responsible enough to yield the primary role to physicians best fitted by temperament and training to care for the incurably ill patient in a humane and competent way.

The advocates of legalized euthanasia are right when they insist that the physician is crucial to any effective social policy permitting patients to be killed on request. Doctors do have the necessary knowledge. They do control the prescription of the necessary lethal agents. They do know when the patient's diagnosis and prognosis portend a painful and inevitable death. These very facts impose an enormous moral responsibility on the profession to resist becoming moral accomplices or society's designated killers.

If euthanasia is legalized, the medical profession will bear a large burden of the blame if it does not educate the public to the dangers and if it fails to refuse to participate. The profession must also work to alleviate the societal conditions that foster euthanasia—the attitude of hopelessness and futility before death and dying, the financial pressures that all too forcibly convince the patient that he is a burden, and the illusion that life must be perfect and that any chance illness is an affront to human dignity. The profession as a whole must make it morally mandatory to make competent use of all measures that relieve pain and suffering.

There is much that physicians, individually and as a profession, can—and must—do short of killing patients to eliminate the problem of suffering. Legalization of euthanasia poses a far deeper moral challenge than the profession may appreciate. It challenges us to define what it really means to be a physician.

All morally responsible, compassionate, and merciful physicians share the same goal when confronted with a suffering, dying, terminally ill human being. They all strive to assist the suffering person to achieve a gentle and good death. They all share the "objective to do good," to use the phrase from Trollope's novel that I have used as an epigraph. What we do not share is the definition of "the good to be done." A good death does not, I have argued, include killing the patient, nor can one be a good physician and do so.

Acknowledgements

I am indebted to Pat McCarrick and Mary Coutts for invaluable bibliographic assistance.

C. Treatments at the End of Life

22

CPR, DNR, and the Patient's Good

G. Kevin Donovan, MD

CARDIOPULMONARY RESUSCITATION (CPR) occupies a unique position in medical practice and in ethical decision making. Its widespread application and now routine expectation has come to be a persistent source of ethical controversy and requests for ethics consultation. What began in the 1950s as a life-saving intervention for hospitalized patients who experienced sudden cardiac arrest became, in the 1970s, the default treatment for all patients whose hearts had stopped.[1] In their 1965 book, *Fundamentals of Cardiopulmonary Resuscitation,*[2] Jude and Elam made it clear that CPR should only be used with patients who experience sudden cardiac arrest and could be successfully revived. They emphasized that it was inappropriate to use with dying patients.

How did we arrive at our present state, where it is seen either as a necessary life-saving intervention, or at least a necessary rite of passage? Many times, patients and their families seem to believe that we cannot die without it, that it is an essential part of the ritual of death. This highlights its unique position in medicine, and especially in end-of-life care. CPR has become a default procedure, one of the few that families and surrogates must consent to withhold, or else it will be undertaken even in the face of poor medical justification and outcomes. This forces us to face the question, when should medical professionals offer something that we wouldn't otherwise recommend?

In the past few decades, the public has learned to marvel at the progress of medicine; its technological and therapeutic achievements at times seem almost miraculous. With high expectations, it comes as no surprise that patients and their families routinely "want everything done" even in those situations where clinicians see a diminishing chance for success. Following the ethical guidelines of shared decision making and informed consent, many physicians have felt compelled to offer CPR for

1. American Heart Association Staff, *Basic Life Support (BLS) For Healthcare Providers—Updated With AED Use For Children* (Dallas, Tex.: American Heart Association, 2001), 297.

2. James R. Jude and James D. Elam, *Fundamentals of Cardiopulmonary Resuscitation* (Philadelphia, Penn.: F.A. Davis Co., 1965).

a patient when, in their best clinical judgment, it will provide little or no benefit. They may do this under the mistaken notion of a legal compulsion; more often, they do so because the patients and families evince a strong preference for it. What shapes this prevalent predilection?

Patients learn about CPR from many sources, including physicians, family, and friends—and even CPR courses. However, a number of patients report that much of their information comes from the media: television, movies, newspapers, or books.[3] What do they learn from these sources? One study of cardiopulmonary resuscitation on television found that the success rate portrayed indicated 75 percent of patients survive the immediate arrest.[4] Contrast this with what medical professionals know to be true: that despite improved outcomes in the past decade, hospitalized adult patients are likely to have successful resuscitation with survival to discharge only about 20 percent of the time.[5] Inevitably, these numbers will be worse for patients in the ICU, on pressors, and with multiorgan failure. This disconnect between the expectation of patients and the reality in the hospital is difficult to bridge. Of course, they cannot be told, nor would they accept without offense, the trite but true observation that the best place to have a cardiac arrest is not in the hospital, not in the ambulance, but on television! Nevertheless, patients do look to their physicians for guidance, and the realistic evaluation of harms and benefits is part of the doctor's professional responsibility.

The anticipated benefits of CPR are more readily apparent. The restoration of cardio-pulmonary function is the hoped-for outcome. This may be a conditional benefit, depending on the potential for restoration to health, mental awareness, and discharge from the hospital. At times, it may be seen as beneficial by a patient if it allows them to extend their life until they can experience a desired event, for example, a wedding, graduation, or birth of a loved one. The harms entailed by cardiopulmonary resuscitation are not solely physical, but this category of harm should not be minimized; broken ribs are not uncommon even in appropriate CPR. However, a false hope created by a useless prolongation of the dying process, or resuscitation to an unwanted permanent state of unawareness, will cause distress not only to the survivors but to the medical caretakers as well. The medical team is also able to foresee the chaotic scene at the time of death rather than the peaceful leave-taking that most would desire. Recent studies suggest that the psychological trauma of CPR for the family may be mitigated by allowing family presence during the procedure.[6] These harms do not even take into account the poor utilization of required resources every

3. Susan J. Diem, John D. Lantos, and James A. Tulsky, "Cardiopulmonary Resuscitation on Television: Miracles and Misinformation," *New England Journal of Medicine* 334 (1996): 1578–82, DOI:10.1056/NEJM199606133342406.

4. Ibid., 1579.

5. Saket Girotra et al., "Trends in Survival after In-Hospital Cardiac Arrest," *New England Journal of Medicine* 367, no. 20 (2012): 1912–20.

6. Patricia Jabre et al., "Family Presence during Cardiopulmonary Resuscitation," *New England Journal of Medicine* 368 (2013): 1008–18, DOI:10.1056/NEJMoa1203366.

time a crash cart is fruitlessly opened, nor do they include exceptional risks to the team, such as found in the resuscitation of patients with high-risk transmissible infectious diseases.[7]

It must be acknowledged that there are potential harms in the placement of a Do Not Resuscitate (DNR) order as well. The harms involved for the medical caretakers are less evident, but have been well documented. Following a DNR order, clinicians often consider this the threshold for the limitation of treatments not specifically related to resuscitation, or even that only comfort measures should now be provided.[8] Fewer routine procedures and fewer and shorter visits to the patient's room may ensue. The problems for a patient's family inherent in the decision to reject CPR may be more easily recognized. When we ask a patient's surrogates to make the decision to place a DNR order, it is an occasion for emotional distress on their part. It is always difficult to face the death of a loved one, but it is made more difficult by the feeling that this decision will both assent to and contribute to their earlier death. A proper approach to the use of DNR orders at the end of life should be framed to mitigate these problems for the survivors as well as the healthcare team.

One approach that has been employed in some hospitals and locales is a policy of unilateral decision making. This would entail a process-based conflict resolution in which commonly two physicians agree that CPR would be useless in a particular case. It may require consultation with an ethics committee or hospital administration, but the decision would not require consent from the patient's surrogates. This is seen by its proponents as an appropriate restriction from providing a foreseeably pointless intervention for many patients who would simply not benefit from attempted CPR, including those with metastatic cancer, major trauma, or end-stage liver disease.[9] Pellegrino might see this as an appropriate application of the concept of futility, but not an appropriate implementation of that concept (see chap. 19 above).[10] As laudable as the physician's intentions may be, having the decision implemented regardless of whether and how much the surrogates may disagree, is an unjustifiable return to paternalism. It may cause great distress to the family involved, and it may impact more heavily those minority populations who lack sufficient resources to successfully object to this physician-imposed decision.

7. Daniel B. Kramer, Bernard Lo, and Neal W. Dickert, "CPR in the Covid-19 Era—An Ethical Framework," *New England Journal of Medicine* 383 (2020): e6, https://www.nejm.org/doi/full/10.1056/nejmp2010758.

8. Amy Sanderson, David Zurakowski, and Joanne Wolfe, "Clinician Perspectives Regarding the Do-Not-Resuscitate Order," *Journal of the American Medical Association Pediatrics* 167, no. 10 (2013): 954–58, DOI:10.1001/jamapediatrics.2013.2204.

9. Gabriel T. Bosslet et al., "An Official ATS/AACN/ACCP/ESICM/SCCM Policy Statement: Responding to Requests for Potentially Inappropriate Treatments in the ICU," *American Journal of Respiratory and Critical Care Medicine* 191, no. 11 (2015): 1318–30, DOI:10.1164/rccm.201505-0924ST.

10. Edmund D. Pellegrino, "Decisions at the End of Life: The Use and Abuse of the Concept of Futility," in *The Dignity of the Dying Person,* ed. Juan De Dios Vial Correa and Elio Sgreccia (Vatican City: Libreria Editrice Vaticano, 2000), 219–41.

We have been advocating a middle path that navigates between the twin failings of inappropriate CPR and paternalism. Some have referred to this same approach as informed non-dissent.[11] It must begin by providing accurate information about expected non-beneficial outcomes of CPR in a particular case as a key to overcoming false expectations. The surrogates are then reassured that we intend to "do everything" that can be expected to provide benefit and ensure comfort to their loved one. This presupposes that the medical team has already ascertained what the patient's values and preferences would be, and have conformed the goals of care to these. The reasons that some interventions such as CPR would be considered inappropriate under the circumstances are then explained, and it is clearly stated that, therefore, there would be no plans to implement it. However, all other indicated interventions that are mutually accepted would continue, especially those providing comfort care. If the surrogate does not reject the plan explicitly, the DNR order is placed. The difference between this approach and paternalism lies in the response to a rejection by the surrogate. Rather than trying to forcefully override such an objection, it can be accepted for the time being, with an understanding that a reevaluation should be made in a specified time as we chart the patient's progress or lack thereof. Given sufficient time and space for such a decision, most surrogates can find a way to accept it without opposition.

The greatest challenge and discomfiture associated with this approach lies in the possibility that a cardiopulmonary arrest may occur before the medical team and the surrogate have reached an accord on the ethically and medically proper response. When the physician is presented with this situation, she is faced with the exact scenario that was to be avoided. There could be three possible responses, but not all would be equally acceptable. At the moment of the arrest, a unilateral declaration could be made to refuse to attempt CPR. The problems with this are manifest: it is precisely the paternalistic approach that was previously deemed inappropriate, it requires the physician to be available for that decision immediately at the time of the arrest, and it would not even allow for the endorsement of a second consulting physician, all of which would put this action on thin ice both medically and legally.

A second approach that some might follow would be to enact a "slow code" or "show code." This has been done by some who are reluctantly forced into a presumed non-beneficial resuscitation attempt. Rather than respond with the typical urgency that is normally required, the team assembles in a deliberately slow manner, making no concerted effort to begin CPR in a timely manner. If this does not result in the obvious demise of the patient before CPR can begin, a halfhearted pantomime of CPR may be done, more for the benefit of any spectators and of the patient. It is difficult to provide any reasonable defense on an ethical basis of this obvious deception. Ultimately, we are likely to be left with only the third option.

11. Alexander A. Kon, "Informed Non-Dissent: A Better Option than Slow Codes when Families Cannot Bear to Say 'Let Her Die,'" *American Journal of Bioethics* 11, no. 11 (2011): 22–33, DOI: 10.1080/15265161.2011.603796.

A patient who is still a full code should have a full and timely response at the moment of their arrest. Unless the surrogate decision maker stops the team at the door, CPR should commence in a normal fashion at a normal pace. All the indicated maneuvers and drugs for resuscitation should be available. However, the duration of such a code, as with any CPR attempts, is a matter of medical judgment. There are no absolute criteria for the proper duration of an attempted CPR, and only general guidelines for the determination that it should not go on too long.[12] The focus and justification for this procedure, as in all medicine, should have as its chief concern the good of the patient. Pellegrino addresses this eloquently in *For the Patient's Good,*[13] wherein he contends that the answer to a misplaced over-emphasis on autonomy is not rampant paternalism. If we restore beneficence as the guiding principle of medical practice, we should be able to determine the patient's good that is in accord with not only their values but good medical practice as well. In this particular case, the process of shared decision making may have failed to adequately address the patient's good. This could be due to a misapprehension by the surrogate of the patient's actual preference in this situation, or a misunderstanding of our ability to provide an achievable medical good. As a result, although medical caretakers might be forced into an unwanted attempt at CPR, good medical practice would not endorse a prolonged attempt in this situation.

A full-fledged, sincere attempt at resuscitation, but limited in the number of cycles or the number of minutes of duration, may be justifiable in finding the right balance for physicians forced to choose between the achievable medical good for the patient and the emotional needs of the family. A limited code under the circumstances tries to find the best compromise between tragic reality and illusory hope. It does not constitute a deception in that adequate compressions and the usual medications are employed—albeit for a limited time. The prudential judgment of the physician is always needed to determine how long a code must be continued for the good of the patient. The advantages of this approach are several: it changes the default of an automatic CPR status. It does not require a determination of absolute physiological futility in order to determine that CPR would be non-beneficial. It does not paternalistically override the objections of the surrogate if they are raised, but leaves further decision making on the table, allowing for further evaluation of the benefit or lack thereof. This is the most promising way of avoiding confrontation with an anguished family.

For this approach to be successful, it requires strong communication skills as well as compassion for the family's distress. It must take place in the larger framework of planning goals of care—it cannot be simply an approach to "get the DNR."

12. American Heart Association, "2020 American Heart Association Guidelines for Cardiopulmonary Resuscitation and Emergency Cardiovascular Care," *Circulation* 142, no. 6, suppl. 2 (2020): S336ff, https://www.ahajournals.org/toc/circ/142/16_suppl_2.

13. Edmund D. Pellegrino and David C. Thomasma, *For the Patient's Good: The Restoration of Beneficence in Healthcare* (New York: Oxford University Press, 1988).

Therefore the most skilled and well-informed physician or ethicist should lead the discussion, not the intern or the student. Such an approach is ethically preferable, because it does not show inappropriate deference to autonomy. It recognizes that not every decisional request supports autonomy. Although autonomy can be defined as self-rule, it is ultimately based in respect for persons. Respect for the patient's dignity does not require that we put them through procedures at the end of life that have no expectation of benefit. This approach avoids the false notion that we must offer the option to choose what we know to be a predictably ineffective intervention. This would be essentially a non-choice. Taking such a choice off the table relieves the decisional burden from the surrogate, avoids a non-beneficial intervention for a patient in the last moments of life, and maintains the professional integrity of the medical caretakers.

23

Ethical Issues in the Provision of Artificially Supplied Nutrition and Hydration: A Catholic Perspective

G. Kevin Donovan, MD

Introduction

Although Dr. Pellegrino did not publish articles on the issues surrounding the provision of nutrition and hydration to impaired patients, he was quite aware and conversant regarding the controversies described here. I know this with certainty from numerous discussions he and I had on this topic during and after my initial sabbatical at Georgetown (1989–90). I was privileged to get his feedback when I also chose this as the topic of a master's thesis. Although the following analysis is mine alone, it was clearly shaped and strengthened by my personal discussions with Edmund Pellegrino and his invaluable input.

Few issues in contemporary bioethics appear to be settled into a consensus position both ethically and legally, yet remain a persistent source of controversy. The provision of nutrition and hydration to patients, particularly through artificial means, is clearly one of them. As pointed out by Seigler and Weisbard over three decades ago, "the idea that fluids and nutriment might be withdrawn, with moral and perhaps legal impunity, from dying patients, was a notion that would have been repudiated, if not condemned, by most health professionals."[1] Even at that time, they felt they were arguing "against the emerging stream." Much has happened since then that has challenged and changed the traditional perspective that they and others championed.

History

It was not a single case but a series of them that worked their way through the courts over the years, highlighting the issues and resulting in evolving ethical analyses and legal justifications that have led to the present *modus operandi*. Beginning with Karen Quinlan,[2] courts recognized the right of patients to refuse life-sustaining med-

1. Mark Seigler and Alan J. Weisbard, "Against the Emerging Stream—Should Fluids and Nutritional Support Be Discontinued?" *Archives of Internal Medicine* 82, no. 1 (1985): 129–31.

2. *In the Matter of Karen Quinlan*, 70 NJ 10 (1976).

ical treatment. In cases such as *Barber*, *Conroy*, and *Bouvia*,[3] courts began to deal with issues of withholding or withdrawing feeding tubes and with a general movement towards allowing their discontinuation and the patient's subsequent death. Even at this stage, there were dissenting voices. Pellegrino, for one, would have allowed a competent Bouvia to have a feeding tube withheld in the first place, "but once the tube was in place, we would not agree with its removal since the clear intent is to cause death by freeing her from her suffering."[4] The trend for legal support being given to the practice of (mostly) withdrawal of feeding tubes in selected patients continued, as did the voices in opposition. The landmark *Cruzan* case in 1990[5] ruled that, as with any other medical treatment, competent patients have the right to refuse artificially supplied nutrition and hydration. It further ruled that states have the right to determine the level of evidence of the patient's preference necessary for a guardian to successfully request this. Even more notoriety was associated with the case of Theresa Schiavo in Florida. Schiavo had suffered a cardiac arrest in 1990, with subsequent central nervous system damage resulting in a diagnosis of persistent vegetative state. She was in a nursing home, breathing without assistance, and being fed through a gastrostomy tube in February 2000, when her husband requested that her feeding tube be removed. This resulted in a five-year process of appeals, counter appeals, legislative action, and medical dissent about her diagnosis, with the controversy joined by ethicists, politicians, legal scholars, theologians, and the general public. The feeding tube was ultimately removed, and she died a very short time later.[6]

Before that case, there had been disagreement about the propriety of withdrawing feeding tubes from patients. That bioethicists and legal scholars disagreed was no great surprise; the public divergence of opinion among theologians and Catholic bishops was somewhat more surprising. This led to a statement in 1992 by the Committee for Pro-Life Activities of the National Conference of Catholic Bishops offering a tentative guideline on this difficult issue.[7] They supported a presumption in favor of providing medically assisted nutrition and hydration; this and subsequent statements from the Vatican led to a revision in the guidance provided in the *Ethical and Religious Directives for Catholic Healthcare Institutions* (ERDs) in the 5th and 6th editions.[8] As help-

3. Edmund D. Pellegrino and David C Thomasma, *For the Patient's Good* (Oxford: Oxford University Press, 1988), 191–97.

4. Ibid., 200–201.

5. *Cruzan v. Director, Missouri Department of Health*, 497 U.S. 261 (1990).

6. Cf. Arthur L. Caplan, James J. McCartney, and Dominic A. Sisti, eds, *The Case of Terri Schiavo: Ethics at the End of Life* (Buffalo, N.Y.: Prometheus Books, 2006).

7. Committee for Pro-Life Activities, National Conference of Catholic Bishops, "Nutrition and Hydration: Moral and Pastoral Reflections: Committee for Pro-Life Activities National Conference of Catholic Bishops April 1992," *Linacre Quarterly* 59, no. 4 (1992): 33–49.

8. Cf. United States Conference of Catholic Bishops, *Ethical and Religious Directives for Catholic Health Care Services*, 5th ed. (Washington, D.C.: United States Conference of Catholic Bishops, 2009); and United States Conference of Catholic Bishops, *Ethical and Religious Directives for Catholic Health Care Services*, 6th ed. (Washington, D.C.: United States Conference of Catholic Bishops, 2018).

ful as this guidance may have been, it was not unchallengeable nor unchallenged. Things changed following an allocution by Pope John Paul II in 2004 regarding the subject of providing artificially supplied nutrition and hydration to persons in a persistent vegetative state.[9] The magisterial teaching authority of his position was augmented by the Congregation for the Doctrine of the Faith in 2007.[10] The arguments for and against these positions will be considered in greater length below.

Persistent Vegetative State

Most, although not all, of the controversies and cases regarding nutrition and hydration have evolved around patients diagnosed with a "persistent vegetative state." Some understanding of this condition, therefore, is in order. Although some still use this term, it is both inexact and considered demeaning to the human condition by likening human persons to vegetables. Current guidelines describe these conditions as "disorders of consciousness" which range from locked-in syndrome (a condition in which full consciousness is retained) to minimally conscious state (MCS), to unresponsive wakefulness syndrome (UWS), the preferred term for vegetative state. A prolonged disorder of consciousness can be recognized at 28 days or longer, but the natural history of these conditions is not well defined, particularly for populations with nontraumatic disorders of consciousness. Individuals can fluctuate between different diagnostic categories, which is particularly common early in the course of recovery. One study suggests a 30-percent probability of observing behaviors suggestive of minimally conscious state in patients diagnosed with UWS when they are examined in the morning.[11] A vegetative state (UWS) is defined as permanent or persistent three months after a nontraumatic injury and twelve months following a traumatic injury, acknowledging that unexpected recoveries rarely will occur after these times. Such recoveries have occurred spontaneously as well as following the use of certain drugs such as zolpidem (Ambien). The long-term prognosis is more favorable in patients with minimally conscious syndrome, and in those who have sustained traumatic versus non-traumatic brain injury. Evidence from a natural-history systemic review identified that individuals with a disorder of consciousness

9. Pope John Paul II, "Care for Patients in a Permanent Vegetative State," *Origins* 33, no. 43, section 4 (April 8, 2004): 737–40.

10. Congregation for the Doctrine of the Faith, "Responses to Certain Questions of the United States Conference of Catholic Bishops concerning Artificial Nutrition and Hydration" (August 1, 2007), https://www.vatican.va/roman_curia/congregations/cfaith/documents/rc_con_cfaith_doc_20070801_risposte-usa_en.html.

11. M.D. Cortese, F. Riganello, F. Arcuri, et al., "Coma Recovery Scale R: Variability in the Disorder of Consciousness," *BMC Neurology* 15 (2015): 186. Referenced in Joseph T. Ciatino, Douglas I. Katz, Nicholas D. Schift, et al., "Practice Guideline Update Recommendations Summary: Disorders of Consciousness Report of the Guideline Development, Dissemination, and Implementation Subcommittee of the American Academy of Neurology; the American Congress of Rehabilitation Medicine; and the National Institute on Disability, Independent Living, and Rehabilitation Research," *Neurology* 91, no. 10 (September 4, 2018): 450–60.

at one month post-injury may still attain functionally significant recovery after one year post-injury, with additional longitudinal studies showing that approximately 20 percent of patients recovered to the level where they could return to work or school. VS/UWS and nontraumatic disorders of consciousness are associated with poor outcomes, but individual outcomes vary and prognosis is not universally poor. Nevertheless, a prolonged recovery may occur over months to years, and many will remain severely disabled. Therefore, they have not been defined as entering a chronic phase until three months after non-traumatic brain injury and twelve months after traumatic brain injury.[12]

Artificial Feedings or Artificially Supplied?

Patients with the above disorders routinely require some sort of tube feeding. To understand the ethical issues that arise from this, it is helpful to understand the medical procedures involved in "artificially supplied nutrition and hydration" (ASNH). There are a variety of ways to deliver nutrients to the gastrointestinal tract, increasing in complexity and invasiveness. At the simplest and most mundane level, plastic bottles and rubber nipples are routinely substituted for infant breast-feeding, and are not typically included under the ASNH rubric; specialized nipples extending further into the hypopharynx, designed for infants with a cleft palate, are admittedly more invasive and artificial but also not under consideration in this discussion. For others, tube feedings can be inserted and employed in a variety of ways. A nasogastric tube, passed through the nose to the stomach, commonly refers to a larger stiffer tube that is designed primarily for suctioning out stomach contents rather than for feedings, but it can be used for either purpose on a short-term basis. A specialized feeding tube is designed to be softer and more comfortable with a smaller diameter. It can be inserted, typically by a nurse, into the stomach and is sometimes advanced into the small intestine. Placement in the small intestine reduces the likelihood of reflux, regurgitation or aspiration, but requires that feedings be controlled by a pump as a continuous drip. Feedings given directly into the stomach can be given as a bolus of fluid on a scheduled basis. These tubes may cause some nasopharyngeal discomfort or irritation as well as possible gastroesophageal reflux resulting in esophageal inflammation if they are kept in place too long. They are suitable for use over several weeks, but for longer durations a gastrostomy tube would be indicated.

Gastrostomy tubes are inserted from the abdominal wall directly into the stomach. This is commonly done through the skin using an endoscope (percutaneous endoscopic gastrostomy, PEG), but can also be placed in a similar fashion by a surgeon or an interventional radiologist. Once again, these tubes typically are placed within the stomach for intermittent feedings, but can be deliberately designed to deliver nutrients past the stomach directly into the small bowel, as described above. The burden of these tubes occurs primarily in their placement, with the majority of potential complications (bleeding, infection, misdirected tubes, or adverse reactions

12. Ciatino et al., "Practice Guideline Update Recommendations Summary."

to sedation/anesthesia) occurring then. Once a PEG tube is in place, some complications can occur including the risk of aspiration, which can be present in these patients with (or without) a feeding tube, or local inflammation at the insertion site, or diarrhea, but these risks should not be over emphasized. "In fact, such complications are most frequently annoyances and irritants rather than major disasters. Many patients tolerate feeding tubes and tube feeding. . . . Feeding tubes are burdens, but the degree of burden varies from patient to patient."[13]

Ethical Arguments For and Against Withdrawal of ASNH

Medical societies, bioethicists, and court rulings have focused on a common set of arguments supporting the withdrawal of nutrition and hydration from patients with an unresponsive wakefulness syndrome (UWS) or vegetative state. They generally follow this pattern:

- The provision of nutrition and hydration through an artificial means is a medical act.
- Medical treatments can be withheld or withdrawn according to the balance of their benefits and burdens as assessed by the patient, based on the principle of autonomy, even when the refusal may lead to the patient's death.
- When the patient is incapacitated, a morally and legally appropriate surrogate can implement that judgment on the patient's behalf.
- ASNH is analogous to other life-sustaining medical treatments such as a ventilator, which supports or replaces normal bodily functions. Therefore, it can be withdrawn following similar evaluations of benefit and burden.
- The benefit of ASNH to a patient with UWS is minimal. The patient cannot engage in life-affirming activity, thus it can be thought that meaningful life is over.
- The feedings cannot reverse the central nervous system damage and therefore should be considered futile.
- The burden of ASNH is both medical and existential. The medical burden includes risks of complications with the placement and continuation of tube feedings, the cost of the specialized feedings, nursing care, supplies, and other associated expenses. The existential cost is found in prolonging such a life, which can presumably experience no pleasure or pain if truly in a vegetative state, and make no contribution to family or society. Some caution is called for in confirming the diagnosis. Others have argued that this is best evaluated using the traditional distinction between extraordinary (i.e., morally optional) and ordinary, obligatory care[14] rather than the typical approach, which appears to rely on the principle of double effect.

13. Myles N. Sheehan, "Feeding Tubes: Sorting Out the Issues," *Health Progress* 82 (November/December, 2001): 22–27.

14. Daniel P. Sulmasy, "End-of-Life Care Revisited," *Health Progress* 87 (July/August, 2006): 50–56.

The arguments countering this perspective have also remained somewhat consistent over the years that this has been debated. They often begin by pointing out that a proper analysis is found both in the action itself as well as its consequences. If the placement of a feeding tube is a medical act, that can be withdrawn or withheld, then the actual *procedure of tube placement* with its known risks and benefits constitutes the medical act. Once the tube is established, the daily feedings do not themselves constitute a medical act; they are normal and ordinary care. For example, if a similar patient had developed an open fistula into the stomach requiring nothing more to be fed than using a pitcher or a funnel, it would be hard to argue that such a feeding itself is an act of medicine.

Furthermore, the morality of the action of withdrawing feedings is properly seen in its inevitable consequence, death. Unlike the withdrawal of a ventilator or dialysis machine, which in exceptional circumstances does not result in the expected death of the patient, all patients will succumb to a withdrawal of nutrition and hydration, no matter how it is supplied. If a deliberate act leads to an inevitable death, it is difficult to contend that the death was not intended. Causing the intentional death of an innocent person is morally impermissible.

The benefit/burden analysis is also seen as flawed. The medical burden of ongoing feedings through a tube is clearly greatest at the time the tube is initiated. Following this, complications are usually easily managed, the costs of the feeding itself can be minimized because specialized formulas can usually be replaced by a blenderized diet, and medical or dietary supervision, although necessary, is not a major expense. Other costs as described above are not the costs of ASNH, but rather the costs of keeping anyone alive who is profoundly disabled. In fact, tube feedings are routinely offered in the home for patients who need them, regardless of whether they have central nervous system damage or some other indication. They are delivered by family members after minimal instruction, and do not require the presence or assistance of a trained medical professional. If this feeding act for the UWS disabled is construed as a negative or worthless act, the same judgment can and has been made whether or not the patient requires a feeding tube. The judgment is not on the cost of the treatment, but on the value of continuing their lives. It could as easily be applied to other mentally or physically disabled persons whose lives are construed to have no value to themselves or others.[15]

The Catholic Position, Theory and Practice

It was against this background that the Catholic position was debated and ultimately resolved. There had been those supporting withdrawal of ASNH among

15. Wesley Smith, "The Deadly 'Quality-of-Life' Ethic," *First Things* (July 6, 2020), https://www.firstthings.com/web-exclusives/2020/07/the-deadly-quality-of-life-ethic.

Catholic bishops and ethicists[16] and those opposed,[17] using similar arguments to those above, as well as focusing on solidarity and compassion with the seriously ill. The balance was strongly tilted in favor of continued feedings in the statement by St. John Paul II.[18] In addressing an international congress at the Vatican on "Life-Sustaining Treatments and Vegetative State: Scientific Advances and Ethical Dilemmas," he made several points. He first warned against diagnostic errors for these patients, decried the use of the term "vegetative" state, and emphasized the human dignity of such patients. He then went on to state definitively that "The sick person in a vegetative state, awaiting recovery or a natural end, still has the right to basic health care (nutrition, hydration, cleanliness, warmth, etc.), and to the prevention of complications related to his confinement to bed." He went on to add, "I should like particularly to underline how the administration of water and food, even when provided by artificial means, always represents a natural means of preserving life, not a medical act. Its use, furthermore, should be considered in principle ordinary and proportionate, and as such morally obligatory insofar as and until it is seen to have attained its proper finality, which in the present case consists in providing nourishment to the patient and alleviation of his suffering." Moreover, "death by starvation or dehydration is in fact the only possible outcome as a result of their withdrawal. In this sense it ends up becoming, if done knowingly and willingly, true and proper euthanasia by omission."

While many thought that this would conclude the debate, there was continued pushback over the pope's intended meaning, any possible exceptions, and the level of

16. Texas Bishops and the Texas Conference of Catholic Health Facilities, "On Withdrawing Artificial Nutrition and Hydration," *Origins* 20 (June 7, 1990): 54; John Paris, "The Catholic Tradition on the Use of Nutrition and Fluids," in *Birth, Suffering, and Death: Catholic Perspectives at the Edges of Life*, ed. Kevin W. Wildes, Francesco Abel, and John Collins Harvey (Dordrecht, the Netherlands: Springer, 1992), 189–208; Walter Shannon, "The PDS Patient and the Foregoing/Withdrawing of Medical Nutrition and Hydration," *Theological Studies* 49 (December 1988): 623–47; and Kevin O'Rourke, "Reflections on the Papal Allocution Concerning Care for Persistent Vegetative State Patients," *Christian Bioethics* 12 (April 2006): 83–97; Kevin O'Rourke and Patrick Norris, "Care of PVS Patients: Catholic Opinion in the United States," *Linacre Quarterly* 68, no. 3 (2001): 201–217; Christopher Tollefsen, ed., *Artificial Nutrition and Hydration: The New Catholic Debate* (Dordrecht, Holland: Springer, 2008), https://doi.org/10.1007/978-1-4020-6207-0; Ronald P. Hamel and James J. Walter, eds., *Artificial Nutrition and Hydration and the Permanently Unconscious Patient: The Catholic Debate* (Washington, D.C.: Georgetown University, 2007).

17. G. Kevin Donovan, *Nutrition/Hydration and PVS* (thesis; Norman, Okla.: University of Oklahoma Graduate College, 1994); Germain Grisez, "Should Nutrition and Hydration Be Provided to Permanently Comatose and Other Mentally Disabled Persons?" *The Linacre Quarterly* 57, no. 2 (1990): 30–43, https://epublications.marquette.edu/lnq/vol57/iss2/6; United States Conference of Catholic Bishops, "Q&A from the USCCB Committee on Doctrine and Committee on Pro-Life Activities regarding the Holy See's Responses on Nutrition and Hydration for Patients in a 'Vegetative State'" (2007), https://www.usccb.org/issues-and-action/human-life-and-dignity/end-of-life/euthanasia/upload/q-a-nutrition-and-hydration-patients-vegetative-state.pdf.

18. St. John Paul II, "Address to an International Congress at the Vatican on Life-Sustaining Treatments and Vegetative State: Scientific Advances and Ethical Dilemmas," April 2004. See also Donovan, *Nutrition/Hydration.*

magisterial authority. Following this and the Schiavo case, the Congregation for the Doctrine of the Faith was queried by the Catholic bishops.[19] They posed two questions:

> (1) Is the administration of food and water (whether by natural or artificial means) to a patient in a "vegetative state" morally obligatory except when they cannot be assimilated by the patient's body or cannot be administered to the patient without causing significant physical discomfort?
> *Response*: Yes. The administration of food and water even by artificial means is, in principle, an ordinary and proportionate means of preserving life. It is therefore obligatory to the extent to which, and for as long as, it is shown to accomplish its proper finality, which is the hydration and nourishment of the patient. In this way suffering and death by starvation and dehydration are prevented.
>
> (2) When nutrition and hydration are being supplied by artificial means to a patient in a "permanent vegetative state," may they be discontinued when competent physicians judge with moral certainty that the patient will never recover consciousness?
> *Response*: No. A patient in a "permanent vegetative state" is a person with fundamental human dignity and must therefore receive ordinary and proportionate care which includes, in principle, the administration of water and food even by artificial means.

This authoritative statement, and the discussion that was included, removed any doubt about what was intended as the official teaching of the Church. As we previously noted, the ERDs were amended to clarify this, with number 58 now reading:

> In principle, there is an obligation to provide patients with food and water, including medically assisted nutrition and hydration for those who cannot take food orally. This obligation extends to patients in chronic and presumably irreversible conditions (e.g. the "persistent vegetative state") who can reasonably be expected to live indefinitely if given such care. Medically assisted nutrition and hydration become morally optional when they cannot reasonably be expected to prolong life or when they would be "excessively burdensome for the patient or would cause significant physical discomfort, for example resulting from complications in the use of the means employed." For instance, as a patient draws close to an inevitable death from an underlying progressive and fatal condition certain measures to provide nutrition and hydration may become excessively burdensome and therefore not obligatory in light of the very limited ability to prolong life or provide comfort.

This is the current position of the Catholic Church, and therefore the directive shaping policy within Catholic healthcare institutions. It should be noted that it was specifically aimed to answer the question about nutrition and hydration supplied via

19. Congregation for the Doctrine of the Faith, "Responses to Certain Questions of the United States Conference of Catholic Bishops concerning Artificial Nutrition and Hydration" (August 1, 2007).

a tube to patients diagnosed with UWS/PVS. It was decided in part on the basis that such patients are *clinically stable and not actively dying*. There are many more patients in clinical situations in which it would not be clearly applicable. Evidence for palliative or life-sustaining benefits for ASNH at the end of life in diseases such as terminal cancer, amyotrophic lateral sclerosis, and dementias does not support either their medical effectiveness or their moral necessity, and therefore continue to require analysis on an individual basis.[20] This has implications for the provision of care in Catholic hospitals and nursing homes. For instance, it has been pointed out that the absence of a feeding tube in the patient who is developing swallowing problems near the end of life would still indicate a moral obligation for hand-feeding when possible.[21] Moreover, those who would refuse nutrition and hydration in situations where it would be medically and morally indicated would need to be informed that these preferences cannot be honored in Catholic institutions when they are contrary to Catholic moral teaching.[22] A clear statement of values and practices within Catholic health care needs to be communicated to every patient and his or her family, without apology. After all, it is these solid values focused on the good of the patient that have driven the creation of Catholic health care in the first place, and still attract many patients to these institutions.

20. Linda Ganzini, "Artificial Nutrition and Hydration at the End of Life: Ethics and Evidence," *Palliative and Supportive Care* 4, no. 2 (2006): 135–43.

21. Carol R. Taylor and Robert C. Barnet, "Hand-Feeding: Moral Obligation or Elective Intervention?" *Healthcare Ethics USA* 22, no. 2 (2014): 12–23.

22. United States Conference of Catholic Bishops, *Ethical and Religious Directives for Catholic Healthcare Services*, 6th ed., no. 59.

24

Brain Death: History, Controversy, and Prognosis

Allen H. Roberts II, MD

IN DECEMBER 1967, there appeared on the cover of *Life Magazine* the stunning image of South African cardiac surgeon Dr. Christiaan Barnard and his smiling patient, Louis Washkansky, at the Groote Schuur Hospital in Cape Town. Earlier that month, Barnard had performed the first successful human heart transplant on Washkansky, who was terminally ill and bedridden with advanced heart disease; the transplanted heart had been procured from Denise Darvall, a 25-year-old woman who had suffered traumatic brain injury and was on life support at the same hospital. According to a fiftieth-anniversary account of that operation, Ms. Darvall had been declared brain dead by neurosurgeons,[1] and her family gave consent for the donation. The image and the event dramatically captured the imagination of the public and the medical profession and signaled a watershed moment—not simply in the domain of transplant surgery—but, more significantly, in that of the philosophy, science, and ethics of the concept and diagnosis of death.[2]

This chapter will review the history of the concept and practice of diagnosing death by neurologic criteria (DDNC), known in common parlance as "brain death," and engage the ethical controversies that have burdened both concept and practice for more than half a century. It will be necessary to provide a technical account of how brain death is currently diagnosed, since much ethics polemic has centered on the very procedural details of the practice itself. I then consider the current state of philosophical, ethical, and scientific debate that surround this paradigm and its foundations. Concerns related to these foundations recently have led some to advocate the abandonment of the prevailing concept and practice of brain death altogether, and one notes that Dr. Edmund Pellegrino, who chaired the President's Council in 2008, never personally embraced the construct of brain death as it was

1. David K.C. Cooper, "Life's Defining Moment: Christiaan Barnard and the First Human Heart Transplant," *Journal of Heart and Lung Transplantation* 36, no. 13 (2017): 1273–75.

2. The irony should not be lost that *Life* helped animate a new ascertainment of death. *Life Magazine* itself is no longer with us.

articulated in that council's report, citing the lack of moral certainty that the construct sufficiently identified the event of death. Therefore, I will conclude the chapter by situating the current status of the brain-death debate within Pellegrino's broader thought regarding the philosophy of medicine and the "ends of medicine." His writing continues to speak to the intersection of the concept of brain death and these very ends.

Claims that the need for transplantable organs was the sole driver of "brain death" in the first place have been labeled as "historically reductive" by De Georgia.[3] Gary Belkin[4] specifically ties the concept's inception to Joseph Fletcher's broader work in situation ethics.[5] To be sure, the development and articulation of the brain-death concept are multi-dimensional; notwithstanding, brain death, as we currently understand and practice it, is *inextricably* tied to the practice of organ procurement and transplantation, and most scholarly consideration of brain death occurs in the context of and literature pertaining to organ transplantation. The relationship between brain death and organ procurement was present from the beginning and abides today. Belkin reports Paul Ramsey's early protest that brain death was an immoral utilitarian program for the purpose of organ procurement.[6] And, as Robert Truog has more recently stated, "The concept of brain death has . . . been essential for organ transplantation to develop and flourish."[7] Much scholarship on these matters is thus heavily influenced by the presuppositions of current transplant practice, reflecting our profession's laudable attentiveness to the deep suffering and abiding need of those awaiting a transplanted organ in order simply to survive. But to separate the consideration of brain death from that of organ procurement is untenable.

Brain Death: Philosophy, History, and Science

A concise definition of "death," and of what it means to be "dead," are matters of deep metaphysical mystery. Virtually all scholars endorse a three-tiered approach to the concept of death: "(1) the philosophical task of determining the definition of death, (2) the philosophical and medical task of determining the best criterion of death, a measurable condition that has been fulfilled by being both necessary and sufficient for death, and (3) the medical-scientific task of determining the tests of death for physicians to employ at the bedside to demonstrate that the criterion has been fulfilled."[8]

3. Michael A. De Georgia, "History of Brain Death as Death: 1968 to the Present," *Journal of Critical Care* 29 (2014): 673–78.

4. Gary S. Belkin, *Death before Dying: History, Medicine, and Brain Death* (Oxford: Oxford University Press. 2014), chap. 2, "The Justification: Beecher's Ethics," 51–91.

5. Ibid., 73.

6. Ibid., 67.

7. Robert Truog, "Brain Death—Too Flawed to Endure, Too Ingrained to Abandon," *Journal of Law, Ethics & Medicine* 35, no. 2 (2007): 273–81, doi: 10.1111/j.1748-720X.2007.00136.x.

8. James Bernat, "The Whole-Brain Concept of Death Remains Optimum Public Policy," *Journal of Law, Medicine & Ethics* 34, no. 1 (Spring 2006): 35–43.

Robert Veatch correctly has observed that the philosophical concept of death must remain distinct from the scientific criteria used to diagnose it. He states,

> There are two more or less separate questions at stake—one normative and one scientific. The normative question is "What change in a human being is so fundamental that we can say the individual is no longer with us as a member of the human community bearing rights such as the right not to be killed?". . . Choosing the point at which we should treat humans as we treat dead people turns out to be a normative issue, not a scientific one. . . . Once we have answered the normative question, we can move on to answer the scientific question, "How do we know that some element critical for human status—the flowing of fluids, the integrative activity of the brain, or the capacity for consciousness—has been lost irreversibly?"[9]

Acknowledging the *normative question* to be metaphysical and mysterious, we will concern ourselves with the scientific questions, although the lines between Veatch's categories become blurred at points because of medical and societal interests; controversies surrounding brain death have occasioned many attempts, over the years, to apply a *scientific* answer to Veatch's normative *philosophical* question.

Michael Nair-Collins articulates what has been presuppositional over the millennia that preceded the modern era of intensive care units and organ transplantation. I endorse as foundational his assertion that

> [d]eath is a biological phenomenon . . . the irreversible cessation of the integrative functioning of the organism as a whole in its capacity to maintain internal physiological stability (homeostasis) and resist entropy and disintegration. On this view, death is an event (rather than a process), something like a thermodynamic point of no return, which separates the processes of homeostatic maintenance and resistance of entropy from increasing entropy and disintegration.[10]

Traditionally, the clinical procedure for pronouncing death had been the assessment for spontaneous breathing, circulation, and pupillary response, and the absence of all of these predicts the onset of rigor mortis and of putrefaction. *This clinical evaluation, then, is a time-honored, scientific, but proxy, determination that the event of death has occurred.*

Historical accounts of brain death in the literature generally begin by citing the description by French neurologists Mollaret and Goulon of the entity *coma depasse* ("beyond coma"),[11] that is, a deep coma condition that is invariably fatal.

9. Robert Veatch and Lainie Ross, *Transplantation Ethics,* 2nd ed. (Washington, D.C.: Georgetown University Press, 2015), 42.

10. Michael Nair-Collins, "Clinical and Ethical Perspectives on Brain Death," *Medicolegal and Bioethics* 2015, no. 5 (September 1, 2015): 69–80.

11. Pierre Mollaret and Maurice Goulon, "Le Coma Depasse (memoire preliminaire)," *Review Neurologique* 101, no. 1 (1959): 3–15.

A number of lifesaving and life-sustaining developments in cardiology and critical care, in the years leading up to this description, had set the stage for the impact of their work. De Georgia has provided a comprehensive review of these developments, and details the nuances of the discussions among scholars as they engaged, even then, the question of whether neurologic criteria could serve as proxy determination of the normative question.[12] Mollaret and Goulon themselves held that "'irreversible coma' was *prognostic* of death but not really *equal* to death."[13] Regardless, their description, within the historical-ethical milieu delineated by Belkin, provided the necessary groundwork for the publication in the *Journal of the American Medical Association* of "A Definition of Irreversible Coma: Report of the Ad Hoc Committee of the Harvard Medical School to Examine the Definition of Brain Death."[14] The stated intent of this paper, which appeared within a year of Barnard's pioneering surgery, was to make much needed hospital beds available, and to increase the number of organs that might be made available for donation; the authors issued a call to the medical profession and society to effect new public policy that would authorize alternative, neurologic, criteria by which a patient might be pronounced dead. It was a seminal publication, the impact of which cannot be overstated.

More than a decade later, the 1981 the President's Commission report on *Defining Death* stated that "death is that moment at which the body's physiological system ceases to constitute an integrated whole. Even if life continues in individual cells or organs, life of the organism as a whole requires complex integration, and without the latter, a person cannot properly be regarded as alive."[15] The commission understood the "proxy" nature of both traditional and neurologic determinations of the normative state of death and affirmed the validity of the "whole brain" formulation, which holds "the centrality of the whole brain, since it is neither revivable nor replaceable . . . as the regulator of the body's integration."[16]

The commission considered, but rejected, the alternate formulation of a "higher brain" criterion (by which a patient might be declared dead when function of the cerebral hemispheres is lost, but the respiratory center remains intact, such as in the so-called persistent vegetative state, also referred to as unresponsive wakefulness syndrome. Importantly, whereas this "partial brain" formulation of brain death was sidelined, we shall see it return again, years later.

12. De Georgia, "History of Brain Death as Death."

13. Ibid., 675.

14. Henry Beecher et al., "A Definition of Irreversible Coma: Report of the Ad Hoc Committee of the Harvard Medical School to Examine the Definition of Brain Death," *Journal of the American Medical Association* 205, no. 6 (1968): 85–88.

15. United States President's Commission for the Study of Ethical Problems in Medicine and Biomedical and Behavioral Research, *Defining Death: Medical, Legal, and Ethical Issues in the Determination of Death* (Washington, D.C.: U.S. Government Printing Office, 1981), https://scholarworks.iupui.edu/bitstream/handle/1805/707/Defininining%20death%20-%201981.pdf.

16. Ibid.

Finally, the commission recommended the creation of a statute incorporating a [whole] brain-based standard, which they predicted would be "accepted by theologians of all backgrounds."[17] The prediction proved to be accurate. Pope John Paul II affirmed the practice in his address to the International Congress of the Organ Transplant Specialists in 2000.[18] It is endorsed by Protestant and Reformed,[19] Eastern Orthodox,[20] and Coptic traditions.[21] Whereas many Orthodox Jews do not recognize whole-brain death, the concept has been accepted in principle by the Rabbinical Council of America in 1991.[22] A survey of Muslim ethical, religious, and legal documents disclosed that the majority of Muslim scholars, organizations, and governments accept the criteria and practice.[23] Both Hindu and Buddhist traditions endorse a whole-brain criterion with some groups favoring a neocortical criterion.[24] Within each religious group, a small percentage of adherents do reject a neurologic criterion for determining death, and continue to accept only a circulatory criterion.[25]

The Uniform Determination of Death Act (UDDA) was passed in 1981 based on the commission's findings. The UDDA is a model statute, created for adoption by state legislatures, which codified that death may be established in an "individual who has sustained either (1) irreversible cessation of circulatory and respiratory functions, or (2) irreversible cessation of all functions of the entire brain, including the brain stem. . . . A determination of death must be made in accordance with accepted

17. Ibid.

18. St. John Paul II, "Address to the 18th International Congress of the Transplantation Society," August 29, 2000, https://www.vatican.va/content/john-paul-ii/en/speeches/2000/jul-sep/documents/hf_jp-ii_spe_20000829_transplants.html.

19. Franklin Payne, *Biblical/Medical Ethics: The Christian and the Practice of Medicine* (Milford, Mich.: Mott Media, Inc., 1985), 198–99.

20. Protodeacon Basil Andruchow, "Medical Bioethics: An Orthodox Christian Perspective for Orthodox Christians," *The Orthodox Church in America,* September 15, 2010, https://www.oca.org/parish-ministry/familylife/medical-bioethics-an-orthodox-christian-perspective-for-orthodox-christians.

21. The Coptic Orthodox Diocese of Los Angeles, Southern California, and Hawaii, "Brain Death," 2009, http://lacopts.org/articles/brain-death.

22. Daniel S. Nevins, "Contemporary Criteria for the Declaration of Death," in *Hakol Kol Yaakov, The Joel Roth Jubilee Volume,* ed. Robert A. Harris and Jonathan S. Milgram, Brill Reference Library of Judaism 61 (Boston, Mass.: Brill, 2021), chap. 8, 202–34; and Gabrielle Loeb, "Judaism's Perspective on Organ Donation After Death," *The Philadelphia Jewish Voice* 38 (September 2008), http://www.pjvoice.com/v38/38700judaism.aspx.

23. Andrew Miller et al., "Brain Death and Islam: The Interface of Religion, Culture, History, Law, and Modern Medicine," *CHEST* 146, no. 4 (2014): 1092–101.

24. Anant Dhanwate, "Brainstem Death: A Comprehensive Review in Indian Perspective," *Indian Journal of Critical Care* 18, no. 9 (2014): 596–605; John-Anderson L. Meyer, "Buddhism and Death: The Brain-Centered Criteria," *Journal of Buddhist Ethics* 12 (2005): 1–24, https://blogs.dickinson.edu/buddhistethics/files/2010/04/meyer402.pdf; Deepak Sarma, "Hinduism, Brain Death, and Organ Transplantation," *Huffington Post,* September 28, 2016, updated September 29, 2017, https://www.huffpost.com/entry/hinduism-brain-death-orga_b_12198832.

25. Robert Truog and Robert Tasker, "Counterpoint: Should Informed Consent Be Required for Apnea Testing in Patients with Suspected Brain Death? No," *Chest* 152, no. 4 (2017): 702–4.

medical standards."[26] All fifty states have since passed legislation based on this model. The "dead donor rule," which followed, is a philosophical synthesis of the UDDA and homicide law and establishes that no organ may be procured from anyone who is not pronounced dead by one or the other of these criteria.[27] That the language of the UDDA did not further specify "acceptable medical standards" acknowledged and provided for the evolution of science and technology to inform practice standards over time.

It was not until 1995 that "acceptable medical standards" for diagnosing "irreversible cessation of all functions of the entire brain, including the brain stem," were more fully articulated by the American Academy of Neurology (AAN).[28] According to these guidelines, the diagnosis of brain death is to be made in a sequence of three well-defined procedures, during which time the patient is sustained on a ventilator: establishing the presence of an irreversible cause for the coma state, establishing the absence of observable brainstem reflexes, and establishing the presence of apnea (that is, the absence of spontaneous breathing). Confounding clinical conditions which might "mimic" brain death (such as hypothermia, paralysis, drug effects, or severe hypoglycemia) must be absent or, if present, medically reversed prior to formal brain-death assessment. Once coma is diagnosed, the etiology for the coma is established, and brainstem reflexes are found to be absent on clinical examination, then the confirmatory test for brain death is the *apnea test,* by which the patient is assessed for respiratory effort under conditions which are believed to provide maximum stimulus to the medullary respiratory center of the brain. A *positive apnea test,* that is, the absence of observable respiratory effort under these conditions, is confirmatory, or diagnostic, for brain death. If the apnea test is not able to be completed for clinical reasons such as hemodynamic instability, a series of "ancillary tests," which assess the electrical or blood-flow activity of the brain, may be employed. The results of these tests may be helpful, but there are instances where they may provide inconclusive or conflicting results.[29]

In 2010, Eelco Wijdicks published an "evidence-based guideline update"[30] to strengthen the guidelines of 1995. In an effort to improve standardization of the practice, these refined the procedural structure for assessing and diagnosing brain death to enhance compliance with the specifics of brain-death testing. Standard

26. The Uniform Definition of Death Act 1980/1981, https://www.uniformlaws.org/viewdocument/final-act-49?CommunityKey=155faf5d-03c2-4027-99ba-ee4c99019d6c&tab=librarydocuments.

27. James Bernat, "Life or Death for the Dead-Donor Rule?" *New England Journal of Medicine* 369, no. 14 (2013): 1289–91.

28. American Academy of Neurology, Quality Standards Subcommittee, report "Practice Parameters for Determining Brain Death in Adults (Summary Statement)," *Neurology* 45 (1995): 1012–14, https://n.neurology.org/content/neurology/45/5/1012.full.pdf.

29. Eelco Wijdicks, "The Case Against Confirmatory Tests for Determining Brain Death in Adults," *Neurology* 75 (2010): 77–83.

30. Eelco F.M. Wijdicks et al., "Evidence-Based Guideline Update: Determining Brain Death in Adults," *Neurology* 74 (2010): 1911–18.

practice holds that once brain death is diagnosed, the patient's family or surrogate is informed of the diagnosis; interventions such as mechanical ventilation, hemodynamic support, and renal replacement therapy are then discontinued, unless consent is given by surrogates for organ donation. The provisions of the 2010 AAN Guidelines Update have constituted the standard of practice until the time of this writing.

The unusual circumstance of brain death occurring in pregnant patients was recently reviewed by Dodaro et al. who studied the clinical course and perinatal outcome of thirty-five such cases. The most notable difference of their recommendations from the 2010 guidelines is the contraindication to performing the apnea test in this setting, given the risk of physiologic compromise to the fetus. Rather, ancillary tests such as radiographic brain perfusion studies are utilized; if brain death is thereby confirmed, somatic support of the mother may be continued for the support of the fetus.[31]

In 2023 the AAN published a comprehensive and updated set of guidelines for determining brain death in pediatric and adult patients.[32] We shall return to these guidelines presently; however, both the philosophy and specifications of these guidelines, which will likely assume 'standard-of-care' status, will best be understood against the backdrop of the ongoing controversies surrounding brain death that have been prevalent in the literature in the decades that preceded their publication. These controversies are centered, essentially, in three broad categories, namely, those pertaining to philosophical/definitional, technical/practical, and integrity/transparency concerns.

Brain Death: Concerns and Controversies

Despite the progressive refinement of methodology set forth in the sequence of published guidelines, controversy persisted over the whole-brain rationale and formula that the 1981 President's Commission established. Sufficient controversy, in fact, that in 2007 a second President's Council on Bioethics, chaired by Edmund D. Pellegrino, convened to address "a number of persistent concerns and novel criticisms . . . both about the meaning of the term 'brain dead' and about its relation to the death of a human being."[33] In 2009 the council released its *White Paper Report on Controversies in the Determination of Death,* which "discarded the ambiguous term *brain death,* replacing it with the philosophically neutral term *total-brain failure* (TBF)."[34]

31. Maria Dodaro et al., "Brain Death in Pregnancy: A Systematic Review Focusing on Perinatal Outcomes," *American Journal of Obstetrics and Gynecology* 224, no. 5 (May 2021): 445–69.

32. David M. Greer et al, "Pediatric and Adult Brain Death/Death by Neurologic Criteria Consensus Guidelines," *Neurology* 101, no. 24 (2023): 1112–32. doi: 10.1212/WNL.0000000000207740. See also correction published in issue 102, no. 3: e208108. doi: 10.1212/WNL.0000000000208108.

33. President's Council on Bioethics, *Controversies in the Determination of Death* (Washington, D.C.: United States Government Printing Office, 2008), 6, https://bioethicsarchive.georgetown.edu/pcbe/reports/death/.

34. De Georgia, "History of Brain Death as Death."

Acknowledging the work of Alan Shewmon and others (*vide infra)* decrying the inaccuracies of the concept of the "integrative unity" function of the brain, the 2009 report asserted that "cessation of the fundamental vital work of a living organism—the work of self-preservation" reflected cessation of the "organism's need-driven commerce with the surrounding world."[35] Total-brain failure, on this premise, would be a sufficient proxy for the death of the individual, since "this organism can no longer engage in the essential work that defines living things."[36] It would provide, effectively, a scientific approach to Veatch's 'normative' question. The council acknowledged the inseparability of the brain-death question from the issue of organ procurement. Like its 1981 predecessor, the council considered, and again rejected, the substitution of neurologic criteria that reflected only a higher brain, or "neocortical," diagnosis of death. Thus did the "partial brain" definition again become dormant—for a time.

At the conclusion of the white paper report, an important contribution appeared by the council's chair, Dr. Edmund Pellegrino. In this "personal statement," Pellegrino attests to the persistent inadequacy of neurologic criteria as a proxy to answer the normative question:

> [T]he reasons that favor the neurologic standard whole-brain criteria are not compelling. . . . [T]here is as much—or more—moral assurance of death of the donor with the cardiopulmonary as with the neurologic standard. In fact, with the cardiopulmonary standard, there is a higher degree of certainty of death than there is with a heart-beating donor, because heart, lung, and brain have all ceased functioning. . . .
>
> . . . Finally, the metaphysical definition of death as the separation of the body from its vital principle is still held as the authoritative definition by many worldwide. Plato put it most bluntly: "Death in my opinion is nothing else but the separation from each other of two things, soul and body." No precise congruence of this concept with any observable set of clinical facts has ever been agreed upon.[37]

It was the absence of moral certainty that *any* neurologic criteria were a sufficient proxy for the metaphysical event of death that caused Pellegrino to stop short of endorsing the council's total-brain failure formula. From a practical standpoint, however, the white paper report's transition from "brain death" to "total-brain failure" has had little impact on how *brain death* (which has remained the operative term at the bedside and in the literature) is practiced. But Pellegrino's caution was quietly prophetic of the controversies which remain to this day.

The requirement for moral certainty in the matter of brain death is pointedly elaborated by Brugger in his critique of the premises and conclusions of both the

35. President's Council on Bioethics, *Controversies in the Determination of Death,* 60.
36. Ibid., 64–65.
37. Ibid., 114, 110.

President's Commission and Council. "I do not think moral certitude is required when the risks posed by some procedure are minor. But bringing about the death of a human being is never a minor matter. . . . those who defend, authorize, and perform transplants from BD [brain-dead] bodies should be morally certain that the subjects are not living human beings."[38] Brugger concludes,

> Since the best evidence does not dispel the doubt that ventilated BD bodies are dead; and since we should not treat as corpses what for all we know might be living human beings, we have an obligation to treat BD individuals *as if* they were living human beings.[39]

Controversy regarding technical aspects of brain-death practice preceded and followed the 2010 Guidelines Update and is embedded within the guidelines themselves, which, on close inspection, seem to contain but little evidence. Wijdicks himself acknowledges that the guidelines are opinion-based.[40] Indeed, in 2008 (prior to the release of the updated guidelines), Greer and coworkers demonstrated wide variability in the application of the existing (1995) guidelines among *US News & World Report* top fifty neurology/neurosurgery centers.[41]

Acknowledging these and similar findings, Wijdicks called for standardization of policies, education of hospital staff on protocols for establishing brain death, checklists, and brain-death determination by designated, experienced physicians who have documented proficiency in such determinations.[42] Subsequent studies in the years following the 2010 update have continued to identify significant variability in hospital brain-death policies and practices among physicians within a single academic institution,[43] as well as among 492 hospitals across the United States,[44] and internationally.[45] In a retrospective review of 226 brain-dead organ donors, wide variability was seen in the degree of adherence to BD testing protocol, with less than fifty percent of tests being substantially adherent to guideline specifications.[46]

38. E. Christian Brugger, "Are Brain Dead Individuals Dead? Grounds for Reasonable Doubt," *Journal of Medicine and Philosophy* 41 (2016): 329–50.

39. Ibid.

40. Wijdicks, "The Case Against Confirmatory Tests."

41. David Greer, Panayiotis N. Varelas, Shamael Haque, and Eelco F M Wijdicks, "Variability of Brain Death Determination Guidelines in Leading US Neurologic Institutions," *Neurology* 70 (2008): 1–6.

42. Eelco Wijdicks, "The Clinical Criteria of Brain Death Throughout the World: Why Has It Come to This?" *Canadian Journal of Anesthesia* 53, no. 6 (2006): 540–43.

43. Ashutosh Pandey, Pradeep Sahota, Premkumar Nattanmai, and Christopher R. Newey, "Variability in Diagnosing Brain Death at an Academic Medical Center," *Neuroscience Journal* 2017 (2017), https://doi.org/10.1155/2017/6017958.

44. David M. Greer, Hilary H. Wang, Jennifer D. Robinson, et al. "Variability of Brain Death Policies in the United States," *JAMA Neurology* 73, no. 2 (2016): 213–18.

45. Sarah Wahlster, Eelco F.M. Wijdicks, Pratik V. Patel, et al., "Brain Death Declaration: Practices and Perceptions Worldwide," *Neurology* 84 (2015): 1870–79.

46. Claire N. Shappell, Jeffrey I. Frank, Khalil Husari, et al., "Practice Variability in Brain Death Determination," *Neurology* 81 (2013): 2009–14.

An additional practical challenge, unsurprisingly, is that the language and parlance surrounding brain death are inherently confusing. In a narrative review of forty-three peer-reviewed articles, representing approximately eighteen hundred participants, Shah found widespread and significant misunderstanding about the biological facts surrounding brain death and about the fact that organs may be procured from brain-dead patients, even while the donor's heart continues to beat.[47] In a retrospective review of brain-dead organ donors and prospective interviews with surviving families three months after the fact, Siminoff found that more than a fourth of participants were unable to provide a correct definition of brain death.[48] Finally, echoing Brugger's dismay, Robert Truog has famously stated that "the root of the problem was not with "the clinicians, but with the concept. . . . brain dead patients share many more features with living patients than they do with patients who have been declared dead by the more traditional criteria. . . . These patients may appear to be completely well."[49]

Conceptual Challenges to Contemporary Brain-Death Practice

Let us now now shift our attention to arguments both for and against the continued use of a total-brain failure paradigm as an acceptable scientific/biological proxy determination of the normative question. It is helpful to frame the discussion against the backdrop of two premises. The first concerns the normative question posed by Veatch, that is, what event signals that the patient is no longer a part of the human community?[50] What Veatch answers philosophically, Pellegrino, appealing to Plato, states metaphysically, that is, the moment at which the soul departs the body.[51] The second premise is that medicine has always been about making biologic, scientific diagnoses via means that are more or less *proxy* in nature, moving along a spectrum from that which may be ascertained by history and physical examination to that which must be ascertained by a tissue biopsy. Diagnosing death as a biological event is no different. However, because of the normative/metaphysical weight of this event, Alan Shewmon, foremost critic of the whole brain/ total-brain failure construct, states, "Every test or diagnostic criterion in medicine has a trade-off between sensitivity and specificity, but the diagnosis of death is unique in that it demands 100 percent specificity and 0 percent chance of a false-positive error."[52]

47. Seema K. Shah, Kenneth Kasper, Franklin G. Miller, et al., "A Narrative Review of the Empirical Evidence on Public Attitudes on Brain Death and Vital Organ Transplantation: The Need for Better Data to Inform Policy," *Journal of Medical Ethics* 41, no. 4 (April 2015): 291–96.

48. Laura Siminoff, Mary Beth Mercer, and Robert Arnold, "Families' Understanding of Brain Death," *Progress in Transplantation* 13, no. 3 (2003): 218–24.

49. Robert Truog, "Brain Death—Too Flawed to Endure, Too Ingrained to Abandon," *Journal of Law, Medicine, and Ethics* (Summer 2007): 273–81.

50. Veatch and Ross, *Transplantation Ethics,* 42.

51. Edmund D. Pellegrino, "Personal Statement," in President's Council on Bioethics, *Controversies in the Determination of Death,* 110.

52. D. Alan Shewmon, "False Positive Diagnosis of Brain Death Following Pediatric Guidelines: Case Report and Discussion," *Journal of Child Neurology* 32, no. 14 (2017): 1104–17.

James Bernat, whose scholarship informed both the President's Commission and the President's Council, has been the foremost advocate of retaining the current total-brain failure formula. Endorsing the premise that death is a biological event, he contends that, in the absence of mechanical ventilation, "traditional tests are absolutely predictive that the brain will be rapidly destroyed by lack of blood flow and oxygen, at which time death will have occurred. . . . The bedside tests satisfying the whole-brain criterion of death have been designed with a sufficiently high degree of concordance to permit the drafting of widely accepted clinical practice guidelines on the determination of brain death."[53] Thus, he concludes that for the biological definition of death, a neurologic criterion is sufficient to make the diagnosis. Bernat also argues for a neurologic criterion for death based on public policy history. "Brain death is widely regarded as . . . a formerly contentious bioethical and bio-philosophical issue that has been resolved to the point of widespread public consensus. Evidence for this consensus is the enactment of effective and well-accepted brain-death laws and policies throughout the world."[54] Robert Truog has also opined that the practice is "too ingrained to abandon."[55]

But Joseph Verheijde and colleagues indict the prevailing brain death testing paradigm on the grounds that "There have been major advances in the field of neuroscience research in human consciousness since the inception of the term 'brain death' 50 years ago"[56] including the finding of latent and covert awareness in the injured brain, the fact that examination of cranial nerves cannot infer higher integrative functions of the cerebral cortex, the absence of medullary abnormalities at autopsy in brain dead patients, and the fact that contemporary neurologic testing does not assay possibly reversible derangements in neurohormonal regulatory functions,[57] that is, such functions as are governed by the hypothalamus. He further cites Bernat's subtle, even shrouded, transition away from the intent of the President's Conference and Council's requirement for *irreversibility* of brain failure with the invocation of the notion of *permanence,* which permits brain death to be established not on grounds of biological testing, but rather on a *decision not to resuscitate*.[58] On Verheijde's argument, both the commitment to scientific advance as provided for by the UDDA, on the one hand, and to a biological criterion, on the other, would seem to have been jettisoned by the AAN in their sequential guidelines. He concludes with the assertion that the

> Abrahamic faiths' acceptability of brain death was conditioned upon (1) equivalency with biological death, (2) clinical determination with scientifically veri-

53. Bernat, "The Whole-Brain Concept of Death."

54. Ibid.

55. Truog, "Brain Death—Too Flawed to Endure."

56. Joseph L. Verheijde, Mohamed Y. Rady, and Michael Potts," Neuroscience and Brain Death Controversies: The Elephant in the Room," *Journal of Religion and Health* 57, no. 5 (2018): 1745–63.

57. Ibid.

58. Ibid.

fiable criteria, and (3) alignment with the theological definition of death, i.e., the separation of the soul from the human body.[59]

On this argument, John Paul II's 2000 affirmation of brain death cannot be seen as a blanket endorsement of subsequent practice—Catholics may rethink "brain death" not on theological grounds but on the requirement that scientific methodology be rightly stewarded in a way that moral certainty is approached. I believe that His Holiness presupposed that very requirement.

No philosophical or ethics polemic showcases the assembled problems of brain death so dramatically as the account of young Jahi McMath, a thirteen-year-old girl who in 2013 underwent a tonsillectomy in Oakland, California for obstructive sleep apnea. The procedure was complicated by hemorrhage, cardiac arrest, and consequent anoxic brain injury; she was pronounced dead by neurologic criteria, and the family was approached regarding organ donation. However, the family rejected the diagnosis on grounds that echo Truog's observations. Despite repeated BD testing that rendered a positive result (i.e., that she was brain dead), the family persisted, and the case made its way to the courts, into the media, and into the arena of religious conscience and exception.[60] McMath was ultimately transferred to a long-term ventilator facility, where, after undergoing puberty and physical growth, she succumbed to cardiac arrest in 2018. When, simply put, did Jahi die? It remains a cogent question.

Eschewing the reasoned arguments of Truog, Verheijde, and others, Lewis et al. in 2019 and Omelianchuk et al. in 2022 called for a revision of the UDDA which would embrace neuro-respiratory criteria and would explicitly exclude consideration of the hypothalamus.[61] The latter authors establish, *de facto*, a novel "bright line" of expediency, enshrining into perpetuity the science of the 2010 AAN Guidelines, in the interest of "enabling confidence in the determination of and the ability to make the distinction between life and death in a timely and efficient manner."[62]

Shewmon was swift to respond to Lewis's article with a comprehensive rebuttal, endorsed by scores of physicians, ethicists, and philosophical scholars, in which he proposes revision of the UDDA, but in a way that permits science to evolve and to inform the process, rather than to enshrine the 2010 Guidelines as a matter of

59. Ibid. Editor's note: Verheijde et al. used the word "acceptability" in their article, but grammatically, the word "acceptance" would have been more correct.

60. John M. Luce, "The Uncommon Case of Jahi McMath," *CHEST* 147, no. 4 (2015): 1144–51.

61. Ariane Lewis, Richard Bonnie, and Thaddeus Pope, "It's Time to Revise the Uniform Determination of Death Act," *Annals of Internal Medicine* 172, no. 2 (2019):143–45; Adam Omelianchuk, James Bernat, Arthur Caplan, David Greer, Christos Lazardis, Ariane Lewis, Thaddeus Pope, Lanie Ross, and David Magnus, "Revise the Uniform Determination of Death Act to Align the Law with Practice through Neurorespiratory Criteria," *Neurology* 98, no. 13 (2022); 532–36. doi: 10.1212/WNL0000000000200024.

62. Omelianchuk et al.

expediency. He contends specifically that the hypothalamus, universally acknowledged to be part of the brain and possessing an as-yet incompletely understood neurohormonal integrative function, be included and its function assayed to the extent possible. Moreover, Shewmon echoes a body of literature that challenges the claim of safety of the apnea test and insists that the apnea test actually may be dispensed with, given the evolving accuracy of what are currently considered to be ancillary tests.[63]

Nair-Collins has observed that when evidence of persistent hypothalamic function is factored into brain death assessment retrospectively, the positive predictive value of currently accepted testing would be only 50%.[64] Accordingly, scholars at Georgetown University's Pellegrino Center for Clinical Bioethics (PCCB) issued a similar critique of Lewis's proposal. They state:

> Clinical criteria for the diagnosis of "cessation of all functions of the entire brain" must include all pertinent functions, *including hypothalamic functions such as hormone release and regulation of temperature and blood pressure,* to avoid the specter of neurologic recovery in those who fulfill the current clinical criteria for the diagnosis of brain death.[65]

On a related note, the McMath case and questions surrounding the safety of apnea testing served as impetus for the publication of two recent debates in *CHEST* regarding the question of whether informed consent by surrogate decision makers should be required prior to performing the apnea test. Collated rationales favoring informed consent include that the apnea test itself incurs the risk of exacerbating intracranial hypertension, that the predefined carbon dioxide target parameter does not guarantee the irreversibility of apnea, that religious objections on the part of the family must be honored, and that apnea testing is unnecessary. Collated arguments against obtaining informed consent include that brain death testing is a recognized exception to informed consent, that U.S. law does not require informed consent in this venue, that apnea testing involves minimal risk, that brain death is legally death, that this procedure on *prima facie* grounds does not require consent, and that cardiorespiratory and neurological death are equivalent in law.[66, 67] To contextualize the

63. D. Alan Shewmon, "Statement in Support of Revising the Uniform Determination of Death Act and in Opposition to a Proposed Revision," *The Journal of Medicine and Philosophy* (May 14, 2021). doi:10.1093/jmp/jhab014.

64. Michale Nair-Collins and Ari R. Joffe, "Frequent Preservation of Neurologic Function in Brain Death and Brainstem Entails False-Positive Misdiagnosis and Cerebral Perfusion," *AJOB Neuroscience* 14, no. 3 (2021) https://doi.org/10.1080/21507740.2021.1973148.

65. G. Kevin Donovan, Myles N. Sheehan, et al. (bioethicists of the Pellegrino Center for Clinical Bioethics), "Proposal for Revising the Uniform Determination of Death Act," *Hastings Center Bioethics Forum,* February 18, 2022, https://www.thehastingscenter.org/defining-brain-death/.

66. Ariane Lewis and David Greer, "Should Informed Consent Be Required for Apnea Testing in Patients with Suspected Brain Death? No." and Robert Truog and Robert Tasker, "Should Informed

"no consent" argument, one must remember that the patient being tested is not yet known to be dead.

Whereas the Uniform Law Commission tabled the movement by Lewis and others toward a revision of the UDDA in the summer of 2023, the AAN quickly published its updated guidelines in October of that year. That which could not be codified in law could be commended to the medical profession in guidelines that are destined to be widely adopted. The merits of this new set of guidelines include the increased rigor of conditions that must be met prior to the performance of the apnea test (which may indeed lend a measure of safety to the testing procedure itself), and the call for a competency qualification for those who would so test.

However, in keeping with the position of Lewis and Omelianchuk, assaying hypothalamic function is summarily and explicitly dismissed. Given the universal acknowledgment that the hypothalamus is indeed part of the brain, a "partial brain" construct has at length made its way, quietly but squarely, into the latest guidelines that purportedly retain the "complete cessation of all functions of the entire brain" definition. Finally, the 2023 AAN Guidelines reject out of hand the requirement for informed consent prior to apnea testing.

Even as the 2023 AAN Guidelines are beginning to make their way into hospital policies across the country, dissent has been vocal. In February 2024, Eble, Camillo, and Colosi issued a statement entitled "Catholics United on Brain Death and Organ Donation." In this white paper report, which boasts more than 100 medical, academic, and ethics signatories, the authors affirm that

> . . . that the [brain death] criteria found in the guidelines and used in current clinical practice do not provide moral certainty that a patient has died. Given that the Catholic Church requires moral certainty of death as a condition for removing organs when this would otherwise kill the patient, that the majority of organ donors are declared dead using the existing BD criteria, and that there is no expectation that the guidelines will be revised any time in the near future in such a way as to establish the moral certainty of death, the endorsers recommend a number of concrete actions for Catholics. . . .[68]

The authors seek to protect the lives of patients who are not known to be dead, to protect practitioners' and institutions' right of conscientious objection to what they regard as immoral or morally uncertain practice, and to promote the development of

---

Consent Be Required for Apnea Testing in Patients with Suspected Brain Death? Yes." *CHEST* 152, no. 4 (2017): 700–706.

67. D. Alan Shewmon, "Whether Informed Consent Should Be Obtained for Apnea Testing in the Determination of Death by Neurologic Criteria? Yes," and Thaddeus M. Pope, "Whether Informed Consent Should Be Obtained for Apnea Testing in the Determination of Death by Neurologic Criteria? No." *CHEST* 161, no. 5 (2022): 1143–49.

68. Joseph Eble, John Di Camillo, and Peter Colosi, "Catholics United on Brain Death and Organ Donation: A Call to Action." *Catholic Social Science Review* 29 (2024): 299–300.

diagnostic procedures for determining death that possess moral certainty. The authors stop short of calling for a moratorium on brain death practice in Catholic healthcare facilities; however, such would seem to be the logical inference of their statement.

Sulmasy and co-authors have suggested a middle ground, affirming brain death to be a real and cogent entity, and that "whole brain" death is the most philosophically defensible construct, but that a disjunction exists between the definition (irreversible cessation of all function of the entire brain) and the AAN 2023 Guidelines. In this respect, these latter guidelines are seriously flawed. He concludes, "Given the state of medical science today, the mismatch between the [whole brain] standard and current clinical testing can best be bridged by expanding clinical testing to include evidence for the loss of functions that typically map to the hypothalamic region of the brain. . . . Medicine owes patients and their families greater certitude, and not legal fictions, both in the articulation of care and in the determination that a human being is dead."[69]

To Sulmasy's point, and within the current and laudable milieu of transparency and integrity in the practice of medicine, we turn now to the most sensitive of points which should be a shared concern of both transplant and critical care communities, namely, the quality of trust that that must exist between those who would donate and those who would procure. As former U.S. Surgeon General Kenneth Moritsugu has stated to the National Academy of Sciences, "Trust underlies the entire organ and tissue donation and transplantation environment, and everyone must act to obtain, sustain, and nurture that trust among those directly and indirectly affected."[70]

Conclusion:

Almost simultaneously with *Life Magazine's* report in 1967, *Newsweek* issued a similar proclamation of Barnard's surgical triumph. Within this piece, an essay-insert entitled "When Are You Really Dead?" reported a comment by the aforementioned situation-ethicist Joseph Fletcher, whose work influenced the Harvard Report of 1968: "Speeding up the donor's death when it is 'positively' inevitable, may be justified if the transplant provides another human with valuable life."[71]

The question is thus raised--is there a life that is, however diminished and frail, *not* valuable? Today, and in our hearing, is Fletcher's prophecy fulfilled?

69. Daniel Sulmasy et al, "A Biophilosophical Approach to the Determination of Brain Death," *CHEST* 165, no. 4 (2024): 959–66.

70. Kenneth Moritsugu, testimony at the February 5, 2021 public workshop, cited in National Research Council. *Realizing the Promise of Equity in the Organ Transplantation System,* National Academies of Sciences, Engineering, and Medicine; Health and Medicine Division; Board on Health Care Services; Board on Health Sciences Policy; Committee on A Fairer and More Equitable, Cost-Effective, and Transparent System of Donor Organ Procurement, Allocation, and Distribution; Hackmann M, English RA, Kizer KW, eds. (Washington, D.C.: The National Academies Press, 2022), 61. https://doi.org/10.17226/26364.

71. *Newsweek,* December 18, 1967, 87.

Brain death in concept and practice remains quintessentially controversial. For the moment, whole-brain death/total brain failure is deeply ingrained in practice, law, and culture, and was recently re-endorsed by the latest AAN Guidelines in 2023. However, that these guidelines will soon enjoy "standard of care" standing should not cloak the abiding, palpable, and high-stakes controversies that have now come to the fore in public discourse, academic polemic, and at the bedside of the potential donor and recipient.

The unequivocal movement and history of those who founded and promulgated the practice of brain death for nearly sixty years have had the sole impetus of making transplantable organs available. And, we affirm, this is a worthy cause. The need for transplantable organs is great, with more than 125,000 patients awaiting a solid organ, and with more than 65,000 patients dying annually while waiting.[72] But Omelianchuk's "bright line" of expediency "to make the distinction between life and death in a timely and efficient manner"[73] shines a "bright light" on the true utilitarian nature of the brain death paradigm, and inevitably of the relativized worthiness (in their estimation) of that *other patient* who may become a donor, and whose prognosis may be abysmal, but *who is not yet known to be dead.* We see, over the course of brain death policy, polemic, and guideline, the move toward expediency in the endorsement of a partial brain criterion and a program of non-disclosure to families and surrogates which is imbedded not in brain death's definition but in the specific details of technique. To exclude the hypothalamus is to render the current iteration disingenuously opaque.

Similarly, within the sustained interdict of informed consent by Greer et al. is the tacit admission that transparency with surrogate decision makers risks missed opportunities for donation. Informed consent is only 'informed' if the proposed procedure is understood by the consenting surrogate. As we have seen, an abundance of literature suggests that a lay public does not grasp the issue, not least, I suggest, because of the inevitable emotional turmoil of this tragic hour of their lives. The dimension of transparency about brain death and the question of informed consent for brain death testing, it would seem, remain open to discussion and public engagement.

Edmund Pellegrino's teaching offers the necessary corrective. Pellegrino would have us be attentive to the plight of those in need of a vital organ transplant—but no less so to that equally sacred donor whose life also hangs in the balance, not only at the bedside, but also in policy, in published guidelines, and in law. Pellegrino taught that the *telos* of medicine is the "right and good healing action and decision"[74]

72. United Network for Organ Sharing (UNOS) Website www.unos.org. American Transplant Foundation, http://americantransplantfoundation.org/about-transplant/facts-and-myths/, accessed 02/21/2023.

73. Adam Omelianchuk et al, "Revise the Uniform Determination of Death Act to Align with the Law with Practice," 535.

74. Edmund Pellegrino, "The Internal Morality of Medicine," in *The Philosophy of Medicine Reborn: A Pellegrino Reader*, ed. H. Tristram Engelhardt and Fabrice Jotterand (Notre Dame, Ind.: University of Notre Dame Press, 2008), 66.

for each patient and in every clinical encounter. He further insisted that the profession of medicine itself is defined by an internal morality[75] which is to shepherd the care of the patient within the context of community and moral law. The right and good healing action and decision for the potential donor is to be equally weighted with that of the potential recipient. The lives of every figural descendant of both Louis Washkansky and Denise Darvall are at stake.

In the tradition of Edmund Pellegrino, the physician and the medical profession are to be about the "visiting" of this *other* patient, the potential donor, who is among the frail and voiceless "least of these" spoken of in Matthew's gospel.[76] In the visiting of these very souls—in hospital, in law, and in guideline—there is unspeakable reward.

75. Ibid., 68.
76. Matthew 25.

IV

Conscience and Truth-Telling

A. Conscience

25

Introduction: Conscience

Daniel P. Sulmasy, MD, PhD

CONSCIENCE IS A POORLY UNDERSTOOD CONCEPT IN ETHICS, especially in medical ethics. Questions about conscience, conscientious decision making, conscientious objection, the possibility of institutional conscience, tolerance, and pluralism have become quite controversial in medicine. Much of this controversy has been swept up in the so-called culture wars that have come to dominate the sociopolitical scene in twenty-first-century America.

In the following chapter, Pellegrino shares his wisdom about this subject, burning off the fog of emotional political posturing to help us see the underlying ethical issues in the clear light of his keen analytical mind. He begins by discussing the concept of conscience and the importance of conscience in living a moral life. He discusses some of the history of conscience as a moral concept.

He then discusses conscience as necessary for the good practice of medicine, since physicians will disagree with patients and each other about questions such as how much risk to undertake in a surgical procedure, whether to perform cosmetic plastic surgery, whether to screen for prostate cancer, or whether to prescribe aspirin for primary prevention of ischemic heart disease. Judgment is inescapable in medicine and what is permissible legally, or even urged by guidelines, does not bind professional judgment. He then discusses the tension that arises when patients and their physicians disagree about what to do. Each is an autonomous moral agent and, ideally, each is respectful of the conscience of the other.

Pellegrino argues that the idea of conscience as a moral phenomenon is not limited to individuals but that institutions undertake corporate moral commitments that bind them conscientiously as well. An Orthodox Jewish hospital, for instance, ought not to be forced to serve pork to its patients or be prohibited from performing male circumcisions.

Finally, noting that "value neutrality" is impossible, he discusses several possible ways of resolving conflicts of conscience in medicine. He prescribes a reasonable course, but, in doing so, casts a nervous gaze towards a future in which he predicts that secularizing forces will push towards the conscription of conscience, setting up a collision course with those whose views differ from those of the prevailing academic elite.

26

The Physician's Conscience, Conscience Clauses, and Religious Belief: A Catholic Perspective

Edmund D. Pellegrino, MD

Introduction

Conscientious persons strive to preserve moral integrity. This requires that their external behavior be congruent with their conscience's internal dictates about what they take to be morally right and feel compelled to do. In our morally diverse world, conscientious persons may come into conflict with each other and with society's moral values. Except for the amoral sociopath, conflicts of conscience are a regular feature of moral life. Even for extreme relativists, resolving these conflicts is a constant challenge.

Any society purporting to serve the good of its members is therefore obliged to protect the exercise of conscience and conscientious objection. However, this involves a serious dilemma for any pluralist, democratic, liberal, or constitutional state. On the one hand, such a society is committed to tolerance of religious diversity, freedom of individual choice, and "neutrality" with respect to religious belief. On the other hand, optimizing freedom of conscience for some individuals may often limit the legal rights, social entitlements, and moral beliefs of others.

This dilemma is most acute for health professionals who hold strong religious beliefs, some of which cannot be compromised in good conscience. Can conscience clauses protect Catholic and other religious health professionals' moral claims to freedom of the exercise of their conscience? To what extent can these legal measures secure rights of conscience in the face of a liberal, democratic, and secular society's commitments to moral relativism, personal freedom of choice, and an implicit social contract with its professionals? Is there some point at which religious believers are

---

Reprinted with permission from Edmund D. Pellegrino, "The Physician's Conscience, Conscience Clauses, and Religious Belief: A Catholic Perspective," *Fordham Urban Law Journal* 30, no. 1 (2002): 221–44.

morally compelled not simply to refrain from participation, but to dissent in the public arena using the processes of a democratic society to change public policy? This essay engages some of these issues in the specific case of Roman Catholic physicians whose religious beliefs are becoming progressively counter-cultural on the so-called human life issues.[1] Roman Catholic physicians serve as paradigm cases for all whose religious beliefs compel them to refuse to participate in certain acts, which are legal and even "required" in their societal roles.[2] Although this essay focuses on physicians, the same issues confront nurses, social workers, allied health workers, and all others who serve any function in our healthcare system. Similarly, although end-of-life issues will be used to illustrate particular conflicts of conscience, similar conflicts arise in other dimensions of modern health care, such as contraception, abortion, various types of assisted reproduction, sterilization, stem cell research, and cloning. This essay will discuss only the ethical dimensions of the conflicts while others at this conference with the requisite legal expertise will discuss the legal aspects of conscience clauses.

Good law should be based on good ethics; in other words, the rights and claims it protects should carry moral weight and justification. Yet, in resolving conflicts of conscience in secular societies, the complexity of the legal issues reflects the complexity of the ethical issues.[3] Often they are extremely difficult to dissect. This is significant because once the ethical issues are expressed in law, the debate may be reduced to instrumental and procedural details that cannot resolve underlying moral sources of controversy.

For this reason, much more debate is required before conscience and exemption clauses can be applied in ethically defensible ways. The existence of a statutory protection does not assure the exercise of freedom of conscience. This essay seeks to examine some of the ethical desiderata behind conscience clauses in the case of Roman Catholic physicians' conflicts of conscience. It does so under five headings: first, why conscientious objection is so important in our day; second, the moral grounding for freedom in the exercise of conscience; third, the components of the physician's conscience; fourth, specific conflicts of conscience for Catholic physicians and institutions; and fifth, competing models of conflict resolution.

I. Why Conscientious Objection Is a Problem

Convictions about the right and wrong conduct, both as a professional and as a person, form the physician's conscience. Conscientious physicians have always had to protect each domain from the demands of tyrants, law, custom, and professional colleagues. Each era has had its own challenges to the physician's conscience. In our

1. U.S. Catholic Conference, Congregation for the Doctrine of the Faith (CDF), Instruction on Respect for Human Life in its Origin, and on the Dignity of Procreation, *Donum Vitae* (1987) [hereinafter: *Donum Vitae*]; John Paul II, *Evangelium Vitae,* encyclical letter, 24 *Origins* 690, 691–700 (1995) [hereinafter *Evangelium Vitae*].

2. *Donum Vitae, supra* note 1, at 14–20, 35–39.

3. Mark R. Wicclair, "Conscientious Objection in Medicine," 14 *Bioethics* 205, 210 (2000).

own time, profound changes in both the physician-patient relationship and society's construction of the ends of medicine, as well as the secularization of American society, have conspired to the physician's claim to freedom of conscience.

Most powerful perhaps is the shift in the locus of decision making from the physician to the patient or her surrogate. Beginning in 1914,[4] extending through both the Karen Ann Quinlan cases[5] and related cases in the 1970s,[6] and the accompanying trend to micromanagement, the right to refuse care has rapidly metamorphosed into a right to demand and dictate the details of care. For some, the ends and goals of medicine are no longer defined solely by physicians, but by social convention or the demands of patients or their families.[7] On this view, the physician practices by virtue of a social contract, which grants her profession the privileges of freedom to practice in return for provision of those services that society requires or demands. What constitutes the practice of medicine is societally determined. In Oregon, for example, assisting suicide is defined as a normal part of the physician practice, whereas it is forbidden in other states.[8]

These trends are exacerbated by the deprofessionalization of medicine, which views health care as a commodity, and its delivery a matter of corporate enterprise, profit, and commercialization.[9] A managed care organization now monitors and controls physicians' decisions.[10] Corporate policy circumscribes the physician's judgments of conscience about the patients' welfare. Recent professional organizations are trying to recapture professional commitment, but it may not be possible given the fact that most physicians are now employees of corporate entities.[11]

In such a society, such profoundly religious issues as the morality of abortion, euthanasia, human cloning, and stem cell research are determined on grounds of

4. See *Schloendorf v. Soc'y of N.Y. Hosp.*, 105 N.E. 92, 93–94 (N.Y. 1914) (stating that a patient has the right to refuse treatment when doctors operate on a patient that explicitly stated that they did not want the operation).

5. See *In re Quinlan*, 355 A.2d 647, 671 (N.J. 1976) (installing father as guardian of comatose daughter "with full power to make decisions with regard to the identity of her treating physicians.").

6. See, e.g., *Rizzo v. United States*, 432 F. Supp. 356, 360 (E.D.N.Y. 1977) (granting patient a preliminary injunction preventing the Food and Drug Administration from prohibiting the importation of a drug not approved for his personal use).

7. Edmund D. Pellegrino, "The Goals of Medicine: How Are They to Be Defined?" in THE GOALS OF MEDICINE, 55, 58–60 (Daniel Callahan and Mark J. Hanson, eds., 1999) [hereinafter Pellegrino, *The Goals of Medicine*].

8. Compare OR. REV. STAT. § 127.630, 127.635, 127.640 (1996), with ARK. CODE ANN. § 5-10-106 (2001); CAL. PENAL CODE § 401 (1999).

9. Edmund D. Pellegrino, "The Commodification of Medical and Health Care: The Moral Consequences of a Paradigm Shift from a Professional to a Market Ethic," 24 J. MED. & PHIL. 243, 244–51 (1999) [hereinafter Pellegrino, "Commodification"].

10. Ibid., 253.

11. "Medical Professionalism in the New Millennium: A Physician Charter Project of the ABIM Foundation, ACP-ASIM Foundation and European Federation of Internal Medicine," 136 ANNALS INTERNAL MED. 243, 244–46 (2002), http://www.annals.org/issues/v136n3/toc.html.

utility, general consensus, or freedom of choice. In the secular philosophy, there is no other world beyond the immediate utopianism of a man-made heaven on earth. This vision determines secular society's decisions about what is permissible and what is not.

All of this is occurring against the recent historical experience of past and present totalitarian governments subverting the uses of medical knowledge to political and economic purposes. We need not recite again the way the Soviet Union distorted the Hippocratic Oath to make it serve the purposes of Communism,[12] the Nazi physicians' acquiescence in using their knowledge in the service of genocide,[13] or the participation of physicians as instruments of torture or terrorism by so many petty dictators and warlords.[14] The laws and social conventions of pathological societies justified all these violations of the ethics of medicine.

Today's societal context poses serious conflicts of conscience for all physicians, but especially for the religious physician. The teachings of the Roman Catholic Church on medical morals and human life issues go back half a millennium.[15] Its present positions on many crucial issues are distinctly and unapologetically ethically counter-cultural.[16] Many Jewish, Protestant, and Muslim physicians share some of the same beliefs and experience equivalent challenges to their moral integrity. Clearly, for all religiously oriented physicians, the question must be addressed—is it possible to maintain moral integrity and remain an active physician in a secular world? Secularists ask the same question, but with different expectations about what would be a morally defensible response.

It was against the background of these changes in the climate of American medicine and its practice that conscience clauses made their appearance. In 1973, when the United States Supreme Court removed the prohibition against abortion, a medical procedure was legalized, which at that time and since, was morally repugnant to many physicians and the public.[17] In recognition of these objections, the United States Congress passed legislation that exempted physicians and others from

12. Edmund D. Pellegrino, "Guarding the Integrity of Medical Ethics: Some Lessons from Soviet Russia," 273 J. AM. MED. ASS'N. 1622, 1622–23 (1995) [hereinafter Pellegrino, "Guarding the Integrity"].

13. Henry Friedlander, THE ORIGINS OF NAZI GENOCIDE: FROM EUTHANASIA TO THE FINAL SOLUTION 216–45 (1995).

14. See generally MAXWELL G. BLOCHE, URUGUAY'S MILITARY PHYSICIANS: COGS IN A SYSTEM OF STATE TERROR (1987).

15. David F. Kelly, THE EMERGENCE OF ROMAN CATHOLIC ETHICS IN NORTH AMERICA 25 (1979).

16. DONUM VITAE, *supra* note 1, at 12–13, 16–20, 23–25, 32–38; *Evangelium Vitae, supra* note 1, at 691–700, 709–16; Pope Paul VI, "Encyclical Letter of Pope Paul VI *Humanae Vitae:* To Our Dearest Sons and Brothers Health and Apostolic Blessing," in HUMANAE VITAE AND THE BISHOPS: THE ENCYCLICAL AND THE STATEMENTS OF THE NATIONAL HIERARCHIES 33, 37–44 (John Horgan, ed., 1972).

17. See generally *Roe v. Wade,* 410 U.S. 113, 152–53 (1973) (holding that abortion is a fundamental right guaranteed by the due process clause of the Fourteenth Amendment).

participation.[18] Most of the states[19] and other countries[20] also enacted exemption legislation, which allowed those who objected to abortion and a variety of other procedures to refrain from participation.

Several decades later, individual states recognized patients' legal right to execute advance directives through a living will or a durable power of attorney for health care.[21] The resulting statutes were designed to guarantee a patient's right to direct the manner and extent of end-of-life care when she had lost the capacity to make her own decisions to accept or refuse treatment.

The Americans with Disabilities Act reaffirmed this right and required hospitals to inquire on admission whether patients had executed an advance directive.[22] If they had, the hospital was bound to respect its requirements.[23] Similarly, in the case of abortion, Congress recognized that some physicians would have moral objections to participation, so they were exempt, provided they transferred care to another physician.[24] In both cases, abortion and advance directives, the moral claim to freedom of conscience was given legal status in "conscience clauses."

II. Moral Foundation for Conscientious Objection and Conscience Clauses

Freedom of conscience, however, is a moral right.[25] The first axiom of a moral life is to do good and avoid evil. This remains true whatever theory of right and wrong one may hold, whether one is a moral absolutist or relativist, a deontologist or utilitarian, or a communitarian or a social constructionist. Every moral system undertakes to determine what is right and wrong, good and evil, and desires that its worshippers act so that one may be done and the other avoided. This remains true whatever substitute term moralists may use for good and bad, and even if they deny their existence. "Values" is now the term in favor. Values are labeled "good" or "bad," morally "wrong" or "right." Values are replacing terms like principles, duties, and virtues; thereby, equating the normative with the subjective.

18. 42 U.S.C. § 300a–7(b) (1999).

19. Wicclair, *supra* note 3, at 207 (citing L.D. Wardle, "Protecting the Rights of Health Care Providers," 14 J. LEGAL MED. 177, 177–230 [1993]).

20. M.L. DiPietro, "Evoluzione Storica dell'istituto, dell'obiezione di coscienza," 6 ITALIA MEDICINA E MORALE 1093, 1093–151 (2001).

21. E.g., ARK. COD. ANN. § 5-10-106(d) (Michie 1987); LA. REV. STAT. ANN. § 40:2233 (West 2001); OKLA. STAT. ANN. Tit. 63, § 3101.4 (West 1997).

22. 42 U.S.C. § 1395cc(f); 28 C.F.R. § 35.130€(2) (2001); see also *Cruzan v. Mo. Dep't of Health,* 497 U.S. 261, 269–70 (1990) (inferring the necessity of inquiry into a patient's desire to refuse treatment as the natural corollary of the duty to obtain consent).

23. 42 U.S.C. § 1395cc(f)(1)(D); see generally *Cruzan,* 497 U.S. at 269–70 (discussing the right of a competent adult to leave express instructions declining consent to life sustaining medical treatment).

24. 42 U.S.C. § 238n.

25. Rev. Donald W. Wuerl, "The Bishop, Conscience and Moral Teaching," in CATHOLIC CONSCIENCE FOUNDATION AND FORMATION, 123, 127–28 (Russell E. Smith, ed., 1991).

All humans, ethicists included, possess an inner conviction of what is right and wrong and feel compelled to act in accord with that judgment. That inner conviction is the result of an act of practical reason applied to the moral status of an act performed in the past, or yet to be performed.[26] In the Catholic tradition, conscience is called "the law of our intellect" because it is a judgment of reason deduced from natural law.[27] In the Catholic tradition, and in other moral traditions as well, the judgments of conscience are morally binding, that is, they must be followed or the moral agent has acted immorally and accountably.[28] The act of conscience may be in error about the facts, it may reason erroneously, and it may misunderstand or misapply the moral precept it is using.[29] However, when it is convinced that it has seized upon the right thing to do, conscience impels the person to act in a certain way.[30] To ignore this "inner voice" is to induce guilt, remorse, and shame. Only the amoral sociopath escapes the grip of conscience.

Errors of conscience occur when an individual misidentifies the good.[31] The person who follows a wrongful conscience may or may not be morally culpable depending upon whether her ignorance of the good is willful or culpable.[32] Roman Catholics are bound to follow the dictates of conscience, but they are also responsible for the formation of a good conscience.[33] This requires serious education and reflection on the content of official Church teachings.[34] Nominal "Catholics," who firmly believe that fidelity to conscience dictates opposition to Church teachings on the issues of human life and sexuality, are arguably examples of wrongly formed conscience.[35] Their "consciences "compel them to oppose official (Magisterial) teaching, which, for Catholics, is a source of authoritative guidance for conscience formation.[36]

To act against the dictate of conscience is to act against natural law—that portion of divine law accessible to human reason. But, for Catholics, conscience is also "said to

26. Thomas Aquinas, "On Conscience. Disputed Questions on Truth 17, in THOMAS AQUINAS SELECTED WRITINGS, 217, 221–23 (Ralph McInerny, ed. and trans., 1998) [hereinafter AQUINAS, "On Conscience"].

27. Ibid. at 224–26; Thomas Aquinas, THE SUMMA THEOLOGICA OF ST. THOMAS AQUINAS. question 79, at art. 13 (Fathers of the English Dominican Province trans., 1920), available at http://www.newadvent.org/summa (last visited September 29, 2002) [hereinafter AQUINAS, SUMMA THEOLOGICA].

28. Thomas Aquinas, "On Conscience," *supra* note 26, at 228–33 (stating "[t]hat conscience binds means that when one does not follow it he incurs sin.").

29. Ibid. at 226–28. Aquinas claims that conscience is a science that never errs, but rather, any errors occur from the application of this science to "some special act." Ibid. at 226.

30. Wuerl, *supra* note 25, at 127–28.

31. Aquinas, "On Conscience," *supra* note 26, at 226–27.

32. See ibid. at 234–36 (claiming that not following an erroneous conscience would constitute a sin, but that actively pursuing this erroneous conscience is also a sin).

33. Rev. Benedict Ashley, "Elements of a Catholic Conscience," in CATHOLIC CONSCIENCE FOUNDATION AND FORMATION, *supra* note 25, at 39, 48–52.

34. Ibid. at 50–52.

35. Frances Kissling, "The Place for Individual Conscience," 27 J. MED. ETHICS ii24, ii25–27 (2001).

36. Ashley, *supra* note 33, at 50–52.

be of divine insertion, in the way in which all knowledge of the truth that is in us is said to be from God, by whom the knowledge of first principles has been placed in our nature."[37] Hence, for the Catholic, to then ignore, repress, or act against conscience for any reason is a violation of philosophical as well as theological ethics, an error in moral agency and a sin against God.[38] In an analogous way, similar gravity would attach to violations of conscience in all moral systems, religious or secular. Though the idea of a "human nature" is in disfavor today, the conclusion is inescapable, that there exists in all but the most morally obtuse, an operative conscience, a sense of moral compulsion to follow its dictates, and a perception of ethical disquiet in not doing so.

In every belief system, fidelity to conscience is closely identified with the preservation of personal moral integrity. To arrive at a conclusion that something must be done or avoided, and to act accordingly, is to exhibit the kind of person one is, and wants to be. That act provides evidence that the individual is the kind of person she says she is. Not acting in accordance with this conclusion is to incur the justifiable charge of hypocrisy. Often, to act against conscience is to violate personal identity so directly as to lead to severe psychosocial and emotional sequelae.

Therefore, conscience clauses are firmly rooted in what it is to be a human person morally, intellectually, and psychologically. Every individual, by virtue of being human, has a moral claim to the free exercise of conscience. The practical question for positive law in framing legal conscience clauses is how to protect this claim in morally pluralistic societies. How can conscience clauses assure the individual right to conscience while at the same time recognizing how widely the content of conscience can vary between and among individuals in their personal, social, and professional roles? Whose conscience is to prevail in the worlds of individual relations and public policy?

This is a question that confronts all individuals in every walk of life. It is of growing significance in the profession of medicine with respect to doctors' professional and personal beliefs and practices.[39] For Roman Catholics it is so crucial that Pope John Paul II has called on physicians to be conscientious objectors with respect to pro-abortion and pro-euthanasia legislation.[40]

37. Aquinas, "On Conscience," *supra* note 26, at 225.

38. Ibid., at 228–33; see also Wuerl, *supra* note 25, at 127–28.

39. Thomas Faunce, "Peri-Gravid Genetic Screening: The Spectre of Eugenics and Medical Conscientious Non-Compliance," 6 J.L. & MED. 147, 152–55, 161–66 (1999); Thomas May, Rights of Conscience in Healthcare, 27 Soc. THEORY & PRAC. 111, 111–21 (2001); Richard S. Myers, "On Laws: On the Need for a Federal Conscience Clause," 1 NAT'L CATH. BIOETHICS Q. 23, 23–25 (2001); Carl A. Osborne, "How Can We Practice Veterinary Medicine Conscientiously?" 215 J. AM. VETERINARY MED. ASS'N. 1238, 1238–39 (1999); Jing-Bao Nie, "Chinese' Conscientious Acceptance of the Birth Control Policy," 13 MED. HUMAN. REV. 82, 82–85 (1999) (reviewing CECILIA NATHANSEN MILWERTZ, ACCEPTING POPULATION CONTROL: URBAN CHINESE WOMEN AND THE ONE-CHILD FAMILY POLICY [1997]).

40. John Paul II, Doctors Protest Conscience Discrimination, the Holy Father's Address to the International Congress of Catholic Obstetricians and Gynecologists, 48 DOLENTIUM HOMINUM 65, 65–66 (2001) [hereinafter John Paul II, "Doctors Protest"]; John Paul II, "The Medical Doctor Should Respond as a Conscientious Objector to Legislation in Favor of the Crimes of Abortion and

III. The Physician's Conscience

Physicians, in the course of their work as healers, must form their consciences in two inseparable dimensions of their lives—the professional and the personal. Professional conscience concerns itself with two facets of the physician's daily work. First, the ethical propriety of her conduct qua physician with references to the moral duties of the physician-patient relationship. Second, is the moral obligation to practice "good" contemporary medicine, that is, medicine that is scientifically competent and humane. Personal conscience deals with the physician's own moral beliefs of a spiritual, philosophical, cultural, and ethnic nature. Both professional and personal conscience are owed protection.

A. Conscience in Physician-Patient Relationships

One set of obligations relates to such matters as the construct one puts on the physician-patient relationship, is it a contract, a covenant, a commodity transaction, a service relationship—or a vocation? How much respect for patient autonomy is obligatory? Do the patients' "values" override the physician's beliefs? What is a just allotment of scarce resources in a given case? How absolute is the obligation to preserve confidences? Should physicians have sexual relationships with patients? Is the patient's good primary, or is society's? Is the physician-patient relationship simply whatever is negotiable between them?

In the past, one might have assumed that a more or less general consensus existed among physicians on these issues even though there always were individual lapses in application. The Hippocratic ethic was the moral *lingua franca* of physicians across history and cultures.[41] Today, consensus on the precepts of the Hippocratic ethic has been seriously eroded. Individual physicians and their professional organizations now hold different positions on the moral status of their relationship with patients.[42] One cannot assume any longer a common formation of professional conscience, or a shared conception of ethical physician behavior. Physicians in our day may act in accord with their consciences with drastically different and even contradictory presuppositions about what is morally permissible in their relations with patients.

Euthanasia," 15 DOLENTIUM HOMINUM 133, 133–35 (2000) [hereinafter John Paul II, "The Medical Doctor"].

41. See generally Edmund D. Pellegrino, "Bioethics at Century's Turn: Can Normative Ethics Be Retrieved?" 25 J. MED. & PHIL. 655, 655 (2000) (discussing the growth of bioethics from traditional medical ethics) [hereinafter Pellegrino, "Bioethics at Century's Turn"].

42. See, e.g., *Mich. V. Kevorkian,* 639 N.W.2d 291, 311–12 (Mich. Ct. App. 2001) (demonstrating the spectrum of physicians' interpretations of the Hippocratic Oath's medical ethics according to their conscience).

B. Conscience in the Practice of Competent Medicine

A lesser, but nonetheless significant element in the physician's conscience, is her perception of what constitutes "good" medicine. This is a subtle combination of personal and professional morality. Its focus is on medicine as a *tekné*, an art based in skills and knowledge of how to heal well.[43] While there are skeptics who would argue that we cannot define what "good" medicine is, there is the fact that we all distinguish between doctors whose judgments and skills we trust, and those we do not. There are also the inescapable differences among physicians in terms of morbidity, mortality, and diagnostic accuracy. Doctors may also differ, as a matter of conscience, in their opinions about the worth of new and old procedures, consultants they do and do not trust, the reliability of clinical data, the use of alternative or complementary medicine, and the role of other health team members. These can be matters of conscience for the physician who wants to be a "good" clinician, surgeon, healer, or counselor; the better to serve the best interests of the patient.

We will consider later whether or not this heuristic dissection of professional and personal ethics is sustainable in actuality or in light of conscience clauses.[44] First, let us turn to a few illustrations of the way these three sectors of the physician's conscience can conflict in her relationship with patients and then with the demands of our secular democratic societies.

C. Conscience in Personal Moral Beliefs

In addition to her perception of professional ethics, each physician brings to her relationship with the patient a personal set of moral beliefs. She bases these moral beliefs in religious affiliation, personal preference, or moral reflection. Here we confront such crucial issues as the licitness of abortion, euthanasia, assisted suicide, in vitro fertilization, and stem cell research—the whole Pandora's Box of "human life" issues, emerging from our unprecedented control of every phase of human life.

These issues center on how we value human life itself, its purposes, quality, destiny, and utility. Conflicts of belief in this realm are more profound and deeply felt in one's conscience than other issues of professional behavior with patients. For religious individuals of many persuasions, these issues bear directly on their personal spiritual destinies and are, therefore, least subject to compromise.

In the last fifty years, secularism has come to dominate much of medical ethics, despite the fact that most Americans personally hold religious beliefs.[45] Secularism is a response to the plurality of moral and religious beliefs in our polyglot society in

43. *Tekné* is an ancient Greek term that characterizes actions and professions as an art or craft rather than just a normal action. OXFORD CONCISE ENGLISH DICTIONARY 1471 (10th ed., 1999) (origin of the prefix "techno-").

44. See *infra* Part VII.

45. See Frank Shakespeare, "A View from Administration and Government," in CATHOLIC CONSCIENCE FOUNDATION AND FORMATION, *supra* note 25, at 259–64 (discussing the evolution of our society into a secular state).

which there is wide disagreement on what a good conscience should dictate. Since no one moral system or religious set of beliefs is universally accepted, society reasons that none can, or should be dominant. All should be free to express themselves, and each should respect the other. So goes the credo of political liberalism. On this view, decisions that must be made as a matter of public policy in areas such as abortion, euthanasia, and stem cell research, should be made democratically, universally, and equally binding. Conscience or exemption clauses presumably are devised to protect the freedom of the dissenter. Without dissecting its merits or demerits, this liberal democratic policy has functioned to avoid civil strife. However, the recent erosion of the number of beliefs held in common, and the increasingly varied demography of our nation, has magnified the complexity and depth of our differences about what is morally right and wrong. The secular solution of moral or value neutrality has generated genuine conflicts of conscience.

Religious exemption laws and conscience clauses have appeared as a device to protect the physician's conscience. Their inadequacy, however, is becoming manifest. Lawmakers have currently drafted some of those clauses so narrowly that they disqualify most religious institutions from exemption, especially if they are involved in providing assistance and social services irrespective of the religious persuasions of those they help.[46] On these grounds, Catholic institutions are simply not religious enough unless they help the sick and needy for distinctly religious purposes. If they are to be classified as "religious," Catholic institutions must serve only Catholics.[47] On the other hand, if they do so, they are disqualified since they would then discriminate against, and take advantage of, the vulnerable, sick, and poor.

Moreover, secular morality, which supposedly tolerates differences, does so only within a narrow range of so-called values that are supposedly "free" of moral or religious taint. But secular morality is itself an orthodoxy. Its "values" are based in democratic procedures, personal preference as the basis for moral choice, commitment to a free market economy, the commodification of health care, and an eschewal of religious belief.[48] To deviate from this notion of moral "neutrality" in public policy is to be "undemocratic," prejudiced, and intolerably sectarian.

This is not the place to challenge these dicta of secularism as a ruling orthodoxy, but to spell out in more detail their implications for Catholic Christian physicians' freedom of conscience. Again, the Catholic physician is the focus, but the same conflicts would apply to any other religious or moral system with a clear and unequivocal set of precepts giving substance to the conscience of its followers. The conflicts can be divided into three groups: first, between patients and physicians;[49]

46. Myers, *supra* note 39, at 23–26; Msgr. Dennis Schnurr, "Mandating Employer Coverage of Contraceptives: Protecting Conscientious Objection," 30 ORIGINS 161, 163–64 (2000).

47. *Cath. Charities of Sacramento, Inc. v. Super. Ct.*, 109 Cal. Rptr. 2d 176, 183 (Cal. Ct. App. 2001), petition for review granted, 31 Cal. Rptr. 2d. 258 (Cal. 2001); "Religious Refusals and Reproductive Rights," REPROD. RTS. UPDATE, 2002, at 8, http://www.aclu.org/issues/reproduct/refusal-report.pdf.

48. Pellegrino, "Bioethics at Century's Turn," *supra* note 41, at 655, 663–64.

49. See *infra* Part IV.A.

second, between physicians and society;[50] and third, between Catholic institutions and society.[51]

IV. Conflicts of Conscience for Catholic Physicians

A. Conflicts Between Patients and Physicians

Physicians and patients may differ sharply in what their consciences tell them about the moral licitness of assisted suicide, euthanasia, the dignity and worth of human life, the relative importance of quality of life, age, or economics as criteria for withholding or withdrawing life-sustaining treatments, terminal sedation, or whether death of the whole brain and partial-brain death are both equivalent to the death of the patient. The same is true of cloning, stem cell research, and fetal tissues transplantation.

Some would argue that the principle of patient autonomy should prevail in such conflicts, and that the physician, irrespective of her own beliefs, should provide whatever secular social convention legally allows.[52] On this view, medicine is bound by a social contract to provide the services patients or society deem worthwhile. This obligation is in return for society permitting them to set their own standards of education and practice. In addition, they would claim that it is a failure of the principle of beneficence not to do what the patient believes to be in her best interests. Still others might reduce the argument to one of commutative justice, holding that the patient is entitled to the same care available to other patients whose doctors do not suffer from the Catholic doctor's scruples against abortion, euthanasia, or other human life issues.

At present, Catholic physicians may withdraw from the care of patients in these circumstances. However, one wonders how long this exemption will survive, as end-of-life and reproductive decisions become so much an individual prerogative that the ethical standard is no longer a determination of what is morally right, but rather, of what can be negotiated to resolve conflicts.[53] One can foresee the day when patients may gain legal rights to demand a full range of death "services" from every licensed physician just as many today feel entitled to a full range of reproductive "services."

Already we hear ethicists suggesting that physicians must separate their personal moral beliefs from their professional lives if they wish to practice in a secular society and remain licensed as fully functioning physicians.[54] If universal health care were to

50. See *infra* Part IV.B.

51. See *infra* Part IV.C.

52. Cf. May, *supra* note 39, at 111–12 (suggesting that there are certain situations in which a healthcare professional could acknowledge her moral concerns with particular treatment choices).

53. Pellegrino, *Bioethics at Century's Turn, supra* note 41, at 656–57.

54. See Jeffrey Blustein and Alan R. Fleishman, "The Pro-Life Maternal-Fetal Medicine Physician: A Problem of Integrity," HASTINGS CENTER REP., Jan.–Feb. 1995, at 22–26 (discussing the dilemma faced by physicians who must perform abortions and maintain their own integrity); Franklin G. Miller and Howard Brody, "Professional Integrity and Physician-Assisted Death," HASTINGS

be instituted and "death care," as well as birth and reproductive care, were to be entitlements, would Catholic physicians be given only limited practice licenses?

This same question could logically be raised, for example, with respect to stem cell therapy. Using stem cells derived from the killing of human embryos is morally offensive to Catholic physicians. If the therapeutic potentials of stem cell research, genetic engineering, and cloning are actualized, could a conscience clause protect Catholic physicians in secular hospitals or in managed care organizations? Would they not be legally required to provide a full range of services despite moral objection? In an HMO, would the commercial gains of services that patients demand trump a conscience clause? Would religious physicians be hired in the first place?

B. Conflicts with Societal Mores

It has already been seriously suggested that Catholic physicians should not become maternal-child specialists, since they cannot, in good conscience, provide the whole range of reproductive, pregnancy, and neonatal "services," such as, selective abortion for genetic defects, or late-term abortion.[55] The logical next step of such proposals is to withhold specialty certification for maternal medicine from Catholic physicians and others who oppose provision of any such services, which are legal. Should anyone who wishes to be a physician be permitted to narrow the range of services to her patients on the basis of moral and religious reservations? A medical education is a socially sanctioned process in which students learn by doing.[56] Some would argue that the physician's social contract requires her to provide what she has learned in accord with whatever society needs because society granted the privileges of a medical education in the first place.

These are not imaginary scenarios. Catholic and other religiously committed applicants to medical school have been asked about their views on the issues of abortion, euthanasia, ending life support, various reproductive technologies, and stem cell research.[57] Evidence that responses consistent with Catholic teaching have militated against admission is hard to come by for obvious legal reasons. Therefore, we do not know how heavily medical schools weigh the "Catholic" responses against a candidate. The fact that they asked the questions in the first place, is sufficient cause for worry given the dominance of the secular viewpoint in academic circles today.

Based on personal experience on admissions committees, there is more than a mere suspicion that "conservative" Catholics, Orthodox Jews, and fundamentalist Christians may be looked upon with disfavor. Much depends on the lottery of interviewers one encounters by chance. Those who hold certain religious beliefs, it is

---

CENTER REP, May–June 1995, at 8–17 (analyzing the relationship between professional integrity and physician-assisted suicide).

55. See Blustein and Fleishman, *supra* note 54, at 25.

56. Edmund D. Pellegrino, "Philosophy and Ethics of Medical Education," in 2 THE ENCYCLOPEDIA OF BIOETHICS, 860, 863–69 (Warren T. Reich, ed., 2nd ed., 1978).

57. Ibid.

argued, cannot provide all the services patients have a right to expect. Whether we shall ever come to the point at which religious believers who insist on following their consciences will be barred from certain specialties or from medicine, itself, is, therefore, a source of more than imaginary concern.

C. Conflicts Between Catholic Institutions and Society

Organizational ethics is the newest addition to the broadening spectrum of ethical issues being subsumed under the term "bioethics."[58] While it may include business ethics, it covers much more, and already embraces such varied aspects of institutional behavior as relationships with employees, advertising, community commitments, quality of care for the poor and uninsured, and mergers with other institutions.

Organizational ethics is a systematic examination of the morality of collective actions in human institutions dedicated to some specific purposes in society.[59] The ethical "code" or commitment of a specific institution is now customarily expressed in its mission statement. This is in a way the "conscience" of the institution.[60] All who work in that institution are in some way accountable for adherence to the organizational mission, which is, in effect, a promise by the institution to behave in a particular way. Catholic institutions in America have for a long time had specific ethical directives that define them as Catholic hospitals.[61] They are also committed to a charitable, even preferential, treatment for the sick and the poor.[62] Catholic hospitals can properly be considered to have a definable institutional "conscience," one, which given the content of the Catholic moral tradition, could and does come into conflict with secular society and its "values."[63]

The ethical content of the institutional conscience of particular hospitals is well known with respect to sterilization, abortion, euthanasia, assisted suicide, contraception, and cooperation through mergers with other institutions that accept these practices.[64] Fidelity to these prohibitions is not negotiable.[65] It applies to all who practice in these hospitals regardless of their personal beliefs. Catholic hospitals, like Catholic physicians, do not have the option of being "value-neutral" or of separating religious from professional ethical precepts.

There is growing evidence that public funding for Catholic health care and social service institutions may be in jeopardy if these institutions do not provide the

58. Susan Dorr Goold, "Trust and the Ethics of Health Care Institutions," HASTINGS CENTER REP., November–December 2001, at 26, 27–28.

59. Ibid. at 32.

60. Ibid. at 28.

61. U.S. CONFERENCE OF CATHOLIC BISHOPS, ETHICAL DIRECTIVES FOR CATHOLIC HEALTH CARE INSTITUTIONS 9–11, 13–16, 18–22, 25–28, 30–33, 36–37 (4th ed., 2001), http://www.nccbuscc.org/bishops/directives.htm.

62. Ibid. at 8–11.

63. Ibid. at 12–16, 23–33.

64. Ibid. at 8–16, 23–37.

65. Ibid.

"full range" of reproductive services. For example, the District of Colombia City Council recently passed a bill mandating that all employers in the city had to provide coverage for contraceptives in their prescription coverage plans.[66] The move to include a conscience clause to exempt Catholic institutions was rejected vitriolically by one member of the council.[67] Fortunately, the mayor, a Catholic himself, gave the bill a pocket veto.[68]

This sort of challenge to institutional conscience is certain to return in one way or another. Some who oppose Catholic moral teaching vehemently and frankly admit to wanting to eliminate the Catholic healthcare system and at the least deny access to public funds.[69] The more moderate alternative is to interpret exemption clauses so narrowly that Catholic institutions cannot be classified as "religious" institutions since they treat all regardless of belief and provide much more than religious services.[70]

The challenge to institutional conscience promises to grow in severity as new, morally questionable therapeutic procedures emerge from the laboratories and research centers. There is a genuine probability that stem cell research of the kind that depends on the death of embryos, human cloning for therapeutic purposes, or cross-species genetic manipulation will eventually be allowed.[71] Should these measures become clinically effective, the public demand for their availability will increase the pressure for conformity by all hospitals regardless of religious affiliation.

Even if Catholic hospitals are allowed the protection of exemption clauses there is still the more subtle threat to the conscience, stemming from cooperation with secular institutions through mergers. These mergers may be dictated by the need for economic survival, but such survival cannot be bought at the cost of even material cooperation of a direct kind with institutions that violate the established ethical directives for Catholic healthcare institutions. Already there is concern that mergers already in existence may involve Catholic hospitals too closely with activities that are morally objectionable.[72] Can these Faustian bargains survive closer moral scrutiny?

Catholic moral tradition contains a carefully nuanced set of conditions under which cooperation with those individuals or institutions that do not share Catholic moral beliefs may be licit.[73] Considerable controversy has already arisen as to whether the interpretation of these conditions in some mergers has been too lax. This is the case even where the non-Catholic partner promises to abide by the ethical directives of the Catholic bishops.[74] Commingling of funds, administrative entangle-

66. Schnurr, *supra* note 46, at 161–63.

67. Myers, *supra* note 39, at 23–26; Schnurr, *supra* note 46, at 164.

68. Schnurr, *supra* note 46, at 161–63.

69. Myers, supra note 39, at 23–26.

70. Ibid.

71. Ibid.

72. U.S. CONFERENCE OF CATHOLIC BISHOPS, *supra* note 61, at 34–37.

73. See generally ORVILLE N. GRIESE, CATHOLIC IDENTITY IN HEALTH CARE: PRINCIPLES AND PRACTICE 373-416 (1987).

74. U.S. CONFERENCE OF CATHOLIC BISHOPS, *supra* note 61, at 34–37.

ments, and other interminglings characteristic of today's complex institutional relationships raise serious questions of illicit cooperation in seemingly "safe" mergers.[75] Exemption and conscience clauses in these mergers or in the relationships with public policy may not be sufficient to permit financial viability for Catholic healthcare systems. Much depends on how well Catholic institutions can maintain their moral integrity and institutional conscience. Catholic healthcare institutions constitute a very significant sector of service for Americans, Catholic and non-Catholic.[76] If these institutions do not survive financially, the loss to the general public will be great. If they survive only by a loss of their institutional Catholic conscience, something even more fundamental will be lost, not just for Catholics, but for that sector of the American public, who, for their own moral integrity, cannot accept the dictates of a secular order.

V. Conflicting Models of Conflict Resolution

How, in a morally pluralist society, can the moral claim to freedom of conscience owed to every person as a human being be sustained? Specifically, how can the Catholic physician or institution sustain freedom of conscience in a secular world whose culture is areligious, if not antireligious? What role might conscience clauses play? Several ways out of the dilemma have been suggested, none of them entirely satisfactory. They include dissociation of the moral and professional life, abandonment of medicine as a profession, or maintaining moral integrity with judicious dissent.

It would be in keeping with secular orthodoxy to allow Catholic physicians and hospitals freedom of conscience, but to limit its overt exercise. Catholic physicians could have the right to participate in medicine as a profession as long as they would do what is allowable by law (for example euthanasia in Oregon, abortion everywhere, sterilization, etc.). The only thing Catholics would need to do is to provide whatever services are defined by social convention as legal medical practice. On this view, all physicians, Catholic and others, whose moral beliefs are at odds with a secular society simply need to take a "value free" stance. In this way, the autonomy of the patient is preserved and the doctor does not "impose" her beliefs.[77] This is the strong version of value neutrality as the litmus test for medical licensure or certification.

In a weaker version, Catholic physicians could be granted the right to participate in medicine as a profession as long as they would agree to what is allowable by law or defined as part of medical practice by the rest of the profession. Catholics, for example, could object to euthanasia in Oregon, sterilization, abortion, reproductive technologies, and stem cell research. They could refuse personally to participate. But,

75. Ibid.

76. See ibid. at 34–36 (discussing how Catholic healthcare institutions are forging partnerships and ventures with many other healthcare providers).

77. Blustein and Fleishman, *supra* note 54, at 22–26; Miller and Brody, *supra* note 54, at 8–17; Edmund D. Pellegrino, "Commentary: Value Neutrality, Moral Integrity, and the Physician," 28 J.L. MED. & ETHICs 78, 78–80 (2000) [hereinafter Pellegrino, "Commentary"].

in practice they would have to be "value neutral" if they entered specialties that require acts to which they had moral objection.

At the very least, they would have to commit themselves to arrange for referral to a physician they know would do what the patient wished and assist the patient in every way to achieve her purposes. On a stronger version of this form of accommodation, Catholic physicians would be compelled to make a clear choice, either fulfill the conditions of the social contract and provide what is legal or socially acceptable or drop out of any specialty which required services of which they took moral exception.

In its strongest form, this model would logically exclude from medical practice, and eventually the study of medicine, all who did not see their social roles as conforming to all the services society felt necessary and required of physicians. Needless to say, the accommodation such a compromise requires, even in its mildest form, would be morally objectionable for several reasons.

It assumes that the Catholic physician and others who hold firm moral beliefs can separate their professional and personal lives when this means cooperation with what is morally objectionable. For a physician with deep religious commitments, a "value free" stance on certain issues is simply unthinkable.[78] Certain matters are so clearly prohibited as inherently wrong that there is no possibility of compromise without compromise of moral integrity and danger to one's spiritual well-being.

For Catholics, Orthodox Jews, and Muslims, the teachings of the Gospel, Torah, or Koran take precedence in their lives and indeed inspire their healing vocations. For these major religions, healing the sick is ultimately a religious act, and it comes ultimately from God.[79] To practice medicine that contravenes religious teaching would be to subvert conscience to secular society and its "values," to act hypocritically, and to violate moral integrity intolerably.

For Catholics this would also apply to the secular demand that those who must refrain from certain practices must refer to physicians who will provide the disputed treatment or procedure which would also be intolerable. To cooperate in an act which is regarded as inherently morally wrong, such as arranging for an abortion or assisted suicide, is to be a moral accomplice.[80] Respectfully, courteously, but definitively the religious physician must inform the patient of her objection while promising to care for the patient until transfer or referral can be arranged by the patient, family, or social services.

Obviously, the patient cannot be abandoned, legally or morally, and must be cared for until a transfer has been effected. The doctrine of cooperation does not forbid transferring information, findings, or records to another physician or hospital.[81] Indeed, this is required in the interests of patient care. What is illicit is active cooperation in finding a physician who will provide the morally objectionable service.

78. Pellegrino, "Commentary," *supra* note 77, at 78–80.

79. 38 Ecclesiasticus (Sirach) (Jerusalem Bible).

80. GRIESE, *supra* note 73, at 386–90.

81. Ibid.

The requirement of a secular society that physicians practice "value neutrality," is impossible to achieve. First, it is a psychological schism that violates the integrity of the person as a unity of body, soul, and psyche. What it amounts to is the elevation of secularism to the level of a social orthodoxy; thereby, violating one of the major tenets of secularism itself—that no ideology would have preference over any other.[82] It also violates a prized precept of the secular, democratic, constitutional social order by discriminating against a significant segment of the population, and the physicians who share certain religious beliefs.

The difficulty of a meaningful compromise between the secular orthodoxy and religious belief is illustrated in those pragmatic attempts to find a way to respect moral integrity and the right of conscientious objection. For example, Wicclair would allow for conscientious objection as long as it corresponded to "one or more core values of medicine."[83] He would use congruence with these core values as the moral test for an acceptable claim to conscientious objection.[84]

Wicclair offers a guide for assigning moral weight within recognized medical norms. For example, he gives more weight to preventing death than protecting confidentiality.[85] He takes his cue here from the integrity of the profession, rather than the integrity of the physician as a moral person.[86] In one of his examples, he clearly states that more moral weight should be given to a physician's request to preserve moral integrity as a physician than to her "moral integrity as an Orthodox Jew who happens to be a physician."[87]

In his guidelines, Wicclair clearly makes religious belief subservient to professional medical belief about what is right and wrong.[88] Its effect is to require the kind of value and belief dichotomy, which is incompatible with moral integrity for a true religious person. The moral values of religious persons transcend the "values" of the profession—especially now that those values have changed so drastically. Where there might have been concurrence in the past between medically held and religiously held beliefs, that congruence has been seriously eroded today.

Indeed, consensus on the moral values of medicine is being progressively reduced to competence, refraining from harm, and the protection of confidentiality. The recent set of commitments advanced by the American Board of Internal Medicine, the American Society for Internal Medicine, and the European Federation

82. See also May, *supra* note 39, at 111 (discussing the conflict that arises when a physician's values and conscience conflict with a patient's values and autonomy rights, and noting that society increasingly and unfairly pressures physicians to disregard their personal consciences in their professional roles).

83. Wicclair, *supra* note 3, at 217.

84. Ibid.

85. Ibid. at 223.

86. Ibid. at 224.

87. Ibid. at 225.

88. Ibid.

attempts to recover the idea of professionalization.[89] However, it omits the prohibitions against abortion and euthanasia, the precepts that are most significant for many religious physicians and especially for Roman Catholics. The "moral integrity" of the profession is thus judged to be morally insufficient to justify overriding the physician's conscience.

The moral authority of professional codes does not derive from their acceptance by the profession or social convention.[90] Rather, the ethics of medicine is grounded in something more fundamental, namely the ethical obligation peculiar to what it means to be ill, to be healed, and to offer oneself as a healer.[91]

Respecting a physician's conscience claims, however, does not mean that the physician is empowered to override the patient's morally valid claim to self-determination. Both the physician and the patient as human beings are entitled to respect for their personal autonomy. Neither one is empowered to override the other. The protection of freedom of conscience is owed to both.

Therefore, patients have an uncontested moral right to informed consent and informed refusal.[92] Wicclair and May spend considerable time defending the patient's moral right to refuse treatment, prolong their life, request palliative care, or "reproductive freedom." This is not the issue here. Conscientious objection implies the physician's right not to participate in what she thinks morally wrong, even if the patient demands it. It does not presume the right to impose her will or conception of the good on the patient.

Therefore, May is correct in stating that, "[r]ights of conscience in health care must be exercised in the context of patients' rights to informed consent."[93] This does not at all imply that we should or must acknowledge limits on the physician's rights of conscience.[94] May agrees that patients do not have a right to demand "anything" they take to be beneficial.[95] Clearly, the patient's moral and legal right to self-determination has limits, even in May's view.[96]

Both May and Wicclair, but especially May, spend much time on examples of conflict in choice of treatment and somewhat miss the point of religious objections, which is not whether a religious believer may impose her beliefs on a patient, but

89. "Medical Professionalism in the New Millennium: A Physician Charter," 136 ANNALS INTERNAL MED. 243, 243–44 (2002).

90. Edmund D. Pellegrino, "Professional Codes," in METHODS IN MEDICAL ETHICS 70, 74–76 (Jeremy Sugarman and Daniel Sulmasy, eds., 2001).

91. Ibid. at 78–79.

92. Wicclair, *supra* note 3, at 208.

93. May, *supra* note 39, at 127.

94. Edmund D. Pellegrino, "Bioethics as an Interdisciplinary Enterprise: Where Does Ethics Fit in the Mosaic of Disciplines?" in *PHILOSOPHY OF MEDICINE AND BIOETHICS: A TWENTY YEAR RETROSPECTIVE & CRITICAL APPRAISAL,* 1–23 (Ronald A. Carson and Chester R. Burns, eds., 1997).

95. May, *supra* note 39, at 127.

96. Ibid.

rather whether she has the moral right to refuse to be an accomplice in an act her religion teaches is wrong.[97]

The only ethically viable course for the religious physician is to maintain fidelity to moral integrity and the dictates of conscience while practicing in a secular world. Catholic physicians and institutions have the same moral claim to exercise of conscience, as all other humans, even when the fruit of conscience is refusal and even resistance to accommodation of secular beliefs or the changing beliefs of their professional colleagues. This moral claim entails the right and obligation to use the methods available in a democratic society to protest morally objectionable practices by persuasion, judicious political action, and public debate, particularly in the most egregious situations.

For such a position to be tenable, Catholic physicians must make their positions publicly known, as in the case with the *Ethical and Religious Directives for Catholic Health Care Institutions*. Individual physicians should prepare a leaflet outlining what they can, and cannot, in good conscience do.[98] Patients should know in advance of a crisis that what they desire and believe to be morally acceptable may not be acceptable to the physicians they may be engaging.

Such advance knowledge will not be possible in emergencies or remote areas where the choice of physicians is limited. Even under these circumstances, the Catholic physician cannot violate her conscience to provide a morally objectionable procedure or treatment. Physicians must know their own belief system well enough to recognize where compromise is possible without loss of moral integrity and where it is not. Parenthetically, the Catholic physician is under serious obligation to know the content of her own faith, so that she does not impose hardship on the patient when alternative routes are morally permissible. Sadly, this is not always the case.

The dissenting physician must always treat her patient with respect, avoid moralizing condemnations, and explain the reasons for her moral objections. She must also be aware that every matter of conscience is not of equal gravity. Choosing when to take a morally dissenting stand is crucial if one's exercise of conscience is to be valid and respected.

The same prescriptions and proscriptions are applicable to the institutional conscience of Catholic healthcare institutions. They cannot compromise on fundamental Catholic moral teachings even if resistance might lead to their extinction. Total extinction is not likely, however, the withdrawal of public funds will probably restrict the number of persons in the community, Catholic or non-Catholic, that can be served by institutions faithful to their religious commitments. Although morally illicit, Catholic hospitals may also "cooperate" more fully with the secular mores.

Conscience clauses for physicians probably have a limited value, although they ought to be sought whenever possible. The likelihood, given the current societal mores, is that conscience clauses will be denied or progressively applied so narrowly

97. Ibid.; Wicclair, *supra* note 3, at 205–11.

98. See *supra* notes 61–65 and accompanying text.

as to be self-defeating. At the least, they provide legal limits that in a democratic society should protect dissenting physicians and institutions from the grosser forms of ostracism.

Catholic institutions will probably find greater difficulty obtaining exemption clauses, especially if they accept public funds or purport to serve community needs. Survival may require formation of a Catholic healthcare system nationally. Mergers with non-Catholic institutions, except those that share Catholic perspectives on the human life issues, will raise an increasing number of questions about cooperation. Given the size, geographic extent, and resources, a Catholic healthcare system faithful to magisterial teachings is not an impossibility. Much would be lost were a secular society's dissonances to require the dissolution or "ghettoization" of Catholic health care.

All the societal and political forces of our day are converging on an actualization of the secular state. As medical technology endows humans with ever greater power over the reproduction, genetic endowment, and dying of our species, crises of conscience will surely increase for those who hold religious beliefs about human life, its creation, and ending. In democratic societies, there is a commitment to protection of the right to hold and exercise individual and institutional conscience.

Conscience clauses are straws in the wind telling all of us that public policy and individual conscience on some of the most important matters of human life may be on a collision course. How individual physicians and institutions preserve their moral integrity in such a socio-political milieu is a matter of significance for both secularists and believers.

Conscience clauses will help at least to establish a right to dissent. However, the conditions under which they will be applicable and their effectiveness are very much at issue. It will be a stringent test both of democracy and religious beliefs to see how these conflicts will be resolved.

B. Truth-Telling and Confidentiality

27

Truth-Telling and *Prima Facie* Principles

David G. Miller, PhD

IN THE ESSAY BELOW, Dr. Pellegrino examines some interesting implications of a nuanced examination of the value of truth-telling. Ordinarily, truth-telling to the patient is understood as an essential element of respect for autonomy insofar as a patient needs true information to be able to make informed decisions about treatment. But some cultures—and some individuals—believe either that decision making is best approached as a shared activity or that the way in which information should be shared with a patient is best accomplished in an oblique or indirect way. Of course, a patient may request that a clinician communicate with the patient's surrogate or family directly and allow the surrogates to communicate with the patient. Pellegrino's examination of the importance of truth-telling leads one to wonder whether respect for a patient's autonomy is perhaps too narrow a conception of what it is we truly want to respect in a person. Pellegrino does not doubt or diminish the value of truth-telling; he does suggest that knowing how to respect a person requires more than merely telling the person the truth. It entails some degree of familiarity with the person's self-understanding, worldview, values, and relationships.

28

Is Truth-Telling to the Patient a Cultural Artifact?

Edmund D. Pellegrino, MD

IN THIS ISSUE OF *JAMA*, Antonella Surbone, MD, describes her dilemma in trying to transfer the ideals of medical ethics she learned in the United States to her native Italy.[1] From her experience in the United States, she found that truth-telling and respect for autonomy have become virtual moral absolutes. On the other hand, in Italy, families and physicians often shield patients from painful truths and difficult decisions. As Dr. Surbone points out, what is beneficent in one country may seem maleficent in another country.

This contrast in moral perspectives, of course, is not unique to the differences between Italy and North America. It has become a worldwide problem as newer models of medical ethics nurtured in the individualistic soil of North America are introduced to other countries with different moral traditions.[2] But similar contrasts may exist within a country, for example, between northern and southern Italy, or between the multiple ethnic groups in the United States.[3] Everywhere, as cultural groups attain freedom of expression, physicians and patients must relate with each other across ethical barriers especially with respect to the importance of autonomy.

These contrasts raise some provocative questions. Is medical ethics a cultural artifact such that a universal medical ethic is not viable? How should physicians who are dedicated to the best interests of their patients conduct therapeutic relationships with patients whose cultural values differ materially from their own?

Reprinted with permission from Edmund D. Pellegrino: "Is Truth Telling to the Patient a Cultural Artifact?" *Journal of the American Medical Association* 268, no. 13 (1992): 1734–35.

1. A. Surbone, "Truth Telling to the Patient," *JAMA* 268 (1992): 1661–62.

2. E.D. Pellegrino, P. Mazzarella, and P. Corsi, eds., *Transcultural Dimensions in Medical Ethics* (Frederick, Md.: University Publishing Group, 1992).

3. H.E. Flack and E.D. Pellegrino, eds., *African American Perspectives in Biomedical Ethics* (Washington, D.C.: Georgetown University Press, 1992).

These questions invite a critical reexamination of the foundation and meaning of autonomy and its relationship with truth-telling. Such reflection suggests that autonomy is not and cannot be unequivocally interpreted. Some of the dilemma is more superficial than real and derives from a narrow interpretation of the concept of autonomy.

Autonomy entered medical ethics as part of a system of *prima facie* principles devised to deal with moral pluralism. *Prima facie* principles, such as autonomy, nonmaleficence, beneficence, and justice, are those that ought to be respected unless powerful reasons for overriding them can be adduced.[4] In this system, each principle is given equal weight. Priorities among principles can be established only when the detailed circumstances of a particular decision are known. No principle, including autonomy, is granted a priori moral hegemony over the others.

The principle of autonomy is grounded in respect for persons and the acknowledgment that as rational beings we have the unique capability to make reasoned choices. Through these choices we plan and live lives for which we are morally accountable. Inhibiting an individual's capability to make these personal choices is a violation of his or her integrity as a person and thus a maleficent act.

Autonomy, therefore, is not in fundamental opposition with beneficence as is too often supposed, but in congruence with it. Problems arise when the content of what is beneficent is defined by others such as family members or physicians. This may be warranted when we are mentally incompetent, when our choices harm others, or when we make choices that contravene good medical practice or harm us seriously. In the absence of these limitations, competent humans are owed the freedom to define beneficence in terms of their own values. This does not mean that all values are morally equivalent or defensible, but only that as humans we are owed respect for the choices we voluntarily make.

In North America, in the United States in particular, autonomy has tended to become a moral absolute. The reasons for this are multiple. They include improved education of the public, a strong tradition of privacy rights and personal liberty, a distrust of authority and the possibilities of medical technology, and the loosening of family and community identification. In this context, truth-telling is a necessary corollary, since human capability for autonomous choices cannot function if truth is withheld, falsified, or otherwise manipulated. Truth-telling is essential to informed consent, the instrument whereby personal autonomy is expressed in concrete decisions.

Respect for autonomy and truth-telling are intrinsic to beneficent medical care for many people in the North American context. It does not follow, however, that this concept of autonomy is always beneficent or that it must be accepted by, or imposed on, everyone living in the United States or elsewhere.

4. T.L. Beauchamp and J.F. Childress, *Principles of Biomedical Ethics,* 3rd ed. (New York: Oxford University Press, 1989); W.D. Ross, *The Right and the Good* (Indianapolis, Ind.: Hackett Publishing Company, 1983): 33–34.

For one thing, autonomous patients are free to use their autonomy as they see fit—even to delegate it when this fits their own concept of beneficence. Some patients need a more authoritative approach than others. This approach is legitimate when, for example, despite efforts to inform and empower patients to make their own decisions, patients find themselves unable or unwilling to cope with choices. Such patients may feel sincerely that a close friend or family member would be able to make decisions that better protect their values than they could make themselves. Such a delegation of decision-making authority may be explicit or implicit depending on the dominant ethos.

In some parts of Italy, among many ethnic groups in the United States, and in large parts of the world, this delegation of authority is culturally implicit. In such contexts, the uniformity of the practice suggests that delegation of decision making is an expectation of the sick person that need not be explicit. To thrust the truth or the decision on a patient who expects to be buffered against news of impending death is a gratuitous and harmful misinterpretation of the moral foundations for respect for autonomy. In many cultures clinicians encounter patients who are fully aware of the gravity of their condition but choose to play out the drama in their own way. This may include not discussing the full or obvious truth. This is a form of autonomy, if it is implicitly and mutually agreed on, between physician and patient. However, autonomy should not be violated by a misconceived attempt to be morally rigorous. Withholding the truth from a patient demands, of course, the utmost care in responding to any occasion when the patient wishes to exert more control.

Among most North American physicians, the withholding of truth, in whole or in part, is a therapeutic privilege accepted as morally licit when we have substantial evidence that offering the patient the truth has a significant probability of causing harm, for example, emotional damage or suicide. This "privilege" must be used rarely and with utmost care since it is so easily abused.

Treating patients within the conception of beneficence defined by their own cultural ideas is a form of therapeutic privilege. However, a palpable risk of harm in knowing the truth must exist that must not be outweighed by the risk of withholding the truth. The amount, manner, and timing of truth-telling or truth-withholding are crucial factors for which there is no ready formula.

Thus, autonomy cannot be an absolute principle with a priori precedence over other *prima facie* principles. Rather, the judicious exercise of respect for autonomy means that health professionals must act in a manner that enables and empowers patients to make decisions and act in a way that is most in accord with their values.

That the patient may draw these values from the circumambient culture does not make autonomy or medical ethics a cultural artifact. Autonomy is still a valid and universal principle because it is based on what it is to be human. The patient must decide how much autonomy he or she wishes to exercise, and this amount can vary from culture to culture.

It seems probable that the democratic ideals that lie behind the contemporary North American concept of autonomy will spread and that something close to it will be the choice of many individuals in other countries. What then does the physician do when a society is in transition, as Dr. Surbone suggests is the case in Italy today, or when many different cultures make up one society as is the case in the United States? To preserve both autonomy and beneficence, physicians must get to know their patients well enough to discern when, and if, those patients wish to contravene the mores of prevailing medical culture. This requires a degree of familiarity and sensitivity increasingly difficult to come by, but morally inescapable for every physician who practices in today's morally and culturally diverse global society.

V

Beyond the Bedside: Systemic and Institutional Concerns

A. Respecting Patients' Decisions

29

Capacity and Informed Consent in Pellegrino's Philosophy of Medicine

Aaron Kheriaty, MD

The backdrop of every clinical decision is a kind of hope.

—Edmund D. Pellegrino and David C. Thomasma[1]

I. Illness and the Doctor-Patient Relationship

My purpose here is to examine the theme of capacity and informed consent in Pellegrino's writings. I will focus specifically on the clinical situation in which the patient has decisional capacity. A sustained examination of surrogate decision making for incapacitated patients is found in chapters 12 and 13 of Pellegrino and Thomasma's book, *For the Patient's Good*.[2] An extended treatment of these issues lies beyond the scope of this chapter.

In his writings, Pellegrino did not (so far as I can tell) single out informed consent as a topic or theme for focused and extended consideration. Rather, his views on this theme are embedded in his philosophy of medicine and his understanding of medical beneficence and patient autonomy. He contrasts this framework with other approaches—including principlism, rights-based ethics, and utilitarianism—each of which prioritizes a particular understanding of autonomy that Pellegrino thought inadequate.

Departing from the typical canons of discourse in academic bioethics, I will use my own recent experience of illness as an example to illustrate this theme. I believe that Dr. Pellegrino would forgive this departure from convention, since the patient's experience of illness was central to his thinking. As a doctor who is also a patient, I will examine this topic from both perspectives, since both were essential in Pellegrino's philosophy of medicine.

1. Edmund D. Pellegrino and David C. Thomasma, *For the Patient's Good: The Restoration of Beneficence in Health Care* (New York: Oxford University Press, 1988), 142.

2. Ibid.

Shortly after receiving the invitation to write this chapter, I sustained an injury and became—for the better part of four months—physically incapacitated. A lumbar disc rupture led to nerve impingement and severe radiculopathy. This involved twelve preoperative weeks of constant pain, muscle spasms, and diminished use of my left leg, followed by eight postoperative weeks of convalescence and recovery.

It was in that convalescent state, where pain was a constant companion, that I revisited Pellegrino's foundational writings. Also in the background here is Dr. Pellegrino's teaching, mentorship, and personal example during my time as a medical student at Georgetown. With gratitude, I attempt here a small contribution to the ongoing examination, assimilation, and critique of his writings.

Returning to our theme, prior to surgery and during my recovery I experienced what Pellegrino described as the existential state of illness. As he explained it, "In illness, the body is interposed between us and reality; it impedes our choices and actions and is no longer fully responsive. The body stands opposite to the self. Instead of serving us, we must serve it. It intrudes on our existence rather than enhancing and enriching it."[3] Because of my injury, I could not work; I could not play; I often could not sleep; at times, although the injury was not in my brain, I could not think very clearly. My life had been unexpectedly interrupted, and my professional and personal plans thrown into disarray.

The question of whether or not to pursue surgery, with its attendant risks and benefits, and the solicitation and provision of informed consent, did not occur in the course of a single clinical encounter or at an identifiable moment in time. It was a gradual process that unfolded in the context of several visits with my surgeon and other consulting physicians, as well as much pondering, research, and discussion with my wife between visits with specialists. I tried various remedies, both mainstream and complementary, in an attempt to avoid surgery. The decision to proceed with the operation required developing a relationship of trust with the surgeon—a man who was, only a short time ago, a complete stranger to me.

The formal surgical informed consent process—during which the risks, benefits, and alternatives were enumerated and my official consent given—occurred after I had been in constant pain, only partially alleviated by medications, for twelve weeks. One night I awoke suddenly with foot pain sufficiently severe to induce dissociative amnesia: an episode that afterwards seemed to my mind to last only a few seconds was actually, as my wife later informed me, several minutes in duration.

In the wake of this, my neurosurgeon elicited my consent to proceed with surgery. I recall his mentioning, at the end of the list of surgical risks, the possibility of "paralysis, coma, or death." My wife, quite understandably, cringed. Yet somehow, without a moment's hesitation, I could only immediately say yes to the surgical intervention. It was as though, after weeks of attempting to manage on my own

3. Edmund D. Pellegrino and David C. Thomasma, *A Philosophical Basis of Medical Practice: Toward a Philosophy and Ethics of the Healing Professions* (New York: Oxford University Press, 1981), 208.

without relief, I had no other choice. "Having no other choice" might seem at first glance to negate the possibility of autonomous action, something contrary to the notion of informed consent. However, for Pellegrino, this was not necessarily so—more on this in a moment.

The surgeon intuitively understood my frame of mind, and the mention of paralysis, coma, or death—though statistically true—was here merely a formality. These terrifying possibilities were reduced to a "risk management" box to check on the consent form. Probably more for my wife's benefit than mine, I shrugged off these rare but dire risks with the tautology, "surgery is surgery"—a phrase which my surgeon repeated with a knowing nod of the head.

This was my autonomy in action, though it may strike us as a strange form of autonomy. Granted, by any modern medical-legal standard, my consent was reasonably free and adequately informed. Indeed, as a physician, I was better informed than most patients could ever expect to be. Yet the exercise of my autonomy did not mean I had a plethora of choices before me, from which to select the most appetizing option.

On the contrary, I was in distress, physically disabled, my life unexpectedly interrupted, my travel plans cancelled, my colleagues at work burdened with my responsibilities, my family life thrown into disarray, and my future functionality uncertain. Pellegrino described this existential state of illness better than any other contemporary bioethicist. He wrote, in *A Philosophical Basis of Medical Practice*:

> This ontological assault of illness is aggravated by the loss of certain specific freedoms which we identify as peculiarly human. The patient is no longer free to make rational choices among alternatives. He lacks the knowledge and the skills necessary to effect a cure or to gain relief from pain and suffering. In many illnesses, the patient is not even free to reject medicine, as in severe trauma or other overwhelming, acute injuries. Voluntarily or not, the patient is forced to place himself under the power of another person, the health professional, who has the knowledge and the skills which can heal—but also harm. This involuntary need grounds the axiom of vulnerability from which follows the obligations of the physician.[4]

What I was experiencing in that moment when I assumed the risks of surgical intervention—this existential state of illness—was, for Pellegrino, the starting point for all medical ethics. The anguishing experience of illness is simply the context within which informed consent occurs. In this vulnerable state, I was not a medical outlier; I was just a typical patient. My neurosurgeon saw twenty patients just like me, some of them worse off, in the course of his clinic that day. This was informed consent at its most routine and mundane.

Pellegrino argued that the informed consent process is not primarily a technical or procedural enterprise, but a fundamentally human and interpersonal engagement. It is grounded in a particular kind of relationship—a fiduciary relationship between

4. Ibid., 208.

one person who suffers the assault of illness, and another person who professes to help and to heal. The sick patient is vulnerable: he cannot exercise freedom and autonomy in the way that was possible when he was well. "The patient presents himself in a state of wounded humanity," as Pellegrino puts it,

> He has lost some of his freedom since he must come to the physician; he must give consent when he is in pain and discomfort, and he does so in the presence of an information gap which can never be closed fully. Medical science, therefore, becomes medicine only when it is modulated and constrained in unique ways by the humanity of physician and patienst.[5]

By way of contrast, Pellegrino points out that, "neither plants nor animals—granting that they become ill as well as humans—can enter into a relationship with the healer in which the patient participates as subject and object simultaneously."[6] And so we arrive at the first feature of informed consent in Pellegrino's thinking: *informed consent always occurs in the context of a distinctive kind of relationship.*

To continue unpacking Pellegrino's understanding of informed consent, we can look at other essential features of the physician-patient relationship. First, it is a relationship entered into only reluctantly by the patient, often after futile attempts at self-administered remedies have been exhausted. After a misdiagnosis of a muscle strain, my wife brought me—wisely, but quite contrary to my preference—to the emergency room for evaluation when my pain had escalated beyond control. The ER physician took one look at me and said, "It's a disk: we're getting an MRI in the morning," to which I replied, "It's not a disk," by which I really meant, "I don't *want* it to be a disk." As a physician, I should have known better. As a patient, however, I was vulnerable—and thereby subject to fears which clouded my judgment.

And so I reluctantly placed myself under the care of this ER doctor. With my nod of consent, I entered into a unique relationship, however brief the clinical encounter in the emergency room. "Medicine is essentially a relationship,"[7] Pellegrino argued, and not primarily a set of techniques or technologies. In a remarkable formulation, citing Plato and Aristotle, Pellegrino argued that the doctor-patient relationship is closer to a friendship than to a contractual relationship.[8] "The physician is a friend insofar as he 'shares your pleasure in what is good and your pain in what is unpleasant, for your own sake.'"[9] This grounds the physician's compassion, understood according to its etymological meaning, "in a very real sense, [the physician] must 'suffer with' in order to heal."[10]

5. Ibid., 24.
6. Ibid., 25.
7. Ibid., xii.
8. Ibid., 86.
9. Ibid., 88.
10. Ibid., 115.

With a balance of compassion and competence, the physician must see her way beyond just the patient's symptoms and into the patient's complete experience of illness; but she must also see her way clear of the patient's experience precisely in order to effect a good and skillful healing act. These virtues are a necessary precondition for an effective informed consent process—for this process requires not only explanation and information, but also understanding and empathy.

Pellegrino also emphasized the inequality of power inherent in the doctor-patient relationship. To cite again my own experience, I was paradoxically at my most powerless precisely when the surgeon was exercising his healing powers to the full, for I was under anesthesia. In those brief hours he was putting his energies, concentration, and skills to work on helping my body to heal. Meanwhile, I was unconscious and completely unaware—temporarily an object rather than a subject.

I had to undergo surgery to regain the powers that the injury had compromised. If the surgery worked as planned, then I would no longer be at the physician's mercy. The patient's autonomy is always, to a greater or lesser degree, compromised by illness. Therefore, the physician has a duty to regard and to engage whatever aspects of the patient's freedom and rationality remain intact. This brings us to the second key feature of informed consent in Pellegrino's view: *respect for patient autonomy constitutes, in large part, the physician's attempt to right the imbalance of power that illness introduces.*

The physician's responsibilities in this context—to help (beneficence) or at least do no harm (nonmaleficence), and to treat the patient with due solicitude and respect (autonomy and justice)—are well-trod themes in bioethics. Pellegrino's distinctive accent here was neither an uncritical acceptance of the typical formulation of these principles, nor a thoroughgoing rejection of principlism. In his view, these principles are not free-floating a priori values; they do not fall out of the sky. Rather, Pellegrino derived them from a careful phenomenological analysis of the doctor-patient relationship. This approach gives the principles a distinctive stamp, and allows us to order them and adjudicate situations in which the principles are in tension with one another.

Pellegrino explains, "Three axioms are developed from an ontology of a living body in need of help. These axioms are Do no harm; respect vulnerability of the patient; treat all patients as equal members of the human race."[11] The norm of health is prior to these principles, and provides their teleological goal: "The first and primary value is that it is good to be healthy," he wrote. "Inasmuch as this value determines the nature of the clinical relationship, it grounds a most ancient [Hippocratic] medical-ethical axiom in the functioning of the living body . . . 'To help, or at least to do no harm.'"[12]

Following naturally from this, he states, "The second principle of medicine is that individuals have intrinsic value. . . . Such a value principle leads to an axiom as well: care must be taken for the susceptible individual."[13] In other words, medicine

11. Ibid., xiii.
12. Ibid., 184.
13. Ibid.

first serves the good of *this particular* sick patient; it serves social purposes only in a secondary and derivative fashion. This is another feature that makes informed consent central to Pellegrino's thought, even if he did not treat the issue in a separate treatise or article.

Pellegrino also recognized that the doctor-patient relationship is a two-way street: the patient has distinctive responsibilities toward the physician. If informed consent is to proceed appropriately, the patient must exhibit the virtues of honesty and truth telling, respect and civility. My wife, appropriately, chastised me whenever she noticed in my communications with my physician that I minimized my symptoms or forgot to mention some essential point of my clinical history. Although I was impaired, I was still responsible to play my role in this healing enterprise. Without my contribution, we would miss the mark, medicine would not achieve its purpose, and I would suffer as a result. This brings us to the third key feature of informed consent in Pellegrino's view: *the patient, just like the physician, bears considerable responsibility*. Patients can decide well or poorly when it comes to informed consent; and an imprudent patient decision, while it might be legally valid, will impede the healing process.

II. The Dialectics of Clinical Decision Making

Within Pellegrino's corpus, the most detailed discussion of informed consent is found in chapter 9 of his book, *A Philosophical Basis of Medical Practice*. Here he argues that medical ethics must rest upon a sound philosophical anthropology: "There must be some idea of man," he wrote, "to order our definition of what is 'good' for the person and society, and for the optimal relationships between them. Individual and social bioethics both derive from this context of our idea of man."[14]

Pellegrino's anthropology integrates Aristotelian and Thomistic sources with findings from the modern sciences. Pellegrino insists upon the substantial unity of mind and body: "The body is experienced as a centered unity,"[15] he writes; and furthermore, "it appears that the neurosystem is not an organ, but an embodiment of the mind."[16] This means that every medical act, including the process of informed consent, involves the patient as a whole person, and engages the patient's entire experience of illness—mental, physical, social, and spiritual.

Informed consent is intrinsically teleological: it aims at an already given end, namely, health and healing. The choices presented by the physician are therefore never arbitrary but are always aimed at a particular kind of good. Unlike the diversity of breakfast cereals in a grocery store, having more treatment options is not necessarily better than having fewer options. The menu of medical selections is not made to order by the patient. These end-oriented constraints mean that medicine cannot

14. Ibid., 32.
15. Ibid., 115.
16. Ibid., 103.

be reduced to a set of techniques for somatic manipulation, offered in an unconstrained medical marketplace. Precisely because he was a good physician, my surgeon offered me very few options. Other circumstances may have permitted more avenues toward healing, but my particular circumstances narrowed the range of medically appropriate choices.

Patients cannot simply demand whatever interventions they want from the physician. They must collaborate with the doctor to select the treatments most conducive to health and healing. From this, and going beyond his explicit writings on the subject, we can deduce that Pellegrino would likely support the right of physicians to deny patient or surrogate requests for interventions that the physician deems medically nonbeneficial—so-called futile care. In cases where differences between doctor and patient regarding a course of action are irreconcilable, Pellegrino would recommend that the doctor recuse himself from the case, though without abandoning the patient.

We can now see that the process of informed consent is not simply the operationalizing of an abstract, a priori, disembodied principle of autonomy—whether a Kantian categorical imperative or a libertarian right to privacy. Informed consent operationalizes the principle of medical beneficence and works harmoniously with a rightly understood principle of autonomy. Pellegrino explains,

> The patient is not a passive object to which a technique is applied, since he or she seeks advice and help, and modifies his or her behavior in conformity with the physician's advice. This dimension impinges directly on the person and his or her values. It involves two persons interacting, each in his own socio-historical moment. The intersection of their values, together with those of medicine, science, and society, creates a nexus of choices and priorities. It is the unraveling of that nexus for this patient, here and now, that constitutes medicine.[17]

Both physician and patient bring something essential to this encounter. "The patient has a richer, more personal knowledge of the disorder because it is happening to him or her. The physician only has a more theoretical knowledge of the disorder due to training."[18] I knew the pain and debility of my injury firsthand; my physician knew how to identify the injury and where to apply the scalpel. Both were necessary for the healing action to occur.

In this context, Pellegrino clarifies that, "the term 'patient' does not necessarily imply a passive restoration in which the physician is the sole agent. The patient ideally also participates in the restoration. . . . The patient can yield this moral agency to the physician only by direct mandate. Only in the rarest circumstances will the physician legitimately be the patient's moral agent."[19] In short, except in an emergency, my surgeon could cut only after I gave the green light. And I would only do

17. Ibid., 24.
18. Ibid., 52.
19. Ibid., 123.

so once I was convinced that the surgeon was not only competent but was putting the good of my health above other competing goods (e.g., making money, filling his operating room schedule, practicing a new surgical technique, etc.). So, we arrive at a fourth key feature of informed consent in Pellegrino's thought: *trust between patient and physician is the essential ingredient here—more important, even, than the information conveyed.*[20]

Informed consent requires disclosure of risks, benefits, and alternatives to a proposed treatment. But Pellegrino shows that this process of information disclosure is not reducible to a procedural or technical formula. It requires the physician's keen and personal understanding of *this particular* patient: "The definition of risk is highly personal," he wrote, "and it turns on the patient's estimate of a danger 'worth' running."[21] Informed consent is thus a process of co-reasoning: "Physician and patient together must clarify the relationship of one recommendation with its opposite, and weigh the reasons for each action."[22] In a succinct formulation that Pellegrino would have endorsed, bioethicist Gilbert Meilaender wrote, "Given the facts of patient vulnerability and physician expertise, a patient's autonomy cannot be achieved autonomously. It can be accomplished only with the physician's help."[23]

As previously mentioned, where moral conflict between physician and patient is irreconcilable, Pellegrino would not resolve the conflict by a procedural deference to the patient's autonomy as a trump card. On the contrary, he suggests that in such circumstances, it is better for the physician to recuse himself from the case than for the physician to set aside or ignore his ethical convictions or conscience: "Where these conflicts are fundamental, the medical relationship may even be severed by either party, to be resumed with another physician."[24]

Once the treating physician arrives at a preferred course of action that he intends to recommend, "then the reasoning becomes 'rhetorical'—in the classical sense of artful persuasion," according to Pellegrino. Here he raises the delicate but important question, "how much persuasion is ethically defensible, how vigorously [should] the physician pursue a personal priority system"?[25] The doctor must understand his or her own biases to ensure that tactful persuasion does not shade into subtle coercion. Pellegrino suggests some questions that the physician can ask in this regard: "Does he or she believe that the scientifically recommended action must take priority? Is he or she a therapeutic enthusiast who prefers to 'do something' rather than nothing? Does he or she think people should be 'strong' and ignore minor infirmity?"[26] These sorts of

20. Nana Cecilie Halmsted Kongsholm and Klemens Kappel, "Is Consent Based on Trust Morally Inferior to Consent Based on Information?" *Bioethics* 31, no. 6 (2017): 432–42.

21. Pellegrino and Thomasma, *A Philosophical Basis*, 124.

22. Ibid., 134–35.

23. Gilbert C. Meilaender, *Body, Soul, and Bioethics* (Notre Dame, Ind.: University of Notre Dame Press, 1995), 12.

24. Pellegrino and Thomasma, *A Philosophical Basis*, 135.

25. Ibid.

26. Ibid.

biases are inescapable: the physician does not enjoy a value-free, perspective-less, or "objective" view from nowhere. The point is not that the physician can entirely rid himself of these biases, but that honest self-awareness can mitigate their negative effects in the clinical encounter.

"Likewise, the patient can use artful persuasion to modify the physician's scientific or dialectically secured recommendation to gain an action more congenial to a personal view of what is good." Significant here is the fact that "the patient's vulnerability can easily tip the resolution of conflict in the physician's favor."[27] At the conclusion of this process of deliberating cooperatively, the all-important capstone question—what should be done?—is the most prickly. "Scientific and semiscientific conclusions of varying degrees of certitude are examined under a light strongly tinged with moral hues."[28] Here we arrive at a fifth key feature of his thinking: *informed consent is always a moral, and never simply a procedural or technical, enterprise.*

For Pellegrino, medical ethics applies not just to classic ethical dilemmas or trolley car scenarios. Moral reasoning is present in every clinical treatment decision. From beginning to end, the practice of medicine is shot through with ethical implications. Even routine or apparently trivial clinical encounters involve moral reasoning and require virtuous behavior from both the physician and the patient.

III. Objections and Responses

At this point, a critic may raise several questions about this account. Can't we simply ground informed consent in the principle of autonomy, without recourse to the notion of beneficence? Shouldn't the ends of medicine be determined by patient preference, and not be bound to a natural norm of health? Isn't the restoration of beneficence as the primary ethical principle a return to the bad old days of medical paternalism? Wasn't informed consent designed precisely to remove power from the physician and transfer this power to the patient, in a kind of zero-sum game? Isn't Pellegrino's philosophy of medicine rolling back the gains made by the ascendance in bioethics of patient autonomy? Wouldn't a procedural approach (who decides) be preferable to a normative approach (what is the best decision) in a pluralistic society?

Space does not permit here a full response to each of these questions. However, a few remarks are in order, which may help clarify Pellegrino's position. To begin addressing these objections, we must situate Pellegrino's philosophy of medicine between medical paternalism on the one hand, and untethered patient autonomy on the other. The restoration of beneficence in health care should not be interpreted as the restoration of medical paternalism: "The physician will never again be the sole decision maker, the priestly bearer of technical and moral authority, the dominant figure in all his relationships."[29] He notes that the notion of informed consent does

27. Ibid.
28. Ibid.
29. Ibid., 169.

not even appear in the writings of the Hippocratic physicians of Ancient Greece, where paternalism was the norm.[30] Yet it clearly plays a central role in Pellegrino's philosophy of medicine.

On the other hand, the exercise of patient autonomy must be harmonized with beneficence, which aims at the healthy flourishing of the human organism as an integrated whole. In his and Thomasma's book, *For the Patient's Good*, Pellegrino contrasts Paul Ramsey's covenantal model of the doctor-patient relationship—a forerunner of Pellegrino's own model—with a contractual model in which autonomy is prioritized. "In a Lockean model two or more autonomous entities form a bond for some mutual good. In a covenantal model the bond is formed based on the need of one party,"[31] namely, the vulnerable patient afflicted by disease. While Ramsey's view gets us much closer to the mark, Pellegrino and Thomasma do not fully endorse either the contractual or the covenantal model of this relationship.[32] Instead, he develops a fiduciary model of the doctor-patient relationship. In this account, the relationship is grounded in the virtuous physician acting always in a manner worthy of the patient's trust. For Pellegrino, "fidelity to trust" is a kind of overarching virtue that informs and encompasses the other principles.

Pellegrino and Thomasma steer a course between an outmoded medical paternalism and contemporary ethical theories that prioritize patient autonomy above other goods. This middle course brings our theme of informed consent into focus: "paternalism should be understood to mean a medical decision to benefit a patient without full consent of the latter,"[33] and thus, should be rejected. On the other hand, "the almost automatic assumption of some ethicists and patients that autonomy must always supervene"[34] is also problematic, for "to accord autonomy superiority over beneficence is sometimes to abandon the patient in time of need."[35]

In terms of legal rights, Pellegrino and Thomasma argue that "the right to self-determination may be overridden on grounds of beneficence when one's competence is impaired for some reason."[36] Furthermore, "if the physician made no effort to assess the patient's competence, or to dissuade a competent but noncompliant or obtuse patient from an ill-conceived decision, then the patient's autonomy would block the care the physician should give for the patient. It could result in actual harm to the patient,"[37] which clearly undermines the goals of medicine. For Pellegrino, a procedural appeal to autonomy does not excuse such negligence or the resulting harm.

30. Ibid., 201.
31. Pellegrino and Thomasma, *For the Patient's Good*, 6.
32. Ibid., chap. 4, 51–58.
33. Ibid., 10.
34. Ibid., 6.
35. Ibid.
36. Ibid.
37. Ibid.

We can mention one final point before closing, which highlights an unusual feature of Pellegrino's view. Pellegrino argued that every clinical decision involves some assessment of the patient's capacity to make conscious and reasoned choices regarding medical care. The patient's decisional capacity is a necessary prerequisite to informed consent. But after making this uncontroversial claim, Pellegrino and Thomasma make another claim that many ethicists today would find uncongenial, namely, "Incompetence is a moral warrant for others to take actions they perceive to be in the patient's best interests, even against the patient's expressed wishes."[38] Here, Pellegrino and Thomasma seem to rank the "best interest" criteria for surrogate decision making above the "substituted judgment" criteria. This prioritization arises from his distinctive view that beneficence precedes autonomy, and that unconstrained autonomy can lead to harm of the vulnerable patient. A full examination and defense of this point would take us far afield. I note it here merely to illustrate that, while Pellegrino's account of informed consent contains many conventional features, it also contains a few features that are distinctive, and that challenge the elevation of unfettered patient autonomy in much of contemporary bioethics.

IV. Conclusion

Pellegrino's philosophy of medicine grounds the process of informed consent on a rigorously articulated philosophy of medicine, with the doctor-patient relationship at the center. There are important accents here, and a few significant departures from contemporary bioethical norms that render his account distinctive, and in my view, important for medicine today.

Pellegrino's account, for example, provides a solid ethical grounding for physicians to refuse requests from patients or surrogate decision makers for treatments that physicians deem nonbeneficial or disproportionately harmful. Such refusals would need to include appropriate safeguards to ensure second-opinion consultations, and opportunities for transfer of care when differences are irreconcilable.

Pellegrino has charted a course between overbearing medical paternalism and radical libertarian individualism, neither of which achieves optimal healing. In his account, patient and physician engage in a mutual process that respects the expertise, the experience, and the conscience of each. This account of what medicine can and should be, even in its most mundane and routine decisions, is a remarkable achievement for which we can be grateful.

38. Ibid., 148.

30

Advance Directives and Surrogate Decision Makers

Sarah B. Vittone, DBe, RN

Introduction

Advance care planning is a process by which patients incorporate the goals of current treatment plans with their identified priorities for living and dying. As technological advances in medical treatment push the boundaries of life expectancy in the face of dire diagnosis, obligations for clinicians to engage in meaningful discussion and to support personal decision making have become ever more necessary. Advance directives are the documents that provide written evidence of advance-care planning. They support personal control in end-of-life decision making by indicating which treatments the patient would find acceptable or unacceptable at a future time should cognitive decline or other processes limit the ability to express those preferences. An advance directive (AD) has two components: the living will (LW), and the appointment of a proxy or surrogate decision maker, sometimes referred to as the durable power of attorney for health care (HCPOA). A patient may choose to complete one or both components. The living will focuses on what decisions will be made, and surrogate appointments determine who will make those decisions when the patient is unable to do so. Ideally, if both are utilized, there will be some indication whether the proxy or surrogate will be allowed to override some details of the living will.

The difficulty of decision making for patients at the end-of-life is complicated by their inability to participate in these decisions in real time. According to various studies cited by the Institute of Medicine, significant portions (sometimes, the majority) of adult medical inpatients, nursing home residents, and older adults facing treatment decisions at the end of life were found to be incapable of making those decisions themselves.[1] Over the last thirty years, advance directives have become essential for healthcare facilities and clinicians seeking to make treatment

1. Institute of Medicine, Committee on "Approaching Death: Addressing Key End-of-Life Issues," *Dying in America: Improving Quality and Honoring Individual Preferences near the End of Life* (Washington, D.C.: National Academies Press, 2015), 119.

decisions consistent with the personal preferences of adult patients of all ages who lack decisional capacity and who are in irreversible conditions leading to death.

Advance directives, as described in the Patient Self-Determination Act, allow patients to exercise their fundamental right to state preferences for medical treatment and have those preferences honored as well as to identify proxies, surrogates, or agents to act on their behalf when the patients are incapacitated.[2] Healthcare teams frequently hope that patients will state their desires clearly enough to provide direction for health decisions at the end of life. Yet, these directives for future care need to be interpreted in light of often-unforeseen circumstances. For that reason, Pellegrino suggests that the clinician's respect for patient autonomy or self-determination is embedded in the clinician's commitment to the principle of beneficence. Pellegrino and Thomasma establish beneficence as acting in the patient's biomedical, subjective, personal, and ultimate good.[3] They contend that it is ethically imperative that healthcare teams support advance directives for their intended use—supplementing them, as needed, with additional considerations of the patient's good in real time with an "agent" or proxy decision maker.

When the patient's treatment preferences are unknown, clinicians have historically turned to family members to share critical information. Decisions must focus on the patient's good and on limiting harm via a shared interpretation of the patient's good. The clinician identifies the biomedical good, informed by the clinician's expert judgment, which contributes to the identification of the patient's overall good as the patient (or surrogate) understands it. The use of formal documents such as advance directives, which include living wills and/or appointments of agents or proxies with durable power of attorney for health care, now have a fifty-year history in the United States. This chapter will examine the effectiveness, advantages, and limitations of advance directives in protecting and advancing the good and dignity of patients.

History of Advance Directives

Patients routinely share with family and friends their instructions for their funeral and burial, and for the passing down of property to the next generation after their death. It is the sharing of their treatment wishes at the end of life prior to their death that patients find more daunting for emotional, psychological, and relational reasons. As medical technology advanced in the 1940s and 1950s, conditions such as kidney failure and pneumonia were often transformed from fatal to chronic illnesses. Life support, including mechanical ventilators which were initially used solely in surgical interventions, became therapeutic interventions, bridges to help patients to survive and recover from previously lethal conditions.

2. Omnibus Budget Reconciliation Act of 1990, P.L. 101-508, sec. 4206 and 4751, 104 Stat. 1388, 1388-115, and 1388-204. See also, the Patient Self-Determination Act, Pub. L. No. 101-508, §§ 4206 and 4751, 104 Stat. 1388 (codified at 42 USC§§ 1395cc(f), 1396a(w) [1994]).

3. Edmund D. Pellegrino and David C. Thomasma, *For the Patient's Good: The Restoration of Beneficence in Health Care* (New York: Oxford University Press, 1988).

The first "living wills" were proposed in 1967 by Luis Kutner, a human rights lawyer, representing the Euthanasia Society of America. Kutner began with the common law and constitutional law premise that "the law provides that a patient may not be subjected to treatment without his consent" and that the individual has "the right to refuse to permit a doctor to treat him."[4] These first living-will proposals were conceived of as a form of legal trust among the patient, family, proxy, and healthcare provider. In 1968, Dr. Walter F. Sackett introduced the first bill in the Florida legislature proposing that patients be able to make decisions regarding the use of life-sustaining hospital equipment. California became the first state to pass living will legislation in 1976.[5] By the end of the next year, forty-three states had similar legislation.[6] In the famous case of Karen Ann Quinlan, the patient's parents received recognition from the court to represent their adult daughter as proxy decision makers. They claimed that this preserved their daughter's right to self-determination, as she was unable to speak for herself, and included specific indications to weigh benefits and burdens and to make medical decisions, including end-of-life decisions.[7]

State statutes addressing advance directives and surrogates developed over the course of the next twenty years. Medical technology continued to advance, to cure and care for critically ill patients and to save lives that previously would have been lost. At times, however, extraordinary treatments and medical technology were used beyond their ability to improve the patient's clinical condition. By the early 1990s, most states had some form of advance directive legislation. Concerns arose over the insistence by patients (or their surrogates) that clinicians use such treatments and procedures to preserve life when such use was unlikely to achieve the desired end.

In another famous case, in 1990, the parents of Nancy Ann Cruzan were not able to act as surrogate decision makers for their daughter.[8] The court required clear and convincing evidence that Cruzan would have requested the withdrawal of treatment if she had known such withdrawal would be life-ending. Cruzan had not left a written advance directive, and her parents could not produce evidence of such a wish. The Missouri courts permitted the withdrawal of artificial feeding based on the testimony of her friend. This led to important legislation in 1991, the Patient Self-Determination Act, which requires that all hospitals receiving Medicaid or Medicare reimbursement provide written information to all patients regarding their rights under the act and determine whether patients have, or wish to have, written advance directives on file with the hospital. It is important to note that this legislation did

4. Luis Kutner, "Due Process of Euthanasia: The Living Will, a Proposal," *Indiana Law Journal* 44 (1969): 550–51, as cited in Charles P. Sabatino, "Advance Directives and Advance Care Planning: Legal Policy and Issues," U.S. Department of Health and Human Services, Office of the Assistant Secretary for Planning and Evaluation, September 30, 2007, https://aspe.hhs.gov/reports/advance-directives-advance-care-planning-legal-policy-issues-0.

5. Sabatino, "Advance Directives," 213.

6. Ibid., 215.

7. *In re Quinlan,* 70 N.J. 10, 355 A.2d 647 (1976).

8. *Cruzan v. Director, Missouri Department of Health* (88-1503), 497 U.S. 261 (1990).

not create or legalize advance directives; rather, it validated their existence and use in each of the states that adopted them. All states and the District of Columbia now have advance directive statutes and forms that combine living wills and the assignment of durable power of attorney for health care to a proxy.

Over the last thirty years, the federal government has joined efforts to educate the public and healthcare professionals about advance directives and surrogates. The 2010 Affordable Care Act originally included language to support discussions of this nature between physicians and their patients, permitting physician compensation for the time it takes to have such discussions. This was met with criticism in the media and public, which resulted in its removal in the final version of the act.[9] In 2016, Medicare began compensating physicians for counseling patients over sixty-five years of age on advance care planning, signaling the recognition of its importance.[10]

Patient Self-Determination through Advance Directives

Advance directives can express the patient's wishes or provide specific instructions regarding health care. While ADs or LWs in some circumstances may be oral declarations, documenting them is important. Documentation ensures that patient preferences are known, and they incur a professional obligation to follow them. The duty owed by the clinicians to adhere to these directives requires faith in the soundness of the document as well as confidence in the clinician's diagnosis. The living will section of the AD may be in narrative format, or it may indicate preferences through a series of choices indicated by check boxes. The LW allows the incapacitated patient to express preferences in advance with respect to providing or withholding specific treatments. One valid criticism is that the LW may not make sufficient allowance for those patients who would want to receive specific life-sustaining treatments—such as cardiopulmonary resuscitation (CPR), mechanical ventilation (and other medical technology), or artificial feeding and hydration and in what context. Finally, the LW section of an AD may also indicate the level of comfort measures the patient desires. Comfort-care language may include pain management and continued use of medical testing such as phlebotomy, turning and repositioning, etc. Clinicians may act upon those directives directly if the patient so indicates. The most common AD provides for the appointment of an optional "agent" or proxy to enact directions on the patient's behalf.

Advance directives are valid for use at a future time when the patient meets both of two required conditions. First, the AD is only valid when the patient's inability to participate in medical decisions is a permanent condition. Decisional incapacity,

9. Mary E. Tinetti, "The Retreat from Advanced Care Planning," *Journal of the American Medical Association* 307, no. 9 (2012): 915–16, doi:10.1001/jama.2012.229.

10. Office of the Assistant Secretary for Evaluation and Planning (ASPE), *Advance Care Planning among Medicare Fee-for-Service Beneficiaries and Practitioners: Final Report* (Washington, D.C.: Office of the Assistant Secretary for Evaluation and Planning, 2020), https://aspe.hhs.gov/reports/advance-care-planning-among-medicare-fee-service-beneficiaries-practitioners-final-report.

itself, may be a temporary, fluctuating, or permanent condition due to cognitive decline from physiologic or psychological disease progression or treatment side effects. For the purpose of the AD, the condition must be permanent. Second, these directives are only valid to be executed at a time when one of three clinical conditions apply: the patient is suffering from a terminal illness, is imminently dying, or is in a persistent vegetative state.

Advance directives are also intended to decrease the burden placed on family and clinicians at the end of life. Each state and the District of Columbia legislate a framework or document for this purpose.[11] Religious organizations may provide guidance to their faithful on the use of advance directives.[12] The AD includes sections for specific directions such as the naming of a "proxy" or "agent" to act on a patient's behalf. This person may also be referred to as a durable power of attorney for health care or healthcare proxy (HCP). In some states, these directives may include directions by the patient for the transfer of the body for science or for organ donation at time of death. Importantly, states may or may not recognize advance directives prepared outside their jurisdiction. The ethical importance of the AD may withstand legal scrutiny and as such should be shared with the organization administration and/or ethics consultation service. Clinicians are encouraged to review and make note of an AD including both an LW and identified HCP in their own patient medical record entry.

Preparation of the Advance Directive Document

Advance directives may be prepared at home or in medical offices, and it is best when an identified HCP and the patient's physician or primary care provider are included in the discussion with the patient. Education for providers on the complexities of this discussion is readily available. Attempts to encourage the preparation of advance directives may be met with reluctance. An individual approach with intentional listening and facilitation is best practice.

Advance directive documents must also be executed in the presence of witnesses, as required by the relevant legislation. The named HCP should not sign as a witness, and neither should a physician or other employee of a healthcare facility, out of concern over conflict of interest. While providers in outpatient settings are to be encouraged to have these conversations with their patients, since long-standing relationships can be of benefit, advance directives are frequently discussed within inpatient settings at a time when decisions are imminent. All inpatient hospitals and skilled-care facilities have a process for documenting an AD as well as addressing AD requests by patients. The social workers or nurses are likely to be the first responders to these requests. It is important for members of the healthcare team to be ready to explain to patients the details of an AD form, and to include the physician and other clinical team members when ADs are made available or are prepared on site.

11. Charles Sabatino, "Myths and Facts about Health Care Advance Directives," *Bifocal* 37, no. 1 (2015): 6–9.

12. See the resources at the end of this chapter.

The process regarding ADs might best be compared to an informed-consent discussion. The patient must be a capable decision maker who understands, is able to reason through, and appreciates the decisions being made and participates voluntarily. The process of providing witnesses should be timely so that these documents may be executed expeditiously. In an attempt to increase the use of ADs, lawyers often include them as part of the estate-planning process; however, legal preparation or notarization of an AD is not required.

Advance directives must be available to the healthcare team in order to implement the patient's instructions. Many times, the location or availability of these documents are unknown, and further, the documents may have been updated over the years, so the most recent version of any advance directive needs to be available. ADs are best prepared at a time when the patient is in a well state and is capable of discerning decisions for their future self at a time when traumatic or terminal illness is predicted to be life-ending. Advance directives may be revised at any time as long as the patient is capable of doing so. Certain life events may bring important changes in future decision making or perceptions about the end of life. It is usually worthwhile to review ADs—and named proxies—at times of significant diagnosis, death of a loved one, divorce or other separation, and at a decline in health or advance in age.[13]

Difficulties in Establishing Meaningful Living Wills in Advance Directives

Unfortunately, the language used in these legislated documents may be too general, vague, or limiting. The nature of end-of-life decisions is often technically unfamiliar for laypersons, who usually have limited life experience in making such critical decisions. The language in the LW may be difficult to interpret. Further, patients may have a difficult time emotionally discussing their own demise. Understanding the implications of a medical intervention, such as dialysis, mechanical ventilation, tracheostomy, vasopressors, cardiopulmonary resuscitation, etc., for an individual with a life-ending diagnosis is markedly complex. The nature of feeding and hydration at the end of life is also complex in the face of some illnesses where such treatments might add to the suffering. For example, a patient might generally indicate a desire for feeding as a good to continue even at the end of life; however, feeding itself might add to their suffering in specific instances such as the case of end-stage gastrointestinal cancer or hepatic-renal failure with ascites. To add to the confusion, the idea of an LW suggests that choices for such life-sustaining medical interventions in the face of terminal or dying conditions may influence the medical outcome. In reality, these decisions may only influence the length of life, extending the dying process, as the patient's clinical conditions will not resolve. Further, the living will is best prepared with the assistance of a clinician with whom the person has a longstanding relationship; this, in fact, rarely happens. Decision making involving clinicians whom the patient has just met is less than ideal. Further, the conditions identified

13. Sabatino, "Myths and Facts."

in the LW when the document was executed may not include the specific condition the patient experiences at the time the document is invoked. Family members may suggest that the written document reflected the patient's understanding at the time of its preparation but not as that understanding subsequently evolved, and, for that reason, family members may express dissent and a desire to override the LW. These issues add to the confounding limitations for the interpretation of these documents. This has resulted in a less-than-optimal experience and less-than-effective implementation of advance directives over the last thirty years. On the positive side, the use of advance directives has clarified the appointment of a HCPOA or HCP. Also, the LW, itself, may be used to discern the patient's preferences and inform substituted judgment (such as when specific indications for withholding CPR in one instance are used to infer the patient's preference for withholding of CPR in other clinical conditions not mentioned in the LW).

Surrogate Decision Maker

Surrogate decision makers for health care should represent the values, beliefs, and specific directions of persons who are now incapable of making decisions. The surrogate's authority is derived from the patient, either as a named durable power of attorney for health care or healthcare proxy in a duly executed AD or as named by state statute. It is of benefit to the patient to name more than one HCP in the event the primary is unable to serve. This person is designated to act on behalf of the patient representing his or her specific wishes in the AD. If there is no LW, the named surrogate is obligated to act on behalf of the patient. Court-appointed guardians are decision makers for all issues, both health and legal. Importantly, guardians must have specified authority from the court to make end-of-life decisions in some states and the District of Columbia. A parent of a minor, as the legally recognized decision maker, makes health decisions including end-of-life decisions in his or her role and authority as the parent. Legislation in some states does allow emancipated minors to execute advance directives and name a healthcare proxy.[14]

Over twenty of the forty-four states with statutes with default surrogate consent laws, now specify that a "close friend" familiar with the person's values can make the decision if none of the listed family members exists or is available. Moreover, approximately eleven states have developed a mechanism for unrepresented patients, usually involving choices by designated physicians often in conjunction with other physicians or ethics committees.[15] Surrogates who are named by state statute, as "next of kin" are held to the same standards for representing patient values, beliefs, and any shared directions for health decision making.

14. Cf. Justia, "Advance Directives and Living Wills Legal Forms: 50-State Survey," October 2022, https://www.justia.com/estate-planning/estate-planning-probate-forms-50-state-resources/forms-for-advance-directives-living-wills/.

15. Erica Wood, "If There Is No Advance Directive or Guardian, Who Makes Medical Treatment Choices?" *Bifocal* 37, no. 1 (September–October 2015): 10–12.

All persons acting in this role (HCPs) have obligations to attend to the patient and to be available to participate in shared decision making with the healthcare team. The HCP must represent the patient's preferences and accept or decline treatments as outlined in the LW or as known to the HCP. When specific indications are not known, the HCP is asked to estimate the choice the patient would have made, using a substituted judgment approach to decisions. The standard of a substituted judgment using the best evidence available of a patient's indication was first established in 1985 *In Re Conroy,* when Judge Schreiber appealed to a "subjective standard."[16] This subjective determination was one in which the question of what the patient would have chosen was specific to the particular patient, not just what a typical patient might choose. Further, HCPs are obligated to advocate for the patient, to seek to relieve suffering, to preserve and restore function, and to influence both the quality and length of life through their decisions on goals of care.

When the patient is a minor—or an incapacitated adult where values, beliefs, and indications are unknown or never known—the use of a "best interest" approach to decision making is favored. This is weighed heavily in a benefits/burdens approach considering the biomedical good of the patient. The HCP adds any cultural, social, spiritual, and familial values which are helpful to the decision at hand. Due to the nature of the HCP, his or her influence or independent judgment may come into any decision. The use of first-person phrases, such as "I want mother to get dialysis" indicates an inappropriately personal decision. The use of third-person phrases, such as "Mother would want dialysis" indicates the HCP is representing the patient. That being said, the HCP cannot be expected to act in isolation. The healthcare team should provide extended support to the HCP validating his or her role and acknowledging the effort it takes to represent the decisions that patients would have made if they were able.[17]

Legal Standing of Advance Directives

Each state and the District of Columbia has AD legislation and some states have a specific document designed for ADs. The documents themselves may be recognized across state lines, but the witnessing requirements and other signing specifics are what create a legally valid, enacted AD. Clinicians have a legal obligation to respect clearly communicated ADs in any form.[18] However, clinicians are not obligated to follow directives to which they conscientiously object or directives that call for medically inappropriate treatment. For example, the patient should not indicate withholding of pain medication in the presence of suffering insofar as clinicians have an obligation to treat pain and suffering and cannot, in good conscience, intentionally fail to treat severe pain.[19]

16. *In Re Conroy,* 98 N.J. 321, 347, 486 A.2d 1209, 1222 (1985).

17. Shana Wynn, "Decisions by Surrogates: An Overview of Surrogate Consent Laws in the United States," *Bifocal* 36, no. 1 (2014): 10–14.

18. Sabatino, "Myths and Facts."

19. Ibid.

Psychiatric Advance Directives

It is valuable to mention that as a result of the Patient Self-Determination Act, for the last twenty-five years legislation has been growing to support psychiatric advance directives. These documents, currently authorized in half of the United States, are designed to support decision making for persons at risk for future decisional incapacity related to a behavioral or mental-health diagnosis. In addition, the patient may name a HCP for future decision making when the patient may be incapable of making decisions.[20]

Decision Making with Minors

Because parents are the natural and legal guardians of their children, rarely do children in the United States, including those who have chronic illness, have an AD or advance-care plan to communicate preferences for end-of-life care.[21] Parents, in addition to the previous obligations mentioned, are obligated to limit harm, to prevent pain and suffering, and to establish a plan of care using shared decision making respecting both duration and quality of life.

Perinatal Advance Directives

While antepartum admissions most commonly center on the joyous delivery of newborns and the expansion of a family, the impact of an unexpected condition leading to the demise of the mother cannot be understated. The preparation of advance directives with these pregnant women could provide valuable guidance to the family and clinical team in the event of an unexpected and terminal event. Nevertheless, it is important to know the legal restrictions that states place on the advance directive with regard to the pregnancy. DeMartino et al. reviewed the fifty states and District of Columbia statutes related to advance directives. They found that:

> Thirty-nine states identified pregnancy as a condition that influences either an incapacitated woman's advance directive or surrogate decision making. . . . Of the thirty-one states that restricted choices about withholding or withdrawing life-sustaining treatment from pregnant women, the restriction appeared in statutes in 29 states and only in advance directive documents in two states. . . . Of the 31 states with a pregnancy restriction, 26 specifically invalidated a woman's advance directive during pregnancy. Nineteen states prohibited a surrogate decision maker from withdrawing life-sustaining therapies from a pregnant woman. . . . The remaining twelve states required that life-sustaining therapies continue to

20. Cf. Justia, "Advance Directives and Living Wills."

21. Alan D. Lieberson, *Advance Medical Directives* (Deerfield, Ill.: Clark Boardman Callaghan, 1992): 44–53.

be provided to a decisionally incapacitated pregnant woman until her fetus could be safely delivered.[22]

Such provisions in the law may not be readily transparent to a woman or her healthcare provider. Any concern regarding the interpretation of an advance directive in these cases should be addressed in healthcare organization policy as well as with legal counsel.

In a well-publicized 2014 court case, Marlise Munoz, fourteen weeks pregnant, was declared brain dead after a pulmonary embolism, yet withdrawal of intensive care support was denied. The Texas Advance Directive Act invalidated any advance directives made by Ms. Munoz due to her pregnancy. Given the fact that Ms. Munoz was dead, the case required additional judicial review because the law was not intended to include deceased pregnant women. Eventually, her body deteriorated, the fetus was determined to be nonviable, and ICU support was removed eight weeks after she had been declared brain dead.[23]

Advance Directives for Research Participation

Advance research directives are beginning to be discussed in this new age of cognitive research, specifically dementia. Advance research directives are not yet available through legislation but could provide a means by which persons could document their wishes about research participation in the event of future incapacity. The concept of research advance directives has been endorsed in some ethics guidelines and position statements; however, formal legal recognition is limited.[24] As these develop for dementia studies, the legal representative option for dissent will likely continue to override written directives prepared by the participant.[25]

Additional Decisions Associated with Advance Directives

Some states address additional decision-making action within their advance directive forms, including organ donation and donation of one's body to science.

22. Erin S. DeMartino, Beau P. Sperry, Cavan K. Doyle, et al., "US State Regulation of Decisions for Pregnant Women without Decisional Capacity," *Journal of the American Medical Association* 321, no. 16 (2019): 1629–30, doi:10.1001/jama.2019.2587.

23. Krista Pikus, "Life in Death: Addressing the Constitutionality of Banning the Removal of Life Support From Brain-Dead, Pregnant Patients," *Gonzaga Law Review* 51, no. 2 (2015): 417–37.

24. Nola M. Ries, Elise Mansfield, and Rob Sanson-Fisher, "Advance Research Directives: Legal and Ethical Issues and Insights from a National Survey of Dementia Researchers in Australia," *Medical Law Review* 28, no. 2 (2020): 375–400, https://doi.org/10.1093/medlaw/fwaa003; Perla Werner and Silke Schicktanz, "Practical and Ethical Aspects of Advance Research Directives for Research on Healthy Aging: German and Israeli Professionals' Perspectives," *Frontiers in Medicine* 5 (2018): art. 81, https://doi.org/10.3389/fmed.2018.00081.

25. Karin Jongsma and Suzanne van de Vathorst, "Advance Directives in Dementia Research: The Opinions and Arguments of Clinical Researchers—An Empirical Study," *Research Ethics* 11, no. 1 (2015): 4–14, https://doi.org/10.1177/1747016114523422.

Organ donation in the United States is managed through the United Network of Organ Sharing.[26] At the time of death, the patient may be eligible for organ donation, and, if so, the patient's permission for retrieval as registered in an advance directive and/or a state driver's license would be actionable. If the patient makes no indication, the HCP will be asked for consent for organ donation. The HCP may be unprepared for this eventuality, and HCPs have been known to interrupt the donation process even when the patient has indicated his or her wish to donate. In the past, UNOS has been unwilling to pressure families to follow the patient's advance directives over the family's objection, but trends appear to be evolving to honor the patient's wishes.

Persons with interest in the training of medical career professionals may wish to donate their bodies to science for use in anatomy education or medical research. This option is available on some, but not all, advance directive forms. Further, even if such donation is indicated on a form, but the HCP expresses dissent, the patient's body is unlikely to be transferred to a receiving academic facility.

Medical Orders as Advance Directives

Advance directives are not intended to act as real-time medical orders, even if CPR is specifically rejected by the patient, the clinician must provide a medical order such as Do Not Resuscitate (DNR) or Allow Natural Death (AND). In an attempt to provide more efficient and immediately actionable advance-care planning, especially specific to life-sustaining treatment, states began to develop legislation for medical orders for life-sustaining treatment (POLST/MOLST) to incorporate the patient's indications for care. This legislation provides for readily available documents for use in the community setting as well as long-term care facilities. Many long-term care facilities require these forms as part of admission to assure that advance-care planning has occurred. State legislation for POLST/MOLST usually provide for continuity of plan of care across healthcare facilities without review.[27]

Healthcare Organization Policy

Healthcare organizations accredited by the Joint Commission are required to have a policy addressing the application of the specifics of ADs and the Patient Self-Determination Act.[28] Ethics committees are ideal for the creation and review of such policies.

26. https://unos.org/about/.

27. Susan E. Hickman, Elisabeth Keevern, and Bernard J. Hammes, "Use of the Physician Orders for Life-Sustaining Treatment Program in the Clinical Setting: A Systematic Review of the Literature," *Journal of the American Geriatrics Society* 63, no. 2 (2015): 341–50, http://doi.org/10.1111/jgs.13248.

28. House of Representatives, Committee on Ways and Means, Subcommittee on Health, report to the Ranking Minority Member, *Patient Self-Determination Act: Providers Offer Information on Advance Directives but Effectiveness Uncertain* (Washington, D.C.: General Accounting Office, 1995).

Challenges and Limitations

While much attention is placed on ADs, patients' actual use of ADs to provide information to their families and clinicians is much less than expected. According to one study in 2017, 36.7 percent of adults had completed an advance directive, including 29.3 percent with living wills.[29] Where, when, and with whom these documents are created, or whether patients discuss these with named surrogate decision makers or clinicians, is quite variable. Clinicians are often not aware that an individual patient has an AD or who the correct HCP might be. HCPs are at times unaware that they have been appointed and also unaware of the patient's directives. HCPs are often as unsure how to decide as they are uncertain as to how to interpret the LW.[30] This may be an impediment to the AD being used for clinical decisions.

In our society, as in many, there is a reluctance to confront our own mortality. This may result in an unwillingness to consider those decisions that might be needed at the end of life. This unwillingness may be shared by the patient's primary care provider, who may fail to broach this discussion, even when a patient is known to have a chronic or terminal condition.[31] Patients may declare that they will leave it to their families to decide, yet they give their families no clear indication of their preferences on which to base those decisions. Familial conflict may be the unfortunate result.

Conclusion

In the words of Edmund Pellegrino, "Properly used advance directives can be of significant help to patients, families, and physicians in treatment decisions when patients become incompetent."[32] When faced with a need for end-of-life decision making for an incapacitated patient, it is an invaluable benefit to have some clear indication of the patient's preferences and values. The best methods for communicating those preferences are in the form of a documented living will or in the naming of the person the patient would most trust to make decisions for him or her. In Pellegrino's model of shared decision making, the physician has the responsibility for identifying a medically good action for this particular patient under the specific set of circumstances. However, patients carry the responsibility for expressing their judgment of what is right for them according to their own values. When incapacitated, patients can fulfill this responsibility best by preparing an advance directive. Clinicians should help patients in this task by gently supporting them in the decision

29. Kuldeep N. Yadav et al., "Approximately One in Three US Adults Completes Any Type of Advance Directive for End-of-Life Care," *Health Affairs* 36, no. 7 (2017): 1244–51, https://doi.org/10.1377/hlthaff.2017.0175.

30. Sabatino, "Advance Directives."

31. Amanda Brown and Toni P. Miles, "Refusing to Admit Defeat: Physicians' Reluctance to Discuss End of Life Care," *Palliative Medicine & Care* 3, no. 1 (2016): 1–2, http://dx.doi.org/10.15226/2374-8362/3/1/00121.

32. Edmund D. Pellegrino, "Ethics," *Journal of the American Medical Association* 268, no. 3 (1992): 354–55, doi:10.1001/jama.1992.03490030066030.

to confront this need and share their decision making. Clinicians in general should model this behavior by preparing their own advance directives. Hospitals and nursing homes can encourage discussions of advance directives at the time of admission. It is even better when these discussions take place in the outpatient setting with a clinician who is familiar with the patient, the patient's medical condition, and his or her prognosis. Failure to provide this last supportive measure would be sad, indeed.

Addendum: An Exceptional Case

In 2010, Dr. Pellegrino and I provided ethics consultation for a 14-year-old patient with a surgically treatable brain tumor. The patient's parents chose to withhold the diagnosis from the patient and further refused the surgical intervention. Based on both the clinical team's and our own assessment of the teen's cognitive capacity, we advised the parents to make the disclosure to their son before the tumor had a negative impact on the teen's quality of life. The parents continued to refuse to disclose, and the hospital contacted Child Protective Services. Our consultation contended that the disclosure was valuable to the teen's understanding of his deteriorating condition, would add to the teen's trust in the clinical team and his parents, and would support his knowing participation in his care. The clinical team's obligation to fidelity to the patient and veracity was crucial to their providing good care. While everyone was in favor of maintaining the family structure, the parents continued to refuse the surgery after the teen had received disclosure. Child Protective Services pursued court-appointed guardianship in order to consent to surgical intervention. Dr. Pellegrino maintained the teen as the focus of our consultation, and the patient's biomedical good was paramount.

Resources for Clinicians

Belief Net offers links to religious and denominational documents for advance directives, https://www.beliefnet.com/wellness/health/2005/03/living-will-resources.aspx.

Five Wishes is the most well-known set of discussion questions developed for the public, https://fivewishes.org/.

National Hospice and Palliative Care Organization (NHPCO) provides links to all states' advance directives, https://www.caringinfo.org/planning/advance-directives/by-state/.

National POLST, https://polst.org/.

Now and at the Hour of Our Death—Catholic Guide to Advance Directives, https://www.catholicendoflife.org/advance-directives/.

U.S. Department of Health and Human Services, National Institute on Aging, https://www.nia.nih.gov/health/advance-care-planning-health-care-directives.

Vital Talk is a training organization for clinicians seeking to advance their communication skills, https://www.vitaltalk.org/.

B. Ethics Committees and Ethics Consultation

31

Introduction: Clinical Ethics Consultations

Claudia Ruiz Sotomayor, MD, DBe

IN HIS PAPER "CLINICAL ETHICS CONSULTATIONS: Some Reflections on the Report of the SHHV-SBC,"[1] Dr. Pellegrino presents his comments regarding the report of the Society for Health and Human Values-Society for Bioethics Consultation (SHHV-SBC) (currently known as American Society of Bioethics and Humanities [ASBH]) where standards for ethics consultation were established (see chap. 32 below). This paper was written in 1999, and since then, there has been an exponential development in the discipline of ethics consultation.

First, there is evidence that improving healthcare ethics consultation services can have a positive impact on healthcare organizations. Some studies show that healthcare ethics consultation services decrease length of stay and costs, and that increasing ethics consultation volume improves physician and nurse perceptions of quality of care.[2] By the year 2007 the usage of consultation services appeared to be low, with a national average in the United States of three to eight consults per year.[3] In the year 2021, a follow-up study was held which determine that there was a 94 percent increase in the number of consults per year;[4] that increment, however, is attributed mostly to an increased awareness of the ethics consultation services in academic hospitals. Small hospitals appear unchanged. There are many theories behind the cause of this disparity, but the main possible cause is the presence of ethicists on staff in academic hospitals creating awareness about the services

1. Edmund D. Pellegrino, "Clinical Ethics Consultations: Some Reflections on the Report of the SHHV-SBC," *Journal of Clinical Ethics* 10, no. 1 (1999): 5–12.

2. Mary McCabe, "Reenergizing the Hospital Ethics Committee: An Opportunity for Nursing," *Journal of Hospice & Palliative Nursing* 17, no. 6 (2015): 480–87, DOI: 10.1097/NJH.0000000000000190.

3. Ellen Fox, Sarah Myers, and Robert A. Pearlman, "Ethics Consultation in United States Hospitals: A National Survey," *American Journal of Bioethics* 7, no. 2 (2007): 13–25, DOI: 10.1080/15265160601109085.

provided (e.g., education, consultation, rounding). Also, some institutions are developing policies that mandate ethics consultation when specific clinical situations arise such as physician-surrogate disagreements around withdrawing life-support treatment, underrepresented patients, donation, initiation of extracorporeal membrane oxygenation, valve replacement denial for patients with subacute bacterial endocarditis, and organ donation after circulatory death, among others.[5] Mandatory ethics consultations are controversial because even when this modality could involve the ethicist at early stages of the case, it could decrease the patient-centeredness that these consults must have, and that Pellegrino highlighted in the following paper.

Another development is that there has been an effort to educate competent clinical bioethicists and to professionalize bioethics. Many people disagree about this recognition because they think that unlike scientists, engineers, doctors, and other professionals, who have a recognized authority, bioethics has a weak claim given the difficulty of achieving a common morality.[6] Some people think that calling bioethics a profession is a "stretch" or even a mistake because the field would lose the freedom and openness that it currently has. However, the idea of advancing the field into a profession is motivating some to create the canons and standards needed, even when there are many different schools of thought. For example, The American Society for Bioethics and the Humanities published the "Code of Ethics and Professional Responsibilities for Healthcare Ethics Consultants," which consists of seven points: (1) be competent; (2) preserve integrity; (3) manage conflicts of interest and obligation; (4) respect privacy and maintain confidentiality; (5) contribute to the field; (6) communicate responsibly; and (7) promote just health care within healthcare ethics consultation.[7] Although a good effort, some questions about how this code is going to be applied and who will enforce it are up in the air.[8]

In the last section of the essay, Dr. Pellegrino addresses accreditation and credentialing. This topic has advanced as well, and there is now a certification exam that clinical ethics consultants can take. The requirements for certification are "a

4. Ibid.

5. Meghan E. Romano et al., "Mandatory Ethics Consultation Policy," *Mayo Clinic Proceedings* 84, no. 7 (2009): 581–85, DOI: 10.1016/S0025-6196(11)60746-5.

6. Craig M. Klugman, "Is Bioethics a Profession?" *Journal of Health Ethics* 5, no. 2 (2008): 1–29, http://dx.doi.org/10.18785/ojhe.0502.06.

7. Anita J. Tarzian and the ASBH Core Competencies Update Task Force, "Health Care Ethics Consultation: An Update on Core Competencies and Emerging Standards from the American Society for Bioethics and Humanities' Core Competencies Update Task Force," *American Journal of Bioethics* 13, no. 2 (2013): 3–13, https://doi.org/10.1080/15265161.2012.750388.

8. Anita J. Tarzian, Lucia D. Wocial, and the ASBH Clinical Ethics Consultation Affairs Committee, "A Code of Ethics for Health Care Ethics Consultants: Journey to the Present and Implications for the Field," *American Journal of Bioethics* 15, no. 5 (2015): 38–51, https://doi.org/10.1080/15265161.2015.1021966.

minimum of a bachelor's degree; and four hundred hours of clinical ethics experience, related to the major domains of the content outline, within the previous four years."[9] This certification is just a starting point, because with so many master's and doctoral degrees in bioethics available, the positions available to hire clinical ethicists require more than "a minimum of a bachelor's degree; and four hundred hours of clinical experience." It will be interesting to see how the certification process continues to evolve in the upcoming years.

9. American Society for Bioethics and the Humanities, "Eligibility for Healthcare Ethics Consultant-Certified Examination" [HEC-C], https://heccertification.org/exam/eligibility.

32

Clinical Ethics Consultations: Some Reflections on the Report of the SHHV-SBC

Edmund D. Pellegrino, MD

Introduction

The Society for Health and Human Values-Society for Bioethics Consultation (SHHV-SBC) report on standards for bioethics consultations is timely as well as welcome. It introduces order into what has been a burgeoning but diffuse, unregulated, and largely unevaluated practice. The report very adequately addresses the skills and competencies of consultants, the process of consultation, and the potential for misuse and abuse. Given the short life span of ethics consultations and their rapid increase in number, the report is a needed guide that is sure to be revised, corrected, and amended.

My comments are offered with these future revisions in view. They are not, however, a detailed critical commentary on the text or a survey of the relevant literature. Rather, they highlight a few aspects of ethics consultation that, on the basis of my own experience, perhaps need stronger or clearer emphasis in future editions. As the authors of the report have done, I will confine myself to the topic of clinical ethics, with only passing reference to its relationship to organizational ethics.

I have chosen seven topics for comment: (l) the centrality of the patient; (2) the character and conscience of the consultant; (3) the place of religious belief; (4) ethics, law, and psychology; (5) advice and conflict resolution; (6) organizational and clinical ethics consultation; and (7) certification and credentialing.

I would like to make some preliminary comments on the style of the report. As is appropriate and necessary, the report is the work of a consensus panel. It has the virtues of this genre but also some of its afflictions: it is long-winded, it sidesteps some of the thornier issues, it sacrifices clarity in the interests of inclusiveness, and it

Reprinted with permission of the Pellegrino family from Edmund D. Pellegrino, "Clinical Ethics Consultations: Some Reflections on the Report of the SHHV-SBC," *Journal of Clinical Ethics* 10, no. 1 (1999): 5–13.

mutes differing opinions to preserve a conciliatory tone. It is also a little overly generous and inclusive in its use of the word "consultant." Most of these minor lapses may disappear with more experience and better evaluation. None detracts from the value of the report.

I. The Centrality of the Patient

Ethics consultations are often requested to resolve conflicts among and between decision makers, health professionals, families, and patients themselves. The reasons for these disagreements are many: variations in conscience or differences in religious, philosophical, or cultural beliefs. As a result, consultants can become preoccupied with conflict resolution and forget that the ultimate purpose of consultation is to arrive at a decision that is morally optimal for a particular patient.

Consultants may reach "consensus" or negotiated agreements and feel "comfortable" with the outcome of the process, yet not act in the patient's interests. Every consultation, therefore, should begin with a clear statement that the patient's good is its aim. This needs to be repeated throughout the process, which sometimes can become overly focused on conflict and interpersonal relationships. Reiteration is necessary even when it seems that the conflict has been resolved. Refocusing on the patient will often bring a wandering discussion back to its purpose and put differences of personal beliefs among consultants into proper perspective.

These differences in belief are often behind the question, "What is the morally right thing to do?" The differences can be made less personal and put into proper context if the consultants can focus on the patient's welfare. To be sure, the "good" of the patient is a complex, many-layered notion.[1] The locus of disagreement cannot always be identified precisely. Keeping the patient's good in the forefront helps to prevent "consensus" from becoming the major and self-justifying purpose of the consultation.

In the end, ethics consultations do not differ from other clinical consultations in which specialists are engaged to clarify obscure diagnostic or therapeutic questions. In both instances, the primary goal is a better decision or action for the patient, not agreement between health professionals. The moral heart of an ethics consultation, as of any clinical consultation, is the improvement of the care of patients.

Ultimately, the ethics committee's opinions or recommendations must be accepted or rejected by the patient or her surrogate and her physician. The final, common pathway for decisions, even in these days of team care, will be through the physician-patient relationship. In the end, the physician writes the order and cannot escape accountability. The patient may discharge the physician; team members may refuse to carry out the order; but, legally and morally the decision rests with the physician and the patient or surrogate, none of whom can displace responsibility to the committee.

1. E.D. Pellegrino and D.C. Thomasma, *For the Patient's Good: The Restoration of Beneficence in Health Care* (New York: Oxford, 1987).

The focus of an ethics consultation, therefore, is *this patient,* here and now, who is facing, by himself or herself or via proxies, a difficult ethical dilemma. The task of the ethics committee and of the consultant is to enable physicians and other health professionals to make the choices that are ethically optimal, given the complexity of the clinical context. The final decision rests in the moral space somewhere between the physician and the patient or surrogate. This is true whether the consultation follows the authoritarian or facilitative model.

II. Conscience and Character

Ethics consultations are usually, but not exclusively, group processes. How to distribute moral accountability among committee members is an insufficiently examined ethical question.[2] In collective decisions, some responsibility rests with each committee member. What is the individual's responsibility for the moral quality of his or her group's decision?

There is a strong psychological tendency to compromise individual opinion in the interest of resolving or avoiding conflict. This tendency grows in intensity the longer the team works together. Each member, therefore, must have sufficient moral maturity to recognize when it is possible to compromise on a moral issue without loss of personal integrity and when it is not. Consultants must be able to discern when they have a moral obligation to abstain, differ from, or object to either the committee process or the outcome.

Committee members cannot absolve themselves of moral accountability simply because the majority reaches an agreement. The legal or institutional legitimacy of ethics committees as advisory bodies confers no moral authority. Neither does ethical expertise, credentials, or eagerness to serve. There is always an obligation to object, civilly and responsibly, if the committee's advice or decision seems ill-advised.

Some of these points are included in the report's discussion of character traits desirable for ethics consultants. These same character traits would be desirable in any activity in which fidelity to trust is central. They can be taught and nurtured, as the report suggests, in a properly functioning ethics committee. Such a committee provides the community of values needed to sustain the virtuous actions of its members.[3] The "collective virtue" of the committee is an important element in the formation of each member, as it is in the making of morally good decisions.

To this end, the essential collective virtue is a commitment to the right and the good, regardless of exigencies to the contrary. Moral compromises confusing consensus with moral propriety or acceding for the sake of peace are less likely to occur when this commitment is explicit. The external appearance of virtuous action must also be matched by internal virtuous motivation.[4] This is much harder to assure or evaluate.

2. E.D. Pellegrino, "The Ethics of Collective Judgments in Medicine and Health Care," editorial, *Journal of Medicine and Philosophy* 7, no. 1 (February 1982): 3–10.

3. A. MacIntyre, *After Virtue* (South Bend, Ind.: University of Notre Dame Press, 1981), 244–45.

4. J. Wallace, *Virtues and Vices* (Ithaca, N.Y.: Cornell University Press, 1978), 42–44.

Serving as an ethics consultant is intellectually, emotionally, and morally demanding. Those who venture to assist others in making crucial moral choices bear moral obligations. No formula can be written for how to steer between the extremes of facilitating and making authoritative statements, or for how to provide expertise while nurturing independent decisions. In order to safeguard against abuses of human dignity and personhood, the welfare of the person in whose interest the decisions are being made—as well as the character of those who help to make the decisions—must be protected.

III. Religious "Values"

The term "values" is notoriously flexible, hence its popularity and its frailty in moral discourse. The report (in a footnote on page 3) equates values with "various normative dimensions. . . ." To that extent, it seems to confine its use to *moral* values. Later (page 4), reference is made to "societal values which frame the context of the consultation." Here the term seems to include other values as well. This ambiguity is especially relevant in discussion of religious values.

It is not clear where religious values, which may be normative or non-normative fit, nor what their relative weight should be compared to other normative values. For many patients engaged in ethics consultations—and their consultants as well—religion has deep personal meaning, and, for many, religion is the ultimate source of morality. It cannot be sacrificed for purposes of exigency or harmony. Given the religious pluralism present in society today, it is increasingly evident that conflicts often center on differences in religious belief. When these beliefs clash with economics, societal mores, conventional beliefs, or the consultant's own beliefs, special care is necessary to protect the patient's interests.

I believe that consultants require a better knowledge of religious beliefs, practices, and spirituality that are different from their own. This is an obligation, regardless of whether or not the consultant, herself, is a religious believer. Respect for the patient makes respect for his or her religious beliefs a moral obligation. This does not mean automatic accommodation to those beliefs. It does mean that religious belief must be given a very high priority and accommodated unless it presents a very serious harm to others.

The fundamental problem is the clash between society's generally relativistic value system and religious absolutist demands in moral matters. A person in a permanent vegetative state, for example, may be viewed very differently by believers and nonbelievers. Some would think it "reasonable" to urge removal of all life support on grounds of poor quality of life, economics, or aesthetics; others might reject these reasons but permit withdrawal if the burdens of treatment outweighed the benefits; still others might want everything done because they hope for a miracle. Each person believes honestly that he or she is acting in the patient's best interests and cannot compromise his or her position.

The consultant may be equally divided. When religious values or norms are seen as simply one set of values among many others and subject to social construction, conflict tends to be negotiated away. When religious values are "trump" values, however, they cannot be negotiated, since there is no basis for compromise. Here, the absence of a common moral vision in our society becomes critical.

This is far too complex a problem for analysis here. It involves conflict resolution or avoidance between health professionals and patient, between patient and institution, and between patient and society. At each point, the conflict of moral visions can be so fundamental that the facilitative model of ethics consultation becomes unrealistic and ineffective, yet the authoritarian model can be injurious.

Here, the mechanism for avoiding fundamental, substantive moral issues in favor of procedural concerns simply will not suffice. This is not a problem that the next report can be expected to resolve. In the future, the role of religious beliefs must be emphasized more strongly in defining the process of ethics consultation and the criteria for competencies of ethics consultants. The importance of the consultant's acquaintance with the more common belief systems, and the gravity of the obligation to respect those beliefs, must be further emphasized.

IV. Ethics, Law, and Psychology

Ethics consultation often involves an intermingling of legal, psychological, and ethical issues. Too often these are confused in practice. The discussion becomes diffuse and the outcome more distressing than helpful. In order to serve the ends of ethics consultation best, these elements, while interrelated, must be disentangled, since their methodologies, sources of evidence, and implementation are different. Each contributes something different to the consultation. At the end of the consultation, they may not be reconcilable.

First, there is a strong tendency among Americans who respect but also fear the law to worry first about lawsuits. Law and ethics overlap to some extent, but they are far from being synonymous. What is legal may not be moral and what is moral may not be legal. Even for those who recognize this distinction, fear of legal action, often based in an inadequate knowledge of the law, tends to dominate discussions. Here, again, it is necessary to remember that the purpose of an ethics consultation is a morally good decision for the patient, not the well-being of the health professional or institution.

Equally problematic is the widespread psychologization of every aspect of life. People today seek to be "comfortable" with all their decisions and take that feeling as an indicator of being morally right. As a result, ethical conflicts tend to be reduced to emotional or personality differences that can be remedied by group psychodynamics. As a further result, careful ethical analyses may be spurned as too abstract, out of context, or divisive. Simply resolving conflict or attaining agreement may be mistaken as "success" in consultation.

Legal and psychological issues are undeniably important and deserve attention. It is necessary to remind participants at the outset, throughout the consultation, and at its end, however, that the goal is a right and good decision for a particular patient. Whether this decision is "legal" or psychologically satisfying is an important but secondary matter in ethics consultation.

None of this is to deny that ethics consultations increasingly are about the human interactions within the healthcare team and between the team and patients or their surrogates. In some ways, the term "bioethics consultation" may obscure the fact that the problem is often primarily one of disagreement about patient-care management. All too often, the ethics consultation is the only occasion when all the decision makers—patient, surrogate, physicians, nurses, social workers, pastoral counselors—are in the same room at the same time. The ethics consultation provides a rare opportunity for a comprehensive survey of the patient's problems and needs around which a fruitful interdisciplinary and interprofessional dialogue can take place.

An orderly analysis—followed by a re-synthesis—of the ethical, legal, and psychodynamic realities should begin with a discussion of what is the morally right and good thing to do. That should be followed with a delineation of the legal implications of what is morally proper, and end with an assessment of how best to implement the morally proper decision, given the socio-psychological relationships between decision makers. In this process, legal and psychological expertise is relevant to the consultation, but does not determine what is morally right or good.

Ethics consultants, therefore, need to gain familiarity with the differences between and among law, ethics, and psychology and the special contributions each may make to a consultation. This entails some form of initial and continuing education for ethics consultants. It also entails making sure that consultants are available to assist in committee deliberations. They should be consultants, not "tie-breakers," lest they be used to relieve ethics committee members of their collective responsibilities.

V. Advice or Conflict Resolution?

The report compares and contrasts the facilitative and authoritarian models of ethics consultation. In today's pluralistic society, the tendency to favor the facilitative is proper. In actual practice, however, the distinction may not be as clear-cut, especially if the moral conflict is internal, that is, within the moral agents themselves, rather than in the group process. Even if a conflict within a group can be resolved, the patient, his or her family, or the health professionals involved may wish to receive personal advice to settle persistent doubts about the morally appropriate thing to do.

Most people do understand their personal accountability for the morality of acts that they voluntarily undertake or condone. Regardless of what a committee's decision or advice may be, conscientious people wish to examine the issues more carefully within the context of their own values—something they may not wish to share in a group discussion. When this happens, they may ask for directive advice not offered within the formal consultation.

During these private consultations, a facilitative stance might not suffice. Consultants are often asked, "What would you do?" "What do you think I should do?" People involved in ethics consultations will ask these questions if they know, trust, or admire the judgment of the consultant. To be too directive incurs the danger of imposing the consultant's values; not to provide direction can be a species of moral abandonment.

"Private" consultation may occur after the fact, but increasingly we see it before the decision is made. This places a heavy burden on the consultant, who must be especially judicious. Here, the character of the consultant becomes especially relevant since she or he becomes, to some extent, an accomplice if his or her advice is followed.

Private consultations are not meant to substitute for more formal group consultation when conflicts among decision makers are at issue. They can be used to encourage more formal consultation, if that is more appropriate. When that happens, the consultant may detect a conflict of interest in subsequent committee deliberation. This should be recognized and disclosed. If the conflict is of significant proportions, the consultant should withdraw from the more public discussion, since confidentiality must be preserved.

Sometimes an ethics consultation is requested to settle disputes within the healthcare team rather than disputes with the patient, institution, or surrogate. Here, the issues center on inter-professional conflict among team members. Consultation of this type, with the team and without the patient or family, can be both facilitative and authoritarian. It may even be prophylactic, to avoid conflict before engaging patients or their families in a subsequent committee discussion.

VI. Organizational Ethics

As the report recognizes, organizational ethics is in its infancy. Its definition, scope, and the procedure appropriate to effective consultations are as yet matters of debate. The section of the report on organizational ethics, therefore, will undoubtedly undergo significant refinement in the future. I have some significant reservations about this section of the report.

For one thing, the report identifies organizational ethics too closely in terms of financial, billing, contractual, and marketing practices. At least those are the examples used to illustrate important issues. There is a need for a more fundamental definition of organizational ethics, one which sets it apart more clearly from clinical ethics, and, thus, more clearly differentiates clinical and organizational ethics consultations.

The term "organizational ethics" covers the whole range of obligations entailed in collective action by a group of persons who act in the name of an institution—framing and implementing policies that may affect individual patients, professional and nonprofessional staff, the community, and so on. Various levels of responsibility are involved—the board of directors, professional staff, administrators, investors, and so on. Each bears some responsibility for the effects of its actions or policies on the care of patients, whom the institutions purport to benefit and for whom the institutions exist in the first place.

Given the commercialization and commodification of health care today, the organizational issues outlined in the report are, indeed, a source of conflict. More subtle but equally in the domain of organizational ethics are the location of responsibility for overall quality of care, credentialing and policing of professional and nonprofessional staff, humaneness and absence of discrimination in care, care for the poor and needy, allocation of resources, and so on. Despite the economic exigencies of the day, hospitals are, at least, socially essential institutions that deal with vulnerable members of society. They are in a trust relationship to the sick who seek their help. They collectively incur responsibility if their policies or the implementation of those policies cause harm. Those who act with the authority of the organization, or gain from its efforts (such as investors), cannot escape responsibility for harms caused by the structure, operation, or personnel of the institution.

Precisely how moral accountability is to be distributed in an organization remains a debated subject. The need for a clearer theoretical framework for decisions is made more urgent by the intrusion of elements of the market and the corporate ethos into health care. Clinical consultation has benefitted from thirty years of extensive study of the ethical theory that applies to the care of patients. Organizational ethics needs equal attention.

The technique and process of consultation might be quite similar in both clinical and organizational ethics, but the differences are too significant for both types of consultation to be left to the same committees or consultants.

First, clinical ethics consultations involve health professionals, patients, and families who have had relationships before the consultation. The conflicts are interpersonal and are usually centered on the welfare of the sick person. Organizational ethics involves persons who have had minimal or no contact before the conflict is aired. Usually these persons are not health professionals. They are administrators, clerks, trustees, or distant owners of institutions, whom the patient has not seen and who do not see the patient. Organizations are notoriously faceless but are not, on that account, excused from moral complicity.

Second, the consultants and others acting for the collective conscience of the institution are employees. Thus they are in immediate conflict of interest when they must clarify disputes between patients and the institutions that pay their salaries. Some, like in-house lawyers and risk-management personnel, are even paid to protect institutions. Such persons would have automatic conflicts of interest during both clinical and organizational consultations.

Third, as the report recognizes, these issues require a different knowledge base in organizational ethics, which is in the process of development. Helping consultants to improve their skills and knowledge is, therefore, more difficult. Moreover, some of these issues will be in the realm of business ethics, which is less morally demanding than clinical ethics, or at least is built on a different set of beginning assumptions.

These reasons outweigh the putative similarities acknowledged in the report (page 29). Thus, the processes and the persons involved in clinical and organizational ethics are best separated. For the most impartial organizational ethics consultation, most, or preferably all, of the consultants should be outside the employ of the hospital and as free of financial conflicts of interest as possible. A community-based committee, assisted by the expertise of ethicists and the use of hospital personnel as resource persons, is probably the safest alternative.

Consultations by commercial, fee-for-service ethicists, not discussed in the report, are beginning to be used by institutions where the usual sources—universities, medical schools, hospital pastoral-care services—are lacking or insufficient. They underscore the necessity for some kind of credentialing. These ethicists are also subject to very significant conflicts of interest, since they are effectively in the employ of the hospital and are dependent upon renewal of contracts for survival as commercial entities. The dangers of subtle capitulation of ethics to fiscal exigencies are obvious.

Similarly, the ethics committees set up "in-house" by managed-care organizations (such as health-maintenance organizations and hospital networks and corporations) are suspect, no matter how well-intentioned their members may be. Self-interest must be kept at a minimum if ethics consultants are to facilitate agreement and clarify issues involving organizations. Even though the focus is the organization, clinicians should provide chosen vital connections with bedside reality that are too easily lost when law, politics, and/or economics dominate the discussion.

Given the differences in their foci of interest and the inevitable expansion of ethics consultation, two separate committees should be formed. There is no denying their overlap at various points. Joint meetings between the two committees would provide opportunities for each to contribute its expertise to the other. If patients and the public are to have confidence in the consultations and the process, however, they must be assured that those who purport to help them make decisions are as unbiased as fallible human beings can be.

This separation of duties does not preclude assigning those hospital policies that are most directly concerned with patient care to the clinical ethics consulting committees. Do-not-resuscitate (DNR) orders, policies on withholding and withdrawing life support, consultations, and so on, are in the domain of clinical ethics proper and should be developed as close to the bedside as possible. Other policies, relating to billing or financial incentives for physicians, which services to pay for, and so on, are best relegated to the organizational committee.

VII. Accreditation and Credentialing

The report wisely refrains from taking a position on the currently hazardous question of credentials and accreditation of consultants as "experts." Nonetheless, the report does devote fifteen pages and two tables to detail a long list of knowledge, skills, and character traits it deems necessary for consultants. The report is vague on how all of this is to be acquired, such as from self-teaching or preparation for a formal degree. The tables and lists are formidable, yet the report seems to suggest in some places that these can be acquired through practice, self-teaching, or brief courses. As the complexity of the issues increases and the literature expands, it seems certain that serious study of bioethics must be required.

Future reports must confront the problem of standardization more directly. Anyone familiar with actual consultation processes knows how widely the knowledge and skills of those acting as "consultants" today vary. Many have natural talents for conflict resolution but lack ethical expertise; many have the technical knowledge but are woefully lacking in interpersonal skills. The ideal consultant would possess both sets of skills, but reality makes this unlikely for most consultants. Are there not two different roles here?

At the beginning of my own observation of ethics consultations, I was opposed to a certification process. Those who are opposed to credentialing and certification fear several deleterious effects, such as the creation of a group of "experts" who "solve" ethical issues, excusing or displacing the persons who must finally take accountability for the decisions and actions. This is especially dangerous in a society obsessed with specialties. Moreover, a cadre of experts, consciously or not, could impose personal or class beliefs, repress dissent, or even constitute an ethics "police." Most of all, experts would dampen the democratic processes of open discussion and impede the development of the capacity all educated people should possess to make their own moral judgments.

These objections are valid; however, there are trends that make ethics expertise essential. Too many ethics committees are composed of good-intentioned, public-service-minded persons who are self-styled "bioethicists." Too many eschew ethical analysis as too "rationalistic" and focus only on the concrete particulars and human interactions involved in the decision. Some even define the whole of bioethics as simply a sophisticated process for conflict resolution and not a source of ethical clarification or normative guidance.

Both positions have some basis in experience. Both must be accommodated in ethics consultations. In a pluralist society, ethics consultation committees and teams should include a wide range of persons from a variety of fields and interests. These persons make up the core consultative group. But each group also needs a resource person who is a trained ethicist—one with an advanced degree in ethics—to provide the expert knowledge base upon which informed opinion can be grounded. In this schema, the "expert" is not a direct consultant to the patient or family but a consultant to the committee, the institution, and the process of consultation.

This does require a re-imaging of the way ethics consultants operate, but it seems a necessary re-imaging. It provides for expertise in a complex field without capitulation to that expertise, and it provides democratic, inclusive, and varied input from a wide variety of community representatives without ignoring the increasing necessity for ethical expertise.

Conclusion

The SHHV-SBC Consensus Panel has produced a very commendable report. It is an excellent working guide that, in the light of experience and evaluation, will assuredly receive serial revision. In anticipation of those revisions, I would make several suggestions. The focus of clinical ethics consultation on the welfare of the patient must be strengthened. The relative weights given to law, ethics, psychology, and religion is a crucial issue that requires continuing assessment. Organizational and clinical consultation must not be conflated. Some standardization of expertise is unavoidable. Finally, the roles of ethics experts and members of the larger health-professional community in the consultation process must be re-imaged.

C. Systemic Challenges and Opportunities

33

Learning Health Systems, Healthcare Delivery Systems, and Systems to Address and Reduce Error

David G. Miller, PhD

THE CONTEXT AND SYSTEMS WITHIN WHICH medicine is practiced can affect clinicians' abilities to foster, respect, and protect the clinician-patient relationship. Dr. Pellegrino recognized at least two ways that healthcare institutions and systems are challenging the practice of medicine today. One is the way in which we deal with medical error, and another is the way in which health care is organized and delivered.

Virginia Ashby Sharpe's introduction to Pellegrino's essay on error situates that threat to medical practice; namely that the diffusion of responsibility in medicine—and, thereby, the potential for diffusion of accountability for error—poses a threat to responsible medical practice. Sharpe points to the ways that those concerns may be addressed.

The context within which medicine is practiced today—which defies the ascription of a healthcare "system"—threatens the centrality of the clinician-patient relationship in numerous ways. The fact that health care is not sufficiently valued as an element of the common good and is, instead, treated as a commodity threatens the clinician-patient relationship in a variety of ways: It envisions clinicians as medical "providers" and patients as "consumers," which diminishes the importance of relationships and the ineradicability of trust for a proper healing act; it encourages a "production-line" mentality with respect to treating patients; it encourages a "service industry" conception of medical care in which consumers are provided with a menu of services to select without respect to their vulnerability and lack of expertise; and it encourages the "proletariatization" of medical professionals such that they are interchangeable and can be pulled into the treatment assembly line as directed by managers. The ways that health care is financed—with for-profit and not-for-profit systems competing for patients, with insurance companies skimming profits by stimulating that competition, with financiers lobbying politicians for regulations that benefit their profitability, etc.—contribute to those pressures. In two essays, Pellegrino offers suggestions for how those pressures might be addressed in morally licit ways—and points to ways that professionals must resist, and work to reform, those systems.

Another aspect of the system of healthcare delivery that may prove to be an opportunity (if approached and finessed properly) or a threat (if approached without careful planning and foresight) resides in the establishment of learning health systems. F. Daniel Davis examines the ways that Pellegrino's approach to medicine and to healthcare systems can provide guidance for the integration of medical research and systems research into medical institutions. Personalized or precision medicine will require massive amounts of data from patients as well as responsive practitioners who learn as they practice. Of course, practitioners are always self-directed learners, but team-based medicine and precision medicine will require them to be self-aware and systems-aware participants in learning health systems.

34

Introduction: Prevention of Medical Error: Where Professional and Organizational Ethics Meet

Virginia Ashby Sharpe, PhD

IN 2001–2002, I led a Hastings Center research project: "Promoting Patient Safety: An Ethical Basis for Policy Deliberation." The project focused on a core premise of the then-new field of patient safety, its ethical implications, and the policy proposals to address it. The core premise is that "the complexity of healthcare delivery systems—rather than flaws in individual performance or character—is the source of most preventable medical errors."[1] The key policy proposal called for a far-reaching, concrete agenda to shift from a "shame and blame" culture in health care to a culture of patient safety.

Dr. Pellegrino was a participant in that project, and his chapter, "Prevention of Medical Error: Where Professional and Organizational Ethics Meet," was a distillation of his thoughts about how to preserve and promote individual moral responsibility in health care even as policy proposals seemed to sweep aside norms of individual accountability in favor of vague references to amorphous "systems" (see chap. 35 below).

In the chapter, Pellegrino is most concerned about what he refers to as a "blame-free" approach and its potential to erode the values and virtues of professionalism. He outlines the moral hazards of "blaming the system," including dulling the moral sensibilities and sense of obligation of health professionals, abdication of individual responsibility for patient care, and abandonment of justice for those harmed. Ultimately, he argues that "[e]mphasis on the system is essential but so too is the retention of moral accountability of the team as a whole and of each of its members [including healthcare executives and trustees]."

In reading the chapter now, my sense is that Pellegrino, like most of us who have thought about responsibility, got hung up on responsibility in the *retrospective* sense,

1. Virginia A. Sharpe, "Mistakes, Medical," in *Encyclopedia of Global Bioethics,* ed. H. ten Have (New York: Springer, 2016), 1923–30, https://link.springer.com/referenceworkentry/10.1007/978-3-319-09483-0_296.

that is, ascription of blame after the fact of harm. Accordingly, he grappled with how healthcare professionalism could be sustained and justice could be done to harmed patients if blame could not be assigned.

Since this paper was written, the "just culture" literature has done a very good job of differentiating between medical harms that are blameworthy and those that are blameless; disclosure of harms to patients and families has become an expectation reinforced by the Joint Commission; states have passed "apology laws" that support providers' honest communication with harmed patients; and many health systems have implemented an "apology and offer" system to fairly compensate harmed patients without the ordeal of the adversarial tort process. These and other steps have helped to address some of Pellegrino's concerns about justice.

That said, Pellegrino is clearly arguing in the paper for responsibility in the *prospective* sense, that is, a resolute commitment to the fundamental norms of healthcare professionalism laid out in his philosophy of medicine. He underscores that the increasing complexity of healthcare systems does not *diminish* responsibility for sustaining the norms of professionalism, rather it *expands* that moral obligation to all actors in healthcare delivery whether they are clinicians, paraprofessionals, administrators, or trustees. If individual choices, actions, and interactions create the complexity of healthcare delivery, then all individuals who have a role also have a responsibility to the primacy of patient welfare, including organizing to prevent patient harm and protect front-line providers so they can deliver safe care. As I write this, the COVID-19 pandemic is overwhelming hospitals across the world. One stark lesson is that the safety of patients depends on a system designed to protect, sustain, and support those who provide direct patient care.

35

Prevention of Medical Error: Where Professional and Organizational Ethics Meet

Edmund D. Pellegrino, MD

Introduction

The report of the Institute of Medicine (IOM) on the prevalence of medical error has engendered widespread attention to a human problem as old as medicine itself.[1] Physicians, patients, and the public have always recognized the fact of medical fallibility. Few physicians can claim that they have never made an error of judgment or procedure. Few have not observed the errors of their colleagues. All are cognizant that the claim of the profession to police itself has never been responsibly actualized. To date, no comprehensive program of error prevention has ever been actualized. It is the possibility of such a program that the IOM report, its shortcomings notwithstanding, has brought into public and professional consciousness.

Past efforts at prevention of error relied exclusively on the moral restraints of medical ethics or the deterrent effect of punitive laws. Like modern tort theory, the Law of Hammurabi (circa 1900 S.C.) and the Roman *Lex Aquilia* and *Lex Cornelia* imposed harsh penalties for medical harm.[2] The Hippocratic Oath and similar codes in many cultures centered on the moral obligation of nonmaleficence. In the last century, autopsy reports, morbidity and mortality reviews, pharmacy and therapeutics committees, as well as quality control and risk management policies have had

---

Reprinted with permission from Edmund D. Pellegrino: "Prevention of Medical Error: Where Professional and Organizational Ethics Meet," in *Accountability: Patient Safety and Policy Reform*, ed. Virginia A. Sharpe (Washington, D.C.: Georgetown University Press, 2004), 83–98. Copyright 2004 by Georgetown University Press, www.press.georgetown.edu.

1. L.T. Kohn, J.M. Corrigan, and M.S. Donaldson, eds., *To Err Is Human: Building a Safer Health System* (Washington, D.C.: National Academy Press, 2000).

2. A. Castiglioni, *A History of Medicine*, trans. E.B. Krumbhaar (New York: Knopf, 1941).

small but definitive effect. But as the IOM report and other studies show, these measures have not on the whole been sufficiently effective.[3]

Against this background of ineffectiveness, the report and many of the chapters in this volume [*Accountability: Patient Safety and Policy Reform*] turn away from blame and moral censure of individual physicians and health workers toward systemic changes in every level of the healthcare system.[4] Models of accident prevention taken from air safety, the nuclear power industry, and occupational safety are being fitted to the organizational context of contemporary health and medical care. Central roles are also given to cooperation between administrators, health professionals, patients, and accrediting agencies as well as individual health professionals. The locus of confidence is being consciously shifted from individuals to systems, institutions, policies, and regulations.

These are obvious responses given the prevalence of medical error and the institutionalized and organized context of today's medical care. Such changes should be encouraged, but with caution. Social crises always spawn enthusiasm for "salvation themes"—new ideas that are expected to revolutionize seemingly decrepit traditional practices. In our enthusiasm for the new, we must remember that system change has its own failings.

The more complex a system of any kind, and the more finely articulated its elements, the more vulnerable it is to failure. Humans are unavoidably elements in a system; individual performance is still crucial. In spite of systemic safety practices in aviation and the nuclear power industry, human error in these areas has not been eliminated.

In this chapter, I argue (1) that a properly organized organizational and systemic context is essential to reduce the prevalence of medical error, (2) that its effectiveness and efficient working depend on a parallel affirmation of the moral duty and accountability of each health professional in the system, (3) that each individual health professional must possess the competence and character crucial to the performance of his or her particular function as well as those of the system as a whole, and (4) that the major function of a system is to reinforce and sustain these individual competencies and virtues.

Systems Are Not Enough

Systems cannot make the professionals within them virtuous, but they can make it possible for virtuous professionals to be virtuous. Correspondingly, a defective system can discourage even the conscientious individual or reduce his or her aspirations to nothing higher than the level of the average. The moral sensibilities of each health professional must be preserved lest the system for error prevention itself become part of the problem.

3. V.A. Sharpe and A. Faden, *Medical Harm: Historical, Conceptual, and Ethical Dimensions of Iatrogenic Illness* (Cambridge: Cambridge University Press, 1998).

4. [Virginia A. Sharpe, ed., *Accountability: Patient Safety and Policy Reform* (Washington, D.C.: Georgetown University Press, 2004)—Ed.]

All of this is especially relevant to healthcare systems. Medical error takes place in a nexus of intricate human relationships. Health "care" is channeled ultimately through persons in unique relationships far less subject to formularization than air travel, nuclear power production, or the manufacture of automobiles or complex pharmaceuticals. No system of error detection or prevention can confine these intricate human relationships within a set of preordained algorithms.

There are also rogue systems, as there are rogue individuals and nations. The ingenuity of humans to distort, bypass, or subvert any system has been demonstrated all too often. Overreliance on a systems approach can be as illusory as past overreliance on the moral integrity of individuals. Even while systemic responses are being designed, it will be necessary to reaffirm the moral nature of medical error, and to retain the notions of blame, accountability, and responsibility. Any system designed to protect patients must sustain and nourish the ethical obligations inherent in professing to be a healer, helper, and caretaker for humans in distress.[5] In fact, the "system" (team, hospital, policy) will need to make the same "profession" of commitment to the welfare of the patient.

This line of argument will require integration of professional and organizational ethics, as well as the relationship between them. Each domain has its own set of obligations. Conflicts between, and among, domains will arise. In an effective "system" of prevention and detection, these conflicts must be put into the right moral order in accord with some order of priorities.

This balancing can only be effected by an architectonic "first" principle fundamental to what medical and health care are all about. That principle, in both the system and the individual clinical encounter, must be the primacy of the good of the patient. The patient is the presumed beneficiary of the individual clinician and the system as well. Detection and prevention of medical error has its ethical foundation in the duty to act for the good of the sick and to avoid harm to patients. Every action of individual professionals and every organizational policy and regulation must be measured by this gold standard of traditional medical morality.[6]

Individuals and Systems—Some Philosophical Notes

The moral interdependence of individuals and societies was a subject of major import to Plato. In the *Republic*, he proposed that a good society was one that acted for the good of its citizens.[7] On the other hand, social good depended upon each person in a society performing that task for which he was best suited and qualified.[8]

5. E. Pellegrino, "The Internal Morality of Clinical Medicine: A Paradigm for the Ethics of the Helping and Healing Professions," *Journal of Medicine and Philosophy* 26, no. 6 (2001): 559–79.

6. E. Pellegrino, "The Healing Relationship: Architectonics of Clinical Medicine," in *The Clinical Encounter*, ed. E. Shelp (Dordrecht: D. Reidel, 1983), 153–72.

7. Plato, "The Republic," in *The Collected Dialogues*, ed. E. Hamilton and H. Cairns (Princeton, N.J.: Princeton University Press, 1983), Book IV, 420b, 519e.

8. Ibid., book II, 369b–e; book IV, 423d.

This interdigitation of individual and collective virtue, according to Plato, makes for a well-ordered and well-functioning republic.

This same ethical reciprocity is applicable in an analogous way to a well-functioning system of medical error prevention. This is different from the utopian social engineering schemata of Charles Fourier or Karl Marx.[9] Their hope was to design social systems that would reshape human moral conduct and perhaps even human nature. The malignant social engineering of totalitarian states had the same goal. Their common failing was to substitute social for personal virtues in violation of the Platonic ideal of a balance between them.[10] This is the danger an overly enthusiastic embrace of any "system" must avoid.

This is not to deny the potential for modifying and sustaining professional virtues in a system whose end is truly the good of the patient. One might then be able to speak of "virtuous" organizations and systems for error prevention. In such organizations the individual and the system reinforce each other in the pursuit of the good each purports to serve. Systems do not have a blank slate on which to write a prescription for human behavior. The "blank slate" conception of human nature is a recurrent fallacy, whether the proposed designing agent is genetic on the one hand or environmental or social engineering on the other.[11]

Individuals can be inspired by the ideals of a morally sound organization. The best always rise above the satisfaction with mediocrity that systems can induce even in ordinarily well-intentioned professionals. Witness the lassitude of even good clinicians in the face of the injustices of today's "healthcare system," or the subversion of professionalism in managed-care organizations. Systems are no more immune to moral corrosion than the individuals within them.

Medical Error Morally Defined

In its report, the Institute of Medicine provides a topography of definitions of medical error.[12] It settles on the following: "the failure of a planned action to be completed as intended (error of execution) or the use of a wrong plan to achieve an aim (error of planning)."[13] The institute focuses for the most part on errors of execution.

Under its preferred definition, the Institute of Medicine includes "adverse events," that is, injuries caused by medical management. When the adverse event is attributable to medical error, it is considered "preventable." A preventable adverse event due to negligent care becomes malpractice when it meets the legal criteria by violation of the prevailing "standard of care."[14] Later the report distinguishes "slips,"

9. Charles Fourier, *Le nouveau monde industriel et sociétaire, Ou invention du procédé d'industrie attrayante et naturelle distribuée en séries passionnées* (Paris: Bossange père, 1829).

10. Y. Simon, *The Definition of Moral Virtue* (New York: Fordham University Press), 3–15.

11. S. Pinker, *The Blank Slate: The Modern Denial of Human Nature* (New York: Viking Press, 2002).

12. Kohn, Corrigan, and Donaldson, *To Err Is Human*, 28, 36, 54.

13. Ibid., 54.

14. Ibid., 28.

"lapses," and "mistakes," as well as latent errors (those at a distance from the "front line") and active errors (those on the "front line").[15]

In each of these definitions, the meaning of "error" much depends upon how error is measured. Methodological limitations and differences in interpretation that may affect the significance of the report have been mentioned already.[16] Whatever more reliable measures of error may later reveal, the report has sounded an alarm that challenges our complacency and commands a critical appraisal of causation, remediation, and prevention of error.

Finally, the very term "medical error" is misleading. It seems to suggest that all error is physician-generated. As the report itself makes clear, this is not the case, particularly if its recommendation of a systems approach is valid.[17] I believe that "patient care error," which includes everyone that participates in health care, is a more suitable term. However, "medical error" is used so widely that I shall retain it in this discussion, despite my preference for the more inclusive term.

Error and Blameworthiness

The ethical desiderata outlined below can apply to all forms of error at the front line, that is, errors attributable to the direct encounter of a patient with the health professional or an organizational policy or procedure. A blameworthy error is one in which there is a failure for which moral accountability can be assigned, one that was preventable if obligations had been fulfilled, if competence had been maintained or properly designed procedures had been consciously followed. Blame can be attached to individuals, teams, organizations, system designers, institutional trustees—to those who act in a system on the "front line" or to those at a distance who authorize others to act in the name of a system or organization.

The degree of blameworthiness will vary with the degree of negligence, the amount of harm done to the patient, the degree of collaboration with the harmful action, and the level of responsibility and authority. With respect to blameworthiness, errors of execution and planning are not as neatly separable as the institute's report suggests. It is helpful to focus on the stages of care at which the error occurs, that is, history taking, diagnosis, prognosis, selection of treatment and laboratory examination, performance of interventional procedures like surgery, cardiac catheterization, prescription writing, etc.

By focusing on specific elements in the patient care process, the relative contributions of the individual healthcare professional, the technical assistant, the system,

15. Ibid., 54–55.

16. E.J. Thomas, D.M. Studdert, and T.A. Brennan, "The Reliability of Medical Record Review for Estimating Adverse Event Rates," *Annals of Internal Medicine* 136 (2002): 812–15; T.A. Brennan, "The Institute of Medicine Report on Medical Errors—Could It Do Harm?" *New England Journal of Medicine* 342 (2000): 1123–25; C.J. McDonald, M. Weiner, and S.L. Hui, "Deaths Due to Medical Errors Are Exaggerated in Institute of Medicine Report," *Journal of the American Medical Association* 284 (2000): 93–95.

17. Kohn, Corrigan, and Donaldson, *To Err Is Human*, 54–55.

the procedures, and the organization to the error can be better differentiated. Any culpability for the error will depend on the degree to which it is preventable in the expected operation of each of these agents, front-line or distant. Errors can occur in what has been termed the "inherent" fallibility of medicine.[18] This term refers to the limitations of medicine as a science of the particular in which the application of general rules and laws may not fit particular cases precisely. Thus, adverse outcomes at certain levels of occurrence can be expected even in the "best hands" following the best procedures. A certain level of morbidity and mortality, for example, can be expected in complicated or delicate procedures. If all has been done properly, error due to the inherent fallibility of a system or the state of the art may be the only acceptable, really "blame-free" error, at least from the ethical point of view.

As the profession's and the professional's knowledge and experience grows, the excusable level of these "inherent" errors may decrease. A preventable adverse action is a moral failure that cannot be exculpated by the "blame-free" approach advocated by proponents of the system dimensions of error. Whether it is committed by an individual, a team, or an institution, culpable error is a violation of the implicit promise each of these entities makes when it offers itself as an instrument of help and healing. By offering to help, the physician and nurse, the healthcare team, the hospital, and even the managed-care organization implicitly promise competence and good judgment in the interest of the person who seeks help. This is their "act of profession," their commitment to the good of the patient.

The professional, the team, or the hospital is morally required to be faithful to this trust. Preventable or negligent harm as the result of error is a moral failure, whatever its legal status may be. Moral blame and culpability are therefore unavoidable. Removing individual blame could conceivably encourage better reporting of errors or more readiness to accept rehabilitative measures. It might also simplify litigation and standardize the tort system. It could mitigate the punitive atmosphere and enhance collegiality and interprofessional communication.[19]

However, despite these beneficial possibilities, there are the associated dangers of complacency and dulling of the moral sensibilities of the humans in the system when either a "blame-free" approach or a "blame-the-system" approach is adopted. The power of individual guilt can be constructive as often as it is destructive. Personal accountability is owed to the person injured. The deterrent effect of fear of shame and blame is not safely ignored.

However well-organized and finely tuned the system may be, it must leave room for discretionary space in so human and uncertain an enterprise as medical

18. S. Gorovitz and A. MacIntyre, "Toward a Theory of Medical Fallibility," *Journal of Medicine and Philosophy* 1 (1976): 51–71.

19. Bryan A. Liang, "Error Disclosure for Quality Improvement: Authenticating a Team of Patients and Providers to Promote Patient Safety," in Sharpe, *Accountability*, 59–82.

20. E. Pellegrino, "The Expansion and Contraction of 'Discretionary Space,'" in *Priorities for the Use of Resources in Medicine*, National Institutes of Health, DHEW Publication No. (NIH) 77-1288 (Washington, D.C.: U.S. Government Printing Office, 1977), 99–112.

care.[20] The health professional needs that space if the care she provides is to be personalized, individualized, and humanized. That space is also necessary to provide for the unforeseen, unique, or unpredictable event in clinical care. It is within that space that the health professional can prevent the systemic shortcomings of the system itself from injuring the patient. It also allows the individual to detect the ill-conceived or poorly designed element in a systems approach. Of course, it is also within the region of discretionary space that the possibility of error may be increased. Within that space, the character of the health professional is the patient's last safeguard.

The blame-free approach advocated by many in the patient-safety movement cannot erase the individual moral onus of preventable error. The physician or other health professional cannot blame the system for his or her failings. Organizational conformity cannot take the place of moral responsibility even if the system follows a no-blame rule. The physician or the nurse, in their respective realms, are the pathways through which harm or good comes to the patient. They are responsible for what they order, or the way they follow an order or procedure. They are responsible for monitoring the healthcare process, at least in the steps immediately preceding and following their personal function in that process.

The blame-free approach is most properly applicable when the harm done is the result of a chain of small errors or the concatenation of such errors inherent in a complex system involving many steps and many persons. This is somewhat parallel to the inherent fallibility notion of Gorovitz and MacIntyre applied to the system of patient care rather than to the procedure itself.[21] It is, in effect, the "inherent fallibility" that may be built into a complex system meant to prevent error that paradoxically produces opportunities for error itself by the way it is structured. This is the case at times in air, rail, or nuclear power industry failures, in which a long history of error prevention has lulled everyone into complacency.

But even here the human factor remains. Just two examples from newspaper stories appearing as this chapter was being written will illustrate this point. Both are from the two industries singled out as models of systemic change of the kind health care should adopt. Neither vitiates the need for system control, but both illustrate the necessity for parallel attention to traditional notions of professional responsibility.

One story tells of an airport control tower in Sweden left unattended because the air traffic controller on duty failed to return from his vacation on time. Nobody noticed until a passenger aircraft sought landing instructions.[22] The plane landed safely at another field. The other story concerns erosion of a nuclear reactor head, reportedly the most extensive erosion found yet in the United States. The Nuclear Regulatory Commission judged that the defect should have been detected at least four years ago. Now there are allegations that the operator of the utility, FirstEnergy

21. Gorovitz and MacIntyre, "Toward a Theory of Medical Fallibility."

22. Associated Press, "SAS Flight Circles Airport after Traffic Controller Fails to Show Up after Vacation," *Associated Press*, August 26, 2002.

Corporation, backdated videotapes, falsified documents, and withheld photographs.[23] The "cover-up" is a classical ethical violation in corporate and government operation.

Who was to blame in these instances? The humans responsible for failure in their duties? The corporation or authority operating the airport or the reactors? On the facts presented it is reasonable to blame both. Professional and organizational ethics were both violated. We need not multiply examples like these. They are all too familiar, not just in systems of air and nuclear reactor safety, but in police departments, fire departments, sanitary systems, and in every walk of life in a society as complex and highly organized as ours. A no-blame system could too often be a travesty of social and commutative justice.

Blameworthiness and the Moral Spectrum of Error

Assigning individual or system responsibility is a difficult task sometimes admittedly impossible to accomplish with any degree of certitude. Bright-line distinctions are also difficult to make with respect to the moral gravity of an error. Yet if individual blame is to be attributed to a certain person or persons, some measure of moral culpability will be required to render justice to those injured. Both punitive and rehabilitative measures will bear some relation to the degree of moral failure in a particular instance.

All negligent error, that is, error that violates generally accepted standards of knowledge and skill, is morally blameworthy. This is true even if an error produces no harm or only slight harm, or is a "near-miss" automatically corrected by some system safeguard. The degree of blame will depend on a number of variables, singly or in combination.

In his study of two surgical services, Charles L. Bosk made a distinction between technical, judgmental, and moral error.[24] Technical error he ascribed to conscientious physicians who met their professional obligations but erred due to lack of knowledge or proficiency. Judgmental errors are the result of a faulty strategic decision despite conscientious attention to professional obligations and requirements. To be blameless, these lapses are expected to be infrequent, not follow a repetitive pattern, and remain within the norms of professional obligation.

When these limits are exceeded, when there is a lack of conscientious attention to expected obligations, or when performance and knowledge deficiencies are allowed to persist, then the error is a moral failure. It is of course difficult to assign a number to the times one may licitly make an "honest" error. Moreover, "conscientiousness" in fulfilling professional obligations is often a subjective evaluation of one's professional peers. The distinctions Bosk suggests are heuristically helpful.

23. Associated Press, "Ohio Atomic Plant Investigated over Acid Leaks," *New York Times*, September 1, 2002.

24. C.L. Bosk, *Forgive and Remember: Managing Medical Failure* (Chicago: University of Chicago Press, 1979).

They are, however, dependent on assessments of the character of one's colleagues. This is a notoriously flexible yardstick often influenced by friendship, kinship, hospital, and school ties.

The central moral issue is the degree of culpability an individual, that is, a truly conscientious physician, may feel about this error. In the realm of conscientiousness, the character of the physician is the only measure that assures that blame is appropriately assumed. This should include steps taken to eliminate the underlying deficiency of information or skill as soon as possible, even if one's colleagues label the error an honest one. Not to do so is to compound the moral onus of further error.

Here again the interaction of a system of error prevention with preservation of a sense of individual responsibility and culpability is essential. Technical and moral error too easily overlap, even with "conscientious" physicians. Ideally, a virtuous physician in a virtuous organization would take even "honest" mistakes to be serious moral issues. Moral conscientiousness can be more demanding than simple fidelity to prevailing professional obligations. The prevailing mores of a blame-free system can all too easily exculpate failures of professional accountability.

The nature and content of professional ethics have been markedly altered in the last thirty years. Its normative content and thrust have become diluted and problematic.[25] One may now fairly ask not only if a physician is conscientious but also what specifically he or she is "conscientious" about. With the gradual erosion of a universal ethic specifically defining professional obligations, professional conscientiousness can cover a wide variety of obligations.

We need only list changes in what is acceptable practice—financial conflicts of interest, limitation of practice to a "boutique" mode,[26] refusal to care for Medicaid patients, unwillingness to see patients at night, etc. In all these instances, error can occur as often as neglect and even as a result of neglect. As a result, the putative line between technical and moral error has become seriously blurred.

In complicated medical procedures and cases, especially in acute intensive care units or when multiple disorders affect the patient, the so-called cascade effect may make detection of causation, responsibility, accountability, and remediation particularly difficult. T. P. Hofer and R. A. Hayward report an illustrative case of a patient subjected to a series of diagnoses and procedures.[27] Different attendings, residents, radiologists, cardiologists, and other specialists sometimes successively, and sometimes simultaneously, treated a series of life-threatening complications. These authors show that in such cases of overlapping judgment and execution, the precise

25. E. Pellegrino, "Bioethics at Century's Turn: Can Normative Ethics Be Retrieved?" *Journal of Medicine and Philosophy* 25 (2000): 655–75.

26. T.A. Brennan, "Luxury Primary Care: Market Innovation or Threat to Access?" *New England Journal of Medicine* 346 (2002): 1165–68.

27. T.P. Hofer and R.A. Hayward, "Are Bad Outcomes from Questionable Clinical Decisions Preventable Medical Errors? A Case of Cascade Iatrogenesis," *Annals of Internal Medicine* 137 (2002): 227–333.

localization of error is uncertain and assignment of responsibility is accordingly difficult. At the end of their paper, Hofer and Hayward state: "Although carefully evaluated error-reducing systems are an important and essential part of healthcare delivery, for many problems in medicine fostering individual professional skills, expertise, values, responsibility, and accountability remain the best available approach to patient safety."[28]

Patient-Care Error and Team Care: Ethics of the Microsystem

Up to this point, I have emphasized the moral accountability of the health professional, principally the physician, though the same spectrum of responsibility can be constructed for all the other health professionals. Modern medical care, of course, takes place in team settings. We must factor in the intricate interdependence of a group of health professionals who are essential to any complex procedure. If the surgeon is to accomplish his task, he must depend on the bypass pump operator, the anesthesiologist, the surgical nurses, the laboratory, the technicians who check, repair, and standardize the physiological monitors, the cardiologist who saw the patient preoperatively, the pharmacist who prepared the preoperative and intraoperative medications, the intensive care unit staff responsible for postoperative care, etc.

All varieties and magnitudes of error may occur between and among team members. Assignment of accountability, and therefore of blame, is more complex than in the one-to-one healer-patient relationship. Ultimately, of course, all group decisions and individual actions do channel back to the individual patient. But the collective nature of the team dynamic opens up the problem of individual versus group moral agency and underscores the need for an ethics of collective human acts as well as an individual ethic.[29]

The team is a microsystem in which adverse actions of execution and judgment may occur—on the part of individual team members or in the articulation of their particular functions with each other. Morally, the microsystem of the team is something more than the numerical sum of the actions of its individual members. Team decisions require a common consent. Team actions share a common purpose, and team error becomes a shared responsibility. In that sense the team is a moral agent—a body of humans collectively capable of communal praise, or of communal blame for their collective errors of judgment or execution. Each member of the team is responsible for the competent performance of his or her own assigned and peculiar function in the team activity. At another level, each member bears some responsibility for the performance of the team as a whole. To varying degrees each team member is therefore implicated in team or system error. This complicity can be expressed in different ways.

28. Ibid., 331.

29. E. Pellegrino, "The Ethics of Collective Judgments in Medicine and Health Care," editorial, *Journal of Medicine and Philosophy* 7, no. 1 (1982): 3–10.

For example, a nurse may perform her part for the care of the patient well but may observe and fail to report, or do anything about, a pharmacist's medication error. The internist may interpret the electrocardiogram correctly but may fail to object to surgery he thinks inadvisable on the basis of that cardiogram. The surgeon may complete his operation successfully but may fail to note the inattention of the anesthesiologist or pump technician at a critical point in the procedure. In each case, the patient might have been endangered, or even injured. Under these circumstances the interdependence of team members imposes a certain measure of complicity in any adverse outcome.

A systems approach to avoidance of patient-care error must attend simultaneously to each member doing her part well and to the shared responsibility of each member for assuring that the team's responsibilities are conscientiously assumed as well. No team member, on this view, could claim innocence if she participated in a wrongful group decision or overlooked the failure of other individual members to perform their specialized tasks well.

Thus, the team member who, for reasons of friendship or a desire to avoid conflict or save face, fails to object to a group judgment she thinks erroneous is complicit in whatever harm may come from that failure. So while the "cascade effect" is a reality, it is a summation of errors both in the system and on the part of individuals in the system. When the relative weight of each item in the cascade might be extremely difficult or impossible to ascertain, the idea of a "blame-free" system gains validity.

But when the loose link in the concatenation of events is clearly identifiable, the collective moral agency of the team as a team is lessened. Thus, the pilot who takes off with assurance from the mechanic that the engine repair is complete cannot be held responsible if that engine fails, nor can the "team" as a collective be held responsible. When the part is defective in manufacture, however, and the mechanic unknowingly uses it in his repair work, neither he nor the pilot is accountable. Rather, the manufacturer incurs the blame. The same is true when the system or policy is at fault.

Whatever safeguards a systems approach may provide and however well they are conceived, someone in team care must be prepared to "throw the switch," so to speak. Emphasis on the system is essential, but so too is the retention of moral accountability of the team as a whole and of each of its members. At some point in every system, there is need to bypass the system if the system's true purposes are not to be frustrated. The responsibility for the discernment and courage necessary for this kind of action is shared by team members. These responsibilities can be preserved only if individual accountability is retained. Those who follow a defective policy may rightfully blame the system if its design is faulty. But it is also necessary to maintain a sense of accountability that will motivate each team member personally to monitor the system and to feel responsible for the way its human components interact with each other.

Social and Organizational Ethics: The Macrosystem

The health professional as an individual functions within the microsystem of the healthcare team that is itself part of two larger "systems": the institution and the society. Any approach to systemic change will inevitably involve the impact of policies, procedures, or laws promulgated in these "macro" systems on the care of patients. Macrosystems are answerable for the way their policies modify the discretionary space, and action, of the individual practitioner, as well as the actual care for the sick.

Even at the macro level, and perhaps especially at this level, the issue of personal accountability remains. Boards of trustees, members of executive committees, and other board members of hospitals or hospital systems can only delegate authority to act; they cannot delegate final responsibility for what is enacted and approved. Administrators and trustees are indeed "entrusted" with the organization, legally, fiscally, and morally. This means they must monitor its fidelity to its mission and those it purports to serve. They act collectively and they share responsibility collectively for what has been done, decided, or approved by those they have appointed as administrators or professionals.[30]

Therefore, organizations as a whole have moral obligations to those they purportedly serve. Hospitals, for example, have obligations to serve the welfare of those they presume to treat. They usually say so in their mission statements. If a procedure or policy ends in harming rather than helping, the institution as well as the responsible individual cannot escape culpability. A "systems" approach to preventing medical error cannot absolve executives, decision makers, or those who carry out their directives. To say, as so many do, "We have a procedure for that. In this case it failed," or simply, "We take responsibility" and "move on," can hardly claim to satisfy any satisfactory criterion of accountability.

The current rash of cases of corporate malfeasance is a cautionary tale for anyone considering systemic approaches to prevention of medical error, or patient-care error. That sad tale need not be repeated here. It is not new, as the history of corporate America so amply testifies. The corporatization, commercialization, and the entrepreneurial cast of health care are already well advanced. We have reason to worry about overreliance on a systems approach and especially a blame-free approach in health care. Despite supposed checks and balances, lines of authority, accounting boundaries, etc., the venal purposes of corporate executives and boards of directors were not prevented. Some if not the majority of these corporations had "ethics" committees. Organizational ethics itself is a dubious enterprise without a cadre of persons of character to detect and correct the tendency of profit to distort even the highest motives.

To acknowledge the failure of corporate ethics and the frailty of the safeguard supposedly built into its systems of surveillance is not to abandon hope for morally responsive organizations. Such organizations do exist and, like morally upright individuals, receive no public notice. Indeed, it is the belief of many virtue ethicists that there can be "virtuous" organizations, those whose habitual disposition is to act in

30. V.A. Sharpe, "Patient Safety and the Role of Hospital Boards," in *The Ethics of Hospital Trustees*, ed. B. Jennings, V.A. Sharpe, B.H. Gray, and A.R. Fleischman (Washington, D.C.: Georgetown University Press, 2004).

such a way as to protect the interests of those they serve. We all know of local businesses, auto repair shops, merchants, and service organizations whom we trust to act well, even while others do not.

Virtuous organizations usually propound mission statements promising ethically responsible conduct. These missions will not become a reality unless they permeate the life of the organization. Needless to say, this must be sustained from the top—from the example and witness of the chief executives and boards of directors. This is not the place to delineate how one generates a sense of ethics and virtuous conduct in organizations.

It is important, however, to emphasize that for all members of a micro- or macro-system to act in a morally responsive way, a sustaining culture of morality is necessary. Of course, there are persons who will be virtuous and upright no matter what milieu they function within. But all persons do not have the same strength of character to be virtuous without a morally responsive work environment to sustain and correct them.

Having said this, it is again important to recognize that even when character formation, honor codes, and fostering a culture of virtue are present, individual virtues are still the bedrock. The military academies in our country emphasize honor codes, character, and virtue. Yet all too sadly, we know they have been the sites of egregious violations of cheating and other reprehensible behavior. Again the necessity of a balance between individual accountability and organizational safeguards is obvious in the prevention of medical error, as it is in every other life arena.

With specific reference to healthcare administrators and executives, we must not forget that the organizations they head are healthcare organizations. They are thus charged with the welfare of sick, dependent, anxious, vulnerable, and exploitable human beings. These humans are patients primarily, not consumers, customers, or subscribers. Nor do they fit any of the other metaphors derived from business, industry, or commerce. Healthcare trustees or administrators, like it or not, must be judged on the impact of their decisions on the patient in the clinic or the bed. Healthcare organizations may be corporate entities, but they cannot be excused if their level of ethical sensitivity does not rise above that of the ordinary corporation.

While the executives of healthcare "systems" may see themselves as primarily corporate officers, they cannot avoid the fact that they too are bound by the norms of clinical ethics. Of course their responsibility is not the immediate responsibility of the doctor or nurse. But the more they recognize the notion of the primacy of the welfare of the sick, the more clearly will they set the proper moral tone of the institutions they direct. We used to hear the claims of certain pharmaceutical companies to be "ethical" and to set a higher standard for themselves. This notion has been dissolved in the corrosive acid of competition and profit so that the pharmaceutical industry is now on the ethical defensive. But the return of the notion of an "ethical" healthcare organization is as important as a new system of rules, algorithms, policies, checks, and monitoring mechanisms in preventing medical error.

The same set of considerations applies to the broadest systemic approach of all, the macro "system" of social and political change. Here the need for restoration and indeed accentuation of personal accountability is ineradicable. Societal legislation and policy are not self-justifying, no matter how great their majority appeal. Certainly democratic societies must elaborate laws, regulations, and the like. But like all the levels of systemic change, however well-intended, we cannot dispense with the concept of personal accountability, or with protection of the whistle blower and the alert individual who detects danger to patients.

By all means, a systems approach should be elaborated. Examples from successful systems of error control outside health care should be translated into the ethics of medicine and health care. These borrowed systems will be most successful if we retain and refine the sense of individual moral responsibility within them as well.

Individual and System Accountability: Synergy, Not Opposition

Some of the defects attributable to the current focus on providers are lack of recognition of the interdependence of team care; suppression of error reporting and information; discouragement of open communication and partnership between patients, families, and providers; and some of the defects of compensation litigation.[31] Yet there is no assurance that a blame-free approach and displacement of responsibility from individuals to systems will remedy these defects. The "system" can be just as reluctant to admit error as the individual. The temptation to cover up, minimize, or deny system error will not be erased. Anyone who has tried to elicit an admission of error or recover compensation for harm done by an institution, business, or corporate entity knows how difficult both justice and satisfaction for harm done can be.

The confidence some bioethicists place in a transfer of trust from physicians to the healthcare organization is wholly without justification in fact. The ethics of organizations are not necessarily more reliable than those of individuals. We need think only of the corporatized impersonality of managed-care organizations to doubt whether a "system" of apology, appeal, or system compensation will be effective in preventing error. We may speak of the ethical obligations of the system to provide quality care, but systems are just as eager to minimize complicity as individuals. In fact, organizational loyalty and complicity all too often make a mockery of appeal systems.

Blaming the system may diffuse accountability so widely that no one feels responsible. Organizations and systems are collectively accountable, to be sure, but this does not assure that each member of the collectivity or team will feel responsible enough to report errors made by friends or associates. It takes as much courage to blow the whistle on the system as it does on an individual in the system. The system is still

31. L.L. Leape, "Error in Medicine," *Journal of the American Medical Association* 272 (1994): 1851–57; L.L. Leape, D.D. Woods, M.J. Hatlie, K.W. Kizer, S.A. Schroeder, and G.D. Lundberg, "Promoting Patient Safety by Preventing Medical Error," *Journal of the American Medical Association* 280 (1998): 1444–47; Liang, "Error Disclosure."

dependent on the ethical sensitivity and courage of each of its members as individuals. Traditional medical ethics remain an essential element of a patient's safety.

Expecting patients and families to be "partners" in a system-oriented error prevention system is a dubious recommendation. This asks people in distress to bear some of the burden in detecting error. The possibility that such an invitation will worsen the situation is not trivial. Patients and families are already involved in micromanagement at the bedside. They may actually hamper the best care by seriously limiting the professional's discretionary space. It takes courage today for physicians to do what they think is right medically when it is thought to violate the autonomy of patients and families to choose otherwise. Rather than being a sign of respect and empowering patients and families, this partnership in error detection asks them to carry a burden in addition to the ordinary burdens of a serious illness. Does this mean that they share the blame for an untoward outcome? Clearly patients and families should be informed, told about lapses in the system or errors made by an individual caregiver, and encouraged to communicate. But this cannot be "sharing" in any genuine sense. We cannot ask the patient and family to beware of medical error as we ask buyers to beware.[32]

The team functions by coordinating the functions of many persons, each of whom has a specific responsibility for an essential part of the team effort. The system can only succeed if each member is faithful to the trust invested in her special function by the patient and the system. Each individual can only function well if the others also do their jobs well. Once this reciprocity is accepted, retention of the ethical accountability of individuals need not be punitive. The system can be designed to emphasize personal accountability yet assist the responsible persons in disclosing error and do so in a supportive environment. The anguish of a conscientious professional for an error, blameworthy or not, is not to be discounted as just punishment for wrongdoing. The emphasis in a system can be on prevention, re-education, and rehabilitation of individual actors.

Overemphasis on the anguish of the caregiver, however, can lead to laxity in removing an inadequate team member. Unfortunately, some cannot be successfully rehabilitated. Others may too readily transfer their personal responsibility to the team. Others may be protected by friends reluctant to hurt them. The primacy of the welfare of the patient must not be diluted in a mistaken effort at compassion for the error-prone caregiver.

In any case, in the processes of error detection, investigation, or disclosure, individuals will unavoidably be identified. That information will often become known. It is unlikely in reality, therefore, that the systems approach will entirely shield the individual from blame. Nor will it assuage the anguish and anger of the patient, who may know or suspect the person responsible. It is wiser, therefore, to acknowledge when an individual is responsible. This will clear away unwarranted suspicion and provide opportunity for personal apology when there is actual fault.

32. R. Goeltz, "In Memory of My Brother," in Sharpe, *Accountability*, 49–58; Liang, "Error Disclosure."

Disclosure: Professional and Ethical Obligation

Principle II of the Code of Ethics of the American Medical Association imposes two duties of disclosure, one based in "honesty with patients and colleagues," and the other based in exposing "physicians deficient in character or who engage in fraud or deception."[33]

The findings of the IOM report underscore the need for a more rigorous application of this principle and the associated guidelines. The urgency for such reaffirmation and further elaboration of these standards of conduct is further underscored by the standards now being set by the Joint Commission on Accreditation of Healthcare Organizations.[34] These standards call for organization-wide disclosure practices with executive involvement and implementation.

If the AMA principles and guidelines and those of the JCAHO were to be cooperatively applied, the desired integration of individual and organizational ethics could be realized. Two "systems"—the AMA and the JCAHO, and the individual physician's professional and ethical obligations—could act synergistically to reduce the prevalence of error.

A fuller exposition of the ethical foundations for disclosure is provided elsewhere in this volume.[35] It suffices to say here that the moral obligation of disclosure lies in the trust relationship that binds patients to physicians and healthcare organizations.[36] Patients submit to care by physicians and hospitals because they must trust them to be competent and to use that competence in the patient's interest. Fidelity to these promises of trust is at the root of professional and institutional ethics. Medical error, however caused, blameworthy or not, is a violation of this trust. That is why patients have a moral claim on disclosure.

Several aspects of this moral claim are linked to the questions of commutative justice that disclosure entails. Patients, in seeking justice, should know when an error is blameless, and thus not a violation of trust, and when it is blameworthy and therefore a violation of trust. Knowledge is needed by the patient to understand and cope with whatever impact the error might have made on the course of his disease. Disclosure also allows patients to change physicians or hospitals. It also permits a more just appraisal of whether or not suspicions of error are based in fact, along with the chance of personal apology, which is more likely and more satisfying than an orga-

33. American Medical Association, *Code of Medical Ethics: Current Opinions with Annotations* (Chicago: American Medical Association, 1999), sec. 8.12

34. Joint Commission on Accreditation of Healthcare Organizations (JCAHO), *Revisions to Joint Commission Standards in Support of Patient Safety and Medical/Health Care Error Reduction* (2001), http://www.dcha.org/JCAHORevision.htm [this link is now obsolete—Ed.].

35. Nancy Berlinger, "'Missing the Mark': Medical Error, Forgiveness, and Justice," in Sharpe, *Accountability*, 119–34; Sandra M. Gilbert, "Writing/Righting Wrong," in Sharpe, *Accountability*, 27–41; Albert W. Wu, "Is There an Obligation to Disclose Near-Misses in Medical Care?" in Sharpe, *Accountability*, 135–42.

36. Pellegrino, "The Ethics of Collective Judgments"; Pellegrino, "The Internal Morality of Clinical Medicine."

nizational apology. Finally, disclosure allows for more genuine and less frivolous litigation for compensation.

In all of this, the need for a conjunction between individual and organizational ethics becomes clear. The Institute of Medicine has alerted the profession and the public to a serious problem. While its data and conclusions have certain shortcomings, they merit the serious response they are receiving.[37] What is clear is that both a systems approach and an approach that reinforces traditional ethical obligations of individual practitioners are necessary. Individual patient-oriented medical ethics and the new domain of organizational ethics are required to act synergistically if the welfare of the patients is to be optimized.

37. Brennan, "The Institute of Medicine Report."

D. Financing Healthcare Delivery

36

Managed Care at the Bedside: How Do We Look in the Moral Mirror?

Edmund D. Pellegrino, MD

Abstract: *Managed care per se is a morally neutral concept; however, as practiced today, it raises serious ethical issues at the clinical, managerial, and social levels. This essay focuses on the ethical issues that arise at the bedside, looking first at the ethical conflicts faced by the physician who is charged with responsibility for care of the patient and then turning to the way in which managed care exacts costs that are measured not in dollars but in compromises in the caring dimensions of the patient-physician relationship.*

THE ETHICAL ISSUES IN MANAGED CARE can be examined at three levels—at the bedside, which centers chiefly on the patient-physician relationship; at the managerial or corporate level, which centers on the relationships of managers, corporate boards and officers, and investors to patients; and at the social level, which centers on the way a society responds to the ill, disabled, and vulnerable in its midst. Managed care as it operates today poses serious ethical questions at all three levels.

This essay confines itself to the ethical issues at the bedside—that is, on the way this particular form of healthcare organization affects the sick and those who profess to help them. Therefore, after some initial remarks and three propaedeutic points, this essay is divided in two parts: the first part deals with ethical conflicts faced by the physician who is charged with responsibility for care of the patient; the second part deals with the way managed care exacts costs measured not in dollars but in compromises in the caring dimensions of the patient-physician relationship.

These "bedside" issues cannot be fully separated from the ethical issues at the managerial and societal levels. If harm results, the moral complicity of corporate officers and investors, as well as society as a whole, cannot be escaped. Those who

---

Reprinted with permission of Johns Hopkins University Press from Edmund D. Pellegrino: "Managed Care at the Bedside: How Do We Look in the Moral Mirror?" *Kennedy Institute of Ethics Journal* 7, no. 4 (1997): 321–30.

manage, invest in, and profit from managed care cannot escape moral complicity for harms that occur at the bedside.

Generically, managed care is morally neutral. It refers simply to any system of health care that aims at constraints on the clinician's management of care to achieve some stated purpose. That purpose may take many forms—the quality of care provided to an individual patient, the personal well-being of the patient, the containment of costs, the welfare of society, or the making of profit. Some of these objectives are morally sound; some are morally reprehensible. Ultimately, the moral status of any system of managed care will depend upon the purpose, the means employed to attain that purpose, and the priorities between and among purposes or means when they conflict with each other.

In all of these variations of managed care, and in actual practice, one ordering principle provides the moral sine qua non; it is the primacy of the moral obligation of healthcare professions to act in the best interests of the person who is ill. This is the moral principle of beneficence contained in traditional codifications of the physician's obligations. This is, has always been, and must remain the telos of the healing relationship, the end built into the nature of medicine and without which it becomes something other than a healing relationship.[1]

The moral test, therefore, of a managed care system is not in the balance sheet of the HMO nor in the dividend reports of the investor nor in the bonus to the "provider" nor even in the costs saved, bed days reduced, efficiency gained, or productivity improved. The moral test of any system of care is its impact on the patient for whom the system is presumably designed and on the physician from whom the patient seeks help.

Three Propaedeutic Points

First of all, none of what follows is a denial of the reality of economics as a constraining force on medicine. The real issue is how to relate the fact of economics to the demands of ethics. Throughout, I shall maintain that when they are in conflict, ethics takes precedence over economics, but cannot ignore it. Important as it is, economics is a science of means. It can help us to attain the purposes we set for ourselves or society by measuring efficiency, productivity, and optimization of the use of scarce resources. But economics cannot set the ends and purposes of the lives of individuals or societies nor determine whether these ends and purposes are morally good or bad.

Thus, an economically sensitive physician, faced with two treatments of equal effectiveness and benefit for her patient, should choose the one that is less costly. This is a moral obligation since the physician acts unjustly if he knowingly wastes the patient's money. In addition, physicians have responsibilities to use not only a

1. Edmund Pellegrino and David C. Thomasma, *For the Patient's Good: The Restoration of Beneficence in Health Care* (Oxford: Oxford University Press, 1987).

patient's resources efficiently, but society's as well. Even when resources are limited, there are morally sound and morally unsound ways of confronting the dilemmas of clinical choice. Under certain conditions rationing can be morally justified so long as these conditions are set by ethics and not economics.

A second important propaedeutic point is that physicians are under a moral obligation not to offer or provide treatments that are unnecessary, especially when treatment is futile.[2] This is not simply for economic reasons. Unnecessary treatment is morally wrong because it violates the obligatory canons of good medicine, imposes costs without reason, and exposes patients to risks without proportional benefits. If physicians in the past had more assiduously observed the moral duty to avoid unneeded treatment, much of the impetus to managed care would not have existed.

A third, and final, introductory point is that in the past, and at present, care has been "managed," albeit indirectly, by measures aimed at enhancing quality—for example, tissue committees, postmortem examinations, pharmacy and therapeutics committees, drug formularies, morbidity and mortality conferences, patient care committees, and the hospital and residency accreditation processes. These and like measures illustrate that managed care can be used for purposes other than cost containment. It shows, too, that physicians can accept managing care if the patient's welfare is the primary goal, even if such management limits their discretionary latitude in clinical decisions. To be sure, these measures can be expanded and improved upon. It is politically and morally offensive for physicians to resist managing care if it is designed for the patient's good. Anyone familiar with the dark side of unfettered clinical decision making will recognize the value of patient- and quality-oriented managed care. Unfortunately, this is not the driving force of managed care as it operates today in the United States.

Ethical Conflicts for the Physician

In any managed care system, the most fundamental and ubiquitous moral dilemma at the bedside is the dilemma of divided loyalty—that is, pitting the welfare of the patient against the welfare of the organization, the physician, or the other plan subscribers.[3] The physician who works in a managed care organization is financially and factually an employee. She accepts remuneration presumably to serve the needs of the employer. This is a contractual relationship, and it carries responsibilities to the employer and to the investors who make the employer's business possible by risking their capital. This contractual obligation conflicts with a morally determined patient-physician relationship, which gives primacy to the well-being of the patient.

2. Hippocrates, "The Art," in *Hippocrates,* vol. 2, trans. W.H. Jones, Loeb Classical Library (Cambridge, Mass.: Harvard University Press, 1981).

3. Edmund D. Pellegrino, "Rationing Health Care: The Ethics of Medical Gatekeeping," *Journal of Contemporary Health Law and Policy* 2 (1986): 23–45; Susan Wolf, "Health Care Reform and the Future of Physician Ethics," *Hastings Center Report* 24, no. 2 (1994): 28–41; Marcia Angell, "The Doctor as Double Agent," *Kennedy Institute of Ethics Journal* 3 (1993): 279–86.

In its strong form, the ideology of managed care asserts that the person-oriented ethics of traditional medical ethics must give way to population ethics, to a concern for the impact of every clinical decision on the availability of resources to others covered by the same system. In this view, the physician becomes primarily an advocate for society rather than for his patient. This is the model of implicit rationing in which the physician as gatekeeper decides who is to receive scarce resources, with or without socially imposed criteria for choice. "Social ethics" takes the place of the traditional ethic of individual patient care.

A weaker form of this model requires at least equal concern for the good of the population as for the good of the individual patient. Here the physician is expected to balance the moral claims of her own patient against those of other plan participants and societal needs in general. This model also depends upon implicit rationing in which the final decisions on allocation at the bedside belong to physicians. In both versions, strong and weak, the patient is at the mercy of the physician's assessment of her relative worth in the competition for limited research.

Although lately challenged, managed care organizations have sought to further the interests of profit or the population at large by imposing "gag" rules of various degrees of stringency on physician employees. While legislation has been developing to limit specific "gag" clauses, there is still the subtler problem of not telling the whole truth. By this I mean that managed care gatekeepers may not feel obliged or expected to tell a patient that the treatment his plan will pay for, or the surgeon, hospital, or laboratory it contracts with, are not the preferred ones for this patient's illness. The patient's freedom of choice and autonomy are thus violated without his knowing it. The physician fails to fulfill his fiduciary obligations to be honest and to enhance patient autonomy. Simultaneously, he silently sanctions a lesser standard of care than he would if he were free to make the decisions he thought best for his patient.

In addition, to serve his patient best, some physicians are tempted to lie—that is, to ratchet up the severity of an illness or to place it in a diagnostic category more favorably received by the managed care plan of the preadmission auditor. How far may the truth be shaded, or stretched, in the interest of an individual patient? Some would say not at all. Others might hold that the restrictions are unjust in the first place and therefore there is no moral obligation to respect them. Others would argue that we must not deceive others even to serve the good of a patient. However stringently or loosely one regards the obligation of truth telling, any system of care that makes truth telling an obstacle to beneficence is suspect on the face of it.

Some have suggested that these conflicts of loyalty can, for the most part, be avoided if physicians themselves become insurers, or "fund holders."[4] Theoretically, on this view, it is presumed that doctors who have more control over allocation decisions for individual patients, or local communities of patients, will act more

4. Donald W. Light, "Managed Care: False and Real Solutions," *Lancet* 344 (1994): 1197–99; Steffie Woolhandler and David U. Himmelstein, "Extreme Risk: The New Corporate Proposition for Physicians," *New England Journal of Medicine* 333 (1995): 1706–8.

beneficently than managers or bureaucrats.[5] Without disparaging the characters or intentions of physicians, the presumption is a precarious one indeed. If physicians share the risks of fiscal loss, this fact will consciously or unconsciously shape their decisions. When the plans involve profit as well, the danger of loss is accentuated in direct proportion to magnitude of gains or losses.

Moreover, when physicians are stakeholders, they become part of the managed care plan—that is, of a corporate entity. The patient loses his advocate with the system since the physician is unlikely to advocate the patient's case against his own interests. At least, when the physician is an employee rather than an owner of an interest, he can make common cause with the patient against the system. Confident assurances that patients can be protected against injustices by grievance mechanisms are so unrealistic as to verge on the ridiculous. Patients in need of care, patients dissatisfied with decisions being made, or those being denied needed care are hardly in a position to initiate, much less to pursue, a grievance process. Such a process might be of some use retrospectively to patients seeking retribution for harm done, but it is an unrealistic safeguard at the time the injustice is being perpetrated.

Clearly, the managed care system, as it is currently constructed, creates ethical conflicts of such magnitude that conscientious physicians feel forced to compromise their personal integrity to survive. Failure to "accommodate" holds the threat of being stricken from the list of "providers" or penalized financially. When the penetration of managed care is in the range of 40–50 percent of the insured, the threat to survival is not imaginary. The result often is that the more sensitive and humane physicians choose to retire or to move out of direct patient care. This is self-defeating since it leaves the field to those who are willing to compromise.

To this we must add the fact that the youngest physicians are being socialized and professionalized into the managed care ideology. Not having experienced other forms of medicine, it is difficult for them to see why there are objections. They believe the evils of the indemnity and fee-for-service systems of the past were even greater. The issue is not choosing the better of two bad alternatives, but designing a system that is morally defensible by the standard of its impact on the care of sick persons—presumably those who should be the beneficiaries of the system.

It must be clear that the conflicts of loyalty outlined above are built into managed care to achieve its end of cost containment. They may or may not contain costs. There are also non-dollar costs that significantly affect the quality and satisfaction of the "care" provided by managed care plans.

Non-Dollar Costs: The Care of the Patient

So obsessed have planners, the general public, and even physicians become with the promised cost reductions of managed care that they have slighted the non-dollar

5. Frank H. Marsh and Mark Yarborough, *Medicine and Money: A Study of the Role of Beneficence in Health Care Cost Containment* (New York: Greenwood Press, 1990).

costs of achieving these economies. These costs may not reflect themselves immediately in morbidity or mortality rates, but they do affect the more subtle, yet important, aspects of "care." They are the basis for the rising tide of complaints we hear from patients—those who are ill, not healthy—as they try to get the help they need. It is these complaints that generate pressure for legislative relief and litigation. Indeed, those who are sick and can do so, opt out of the HMO when they do need real care and can afford to escape, at least temporarily.[6]

One serious non-dollar cost is the chaotic nature of patient-physician relationships in managed care settings. Discontinuity is more the rule than continuity. The physician one encounters is a matter of random meetings conditioned by physicians' work schedules rather than patient need. The clear assumption is that one physician is as good as any other. The personal "chemistry" of the therapeutic relationship is given short shrift. Over an eight-hour shift, patients may have to report their stories to a succession of nurses and physicians "on call," none of whom seems to have communicated with the other.

The dangers of discontinuity go beyond mere inconvenience. They constitute real dangers to the patient. Vital information may be missing, misinformation and negative attitudes may be transmitted in what passes for "communication" between one physician "passing the torch" to another. Doctors are now shift workers, increasingly less committed to the patient and more to the schedule of both shift and productivity. Fewer physicians see patients as their personal responsibility but rather as the responsibility of the organization.

Moreover, the "productivity" schedules imposed on outpatient visits do not foster confidence nor permit getting to the root of a problem unless it is obvious in the first few minutes. It is difficult indeed to establish an effective patient-physician relationship in 15-minute, random, and intermittent office visits. Increasingly, we encounter patients who have been seen five or six times, given a cursory history and physical examination, and sent on their way with a prescription. Symptoms persist, treatment does not "work." The patient and a succession of doctors become frustrated with each other. Unnecessary visits are multiplied and costs increased rather than decreased by these helter-skelter workups. Much of this scenario is avoidable by one, unrushed, carefully conducted "low tech" history and physical examination. This is now a scarce commodity much more difficult to come by than many complicated procedures.

Another non-dollar cost is the conversion of generalists into marginal specialists and specialists into marginal generalists. To save the costs of referrals, physicians should do more themselves. The generalist is invited to do a little orthopedics here, a dash of office gynecology there, and a soupçon of neurology. Time, after all, will tell whether a specialist is needed. The specialist, on the other hand, can treat a minor

6. R.O. Morgan, Beth Vianig, Carolee DeVito, and Nancy A. Pershing, "The Medicare-HMO Revolving Door—The Healthy Go in and the Sick Go out," *New England Journal of Medicine* 337 (1997): 169–75.

respiratory infection, migraine headache, or diarrhea for a while. If it does not go away, then she can refer the patient to a generalist, who starts all over again. All of this adds up to the creep of incompetence. Marginalizing the expertise of either specialist or generalist is a dangerous legitimation of clinical presumptuousness, the costs of which are yet to be measured in missed or delayed diagnoses and increases in disability, suffering, and patient frustration.

To be sure, some of these hassles and harassments were present with indemnification systems and fee-for-service practice. But in the past, such difficulties were recognized as lapses in good care. The finger of guilt could be pointed directly and legitimately at the physician who had personal responsibility for what happened to the patient. Managed care systems, however, legitimate these harassments. They are justified as necessary "inconveniences" legitimized by presumed dollar savings. This, we are told, is necessary to forestall the national economic disaster of spending too much on health care. We do not seem equally worried about our more egregious overspending on self-indulgence, recreation, costs of sports, entertainment, and the like.

Even more disturbing among the non-dollar costs is the marginalization and gradual disenfranchisement of segments of the population—the aged, the chronically ill, the underinsured. Callahan, who is a proponent of age-related rationing, nonetheless worries about covert rationing in the elderly population.[7] Others are equally worried that rationing of "experimental therapies" are subtle excuses for rationing needed care.[8] As the determination of what is experimental or necessary care moves from physicians to managed care organizations, the "needs" of patients will become ever more mechanically defined.[9]

The same kind of danger lies in the bias of managed care against medical technology. This verges on a new Luddite movement in which research in currently untreatable diseases is implicitly discouraged since it will only prolong lives in people who must ultimately die anyway. Measured in terms of increased longevity and improved quality of life, however, effective technologies may well be "worthwhile." What we as a society consider "worthwhile" reflects on the kind of society we want to be. All technology must not be lumped together, therefore, under the heading of "unnecessary care at the end of life."

Conclusion

Managed care is a morally neutral concept. The challenges it can produce in terms of conflicts of obligations for the physician and serious non-dollar costs for the patient cast serious doubt on its moral status as it is practiced today. This does not

7. Daniel Callahan, "Controlling the Costs of Health Care for the Elderly—Fair Means and Foul," *New England Journal of Medicine* 335 (1996): 744–46.

8. F.P. James, "The Experimental Treatment Exclusion Clause: A Tool for Silent Rationing of Health Care?" *Journal of Legal Medicine* 12 (1991): 359–418.

9. Wendy K. Mariner, "Patient's Rights after Health Care Reform: Who Decides What is Medically Necessary?" *American Journal of Public Health* 84 (1994): 1515–20.

preclude an ethically more satisfactory system of care, but this option is just beginning to receive some attention by ethicists.[10] Instead, as a nation we have been so cowed by the specter of economic disaster that we have permitted economics, business, and commerce to drive us into a system that exhibits its most deleterious effects precisely where it ought to be most beneficial—at the bedside. The beginning of an ethically defensible system is the recognition of the moral costs we are already paying for what may well be one time savings.[11]

The ethical issues at the bedside are not just the physician's problems. What we permit, encourage, or reward at the bedside reflects back on all of us. To be sure, medical care should not be given a blank check. But we do need to decide where on our societal value scale the care of the sick fits in relation to the many other ways in which we disburse our unparalleled affluence. The impact of managed care at the bedside is a moral mirror that tells us much about the moral quality of our society. Managed care is not per se evil. It can be a morally creditable enterprise, but if, and only if, it is designed to serve the needs of those among us who are ill—or will be ill—and that is all of us. As it exists today, it does not meet this moral criterion.

10. Robert M. Veatch, "Allocating Health Resources Ethically: New Roles for Administrators and Clinicians," *Frontiers of Health Services Management* 8 (1991): 3–29; Eli Ginzberg and Miriam Ostow "Managed Care: A Look Back and a Look Ahead," *New England Journal of Medicine* 336 (1997): 1018–20.

11. Ginzberg and Ostow, "Managed Care."

37

The Commodification of Medical and Health Care: The Moral Consequences of a Paradigm Shift from a Professional to a Market Ethic

Edmund D. Pellegrino, MD

Abstract: *Commodification of health care is a central tenet of managed care as it functions in the United States. As a result, price, cost, quality, availability, and distribution of health care are increasingly left to the workings of the competitive marketplace. This essay examines the conceptual, ethical, and practical implications of commodification, particularly as it affects the healing relationship between health professionals and their patients. It concludes that health care is not a commodity, that treating it as such is deleterious to the ethics of patient care, and that health is a human good that a good society has an obligation to protect from the market ethos.*

Key words: commodification, economics, ethics, ethos, health care, human good, managed care, marketplace, physician-patient relationship, professional ethics.

Introduction

In Book I of the *Republic*, Socrates asks Thrasymachus, his pragmatic young interlocutor, this question,

> But tell me, your physician in the precise sense of whom you were just speaking, is he a moneymaker, an earner of fees or a healer of the sick? And remember to speak of the physician who really is such. . . . (Plato[1])

Reprinted with permission of the Society for Health and Human Values from Edmund D. Pellegrino: "The Commodification of Medical and Health Care: The Moral Consequences of a Paradigm Shift from a Professional to a Market Ethic," *Journal of Medicine and Philosophy* 24, no. 3 (1999): 243–66. Permission conveyed through Copyright Clearance Center, Inc.

1. Plato, *Republic*, trans. P. Shorey, in *The Collected Dialogues of Plato*, ed. E. Hamilton and H. Cairns (Princeton, N.J.: Princeton University Press, 1982), 341C.

Even the cynical Thrasymachus, who denies the viability of justice in a world dominated by powerful people, admits that physicians are healers first.

Were Socrates's question to be asked in today's healthcare environment, the answer would be different. We would be told that physicians should be encouraged—indeed, impelled—by financial incentives and disincentives to become moneymakers for themselves and money-savers for their corporate employers or investors. Physicians, in short, are asked to be primarily purveyors of a commodity and not the healers Plato said they should be if they were to be *true* physicians.

In this essay, I wish to examine the ethical consequences of commodification of health and medical care on the relations of physicians with patients, with each other, and with society. I will do so by posing three questions and offering three conclusions.

The questions are these: (1) Is health care a commodity like any other, subject to distribution by the operations of the competitive marketplace? (2) What is the ethical impact of commodification on professional ethics and the care of the sick? (3) Does commodification work? (4) If health care is not a commodity, then what is it, and how should it be treated in a just society?

To each question, I shall reply, in turn, as follows: (1) health and medical care are not, cannot be, and should not be commodities; (2) the ethical consequences of commodification are ethically unsustainable and deleterious to patients, physicians, and society; (3) commodification does not fulfill its economic promises; and (4) health care is a universal human need and a common good that a good society should provide in some measure to its citizens.

Is Health Care a Commodity?

It is a fundamental dictum of managed care that health care should be treated like any other commodity, that is, its cost, price, availability, and distribution should be left to the free workings of a free marketplace constrained by a minimum of governmental regulation.[2] Through the usual mechanisms of competition, a "quality product" should emerge since providers will compete with each other in quality, price, and satisfaction of consumers to keep their market share and/or profits.

For their part, "consumers" and "purchasers" will be free to choose among providers selecting the best "buy" suited to their individual needs. Costs will decline, and quality will be maintained or will improve. More care will be accessible to more people on their terms, not the doctor's. The laws of competition will reduce waste, overuse, and error to everyone's advantage. Medicine will be demystified, physicians will become employees, and physicians' decisions will be shaped to conserve society's resources.

In this view, these desirable ends can be facilitated by turning the physician's self-interests to the advantage of the competitive system. By controlling clinicians'

2. R. Hertzlinger, *Market-Driven Health Care* (Reading, Mass.: Addison-Wesley, 1997); [A. Enthoven, "Enthoven Responds," letter to the editor, *Health Affairs* 16, no. 4 (July/August 1997): 282–83—Ed.].

expenditures, employing financial and other rewards and penalties, costs will be kept down, unnecessary care eliminated, and quality outcomes optimized. In this view, there is no objection to physicians being moneymakers so long as this activity is confined to a managed environment and their "product" is traded on the open market. The fundamental question is a deeper one than we can examine here, namely, the issue of the proper relations of ethics to economics.[3]

The underlying ethos of managed care is the theory of Adam Smith's *Wealth of Nations* and *Theory of Moral Sentiments*.[4] If, as Smith taught, everyone pursues his or her own interests, the interests of all will be served. The profit motive as adduced by Smith ultimately works to social benefit. But, for Smith, profit is not the primary end of the market ethos. Rather, it is the best means of attaining the primary end which is "a plentiful revenue of subsistence for the people."[5] As I shall note later in this essay, Smith also felt that some things could not be left to the fortuitous workings of the marketplace and could only be assured by government intervention.[6] For some economists, this reservation would inhibit the liberty only an unrestrained market can provide.[7]

In its most aggressive form, managed care expresses the market ethos in lightly restrained risk selection, high-powered marketing and advertising, strict rules about denial or approval of care, competitive price-cutting, and putting substantial portions of the physician's income at risk. Reward and penalty for doctors depend negatively on the clinical costs they incur and positively on meeting marketing goals and patient satisfaction.[8] Gaining the benefits of market competition for health care when treated as a commodity will, of necessity, lead to a loss of professionalism.[9]

Before discussing the ethical problems consequent on treating health care as a commodity, there is ample reason to question the validity of this line of reasoning even from the purely economic point of view. There is, today, an intense and growing reaction to managed care as it is now practiced. This is of sufficient proportions to cast doubt on the economic theory behind managed care. There is evidence that costs are again rising, services being reduced, young and healthy subscribers being favored over old and sick ones, emergency rooms being closed,

3. M. Friedman, *Capitalism and Freedom* (Chicago: University of Chicago Press, 1967); J.J. Piderit, *The Ethical Foundations of Economics* (Washington, D.C.: Georgetown University Press, 1993); A. Sen, *On Ethics and Economics* (Oxford: Basil Blackwell, 1987).

4. A. Smith, *An Inquiry into the Nature and Causes of the Wealth of Nations*, ed. C.J. Bullock (New York: P.F. Collier, 1909); A. Smith, *The Theory of Moral Sentiments* (Philadelphia: Anthony Finley, 1817).

5. S.T. Worland, "Adam Smith, Economic Justice, and the Founding Father," in *New Directions in Economic Justice*, ed. R. Skurski (Notre Dame, Ind.: University of Notre Dame, 1983), 8.

6. Ibid.

7. Friedman, *Capitalism and Freedom*.

8. R. Kuttner, "Must Good HMOs Go Bad?—Second of Two Parts: The Search for Checks and Balances," *New England Journal of Medicine* 338 (1998): 1635–39; G. Anders, *Health Against Wealth: HMOs and the Breakdown of the Medical Trust* (New York: Houghton Mifflin, 1996).

9. J. McArthur and F.D. Moore, "The Two Cultures and the Health Care Revolution: Commerce and Professionalism in Medical Care," *Journal of the American Medical Association* 277 (1997): 985–1005.

etc.[10] Restrictive legislation is currently before the Congress (Patient Access and Responsibility Act, 1997, H.R. 1415), class action suits are beginning to surface, and legislation has been proposed to remove the protections of managed care organizations against liability provided by the Employee Retirement Security Act.

Economists, healthcare planners, governmental budget officers, industry investors, young and healthy persons who have not made use of the system, and entrepreneurial physicians, nurses, and administrators regard these as transient problems curable by fine-tuning the system. Their faith in democratic capitalism is not without reasonable foundation. After all, they argue, the free market concept has brought America to its position of preeminence and affluence among nations, it has contributed to the superiority of American medicine, and it accords with the spirit of enterprise, freedom, and self-determination Americans rightly cherish. Indeed, the dismal failure of government-operated enterprises and centrally regulated economies in Eastern Europe sharply underlines the superiority of free markets for most commodities.

Much more important than the economic shortcomings of the managed care line of reasoning and operation are the ethical issues encompassed in the paradigm shift from a profession to a market ethic. Before examining these, it is necessary to see whether health and medical care are, in fact, commodities. If they are, then the managed care line of reasoning is essentially correct, and the present shortcomings should yield to instrumental manipulation. But, if they are not commodities, then instrumental manipulation will not cure the present ills. Attention will have to turn to the ends and purposes of medicine, to healing as a special kind of human activity governed by an ethic that serves those ends and not the self-interests of physicians, insurance plans, or investors.

The legitimacy of the marketplace, competition, and democratic capitalism, therefore, are not at issue. Rather, the ethical question commodification raises is whether the marketplace is the proper instrument for the distribution of health care. Specifically, is health care sufficiently different from pantyhose, ocean-front condominiums, or television sets to set it apart from other consumer goods? The answer to this question rests on what we mean by a "commodity."[11]

10. J. Kassirer, "Managed Care's Tarnished Image," *New England Journal of Medicine* 337 (1997): 338–39; M.B. Smith, "Trends in Health Care Coverage and Their Implications for Policy," *New England Journal of Medicine* 337 (1997): 1000–1003; S.A. Flocke et al., "The Impact of Insurance Type and Forced Discontinuity on the Delivery of Primary Care," *Journal of Family Practice* 45 (1997): 129–35; Anders, *Health Against Wealth*; R. Pear, "Government Lags in Steps to Widen Health Coverage," *New York Times* (interactive edition), August 9, 1998, National Desk Section; E. Larson, "The Soul of an HMO," *Time Magazine* 147 (1996), 45–52.

11. L. Zoloth-Dorfman and S. Rubin, "The Patient as Commodity and the Question of Ethics," *Journal of Clinical Ethics* 6, no. 4 (1995): 339–57; E.S. Anderson, "Is Women's Labor a Commodity?" *Philosophy and Public Affairs* 19 (1990): 71–92; A.W. Immershein and C. Lestes, "From Health Services to Medical Markets: The Commodity Transformation of Medical Production and the Non-profit Sector," *International Journal of Health Services* 26 (1996): 221–37; D. Reynolds, "Not Commodity and Not Public Work," *Bioethics Forum* (Winter 1995): 38–39; E.S. Anderson, "The Ethical Limitation of the Market," *Economics and Ethics* 6 (1996), 179–205.

The *Oxford English Dictionary* gives a wide range of meanings to the word "commodity."[12] The definition most relevant to this discussion is the way the word is used in commerce, that is, a thing produced for sale valued for its usefulness to the consumer or its satisfaction of his preferences. Implicit in this definition is the idea of fungibility, namely, that one healthcare encounter is like any other, just like any bag of beans of same weight and quality is like any other bag of beans. It follows also on this view that, providing they are equally competent, any physician is like any other, and any patient like any other. Hospitals, laboratories, and nursing homes are all interchangeable as well.

If health care is a commodity, then it is something we possess and can sell, trade, or give away at our free will. This implies that, like any commodity, ownership of medical knowledge upon which health care depends resides with health professionals or those who employ or invest in them.[13] No one else can lay claim to their medical knowledge or skill unless they were acquired unjustly. In this view, there would be no duty of stewardship over medical knowledge which would require its use on behalf of those who need it but cannot pay for it. Nor can there be any valid moral claim by the sick on society for its allocation or distribution. Whether health care is a commodity in this sense will be discussed below as will the moral implications for the justice of moral claims in the allocation of resources.

The commodification question centers on "health care" not on the facilities, medications, instruments, dressings, and other disposable items used in or necessary to health care. These are, in one sense, commodities since they can be owned, consumed, bought, traded, and donated. They are in a different category from commodities unrelated to health care. To a large extent, they must be subject to market forces. But in the interests of the common good, they become the subject of ethical and public concern when their scarcity threatens human well-being. The sale of body parts, kidneys, or blood is another instance of commodification with serious ethical implications.[14] Granted the importance of all the objects that are essential in health care, the central ethical issue is the quality and nature of personal relationships involved. The materials used are means to the end of healing and helping sick persons.

By "health care," therefore, I mean the provision of assistance to persons in need of care, cure, education, prevention, or help related to trauma, illness, disease, disability, or dysfunction by other persons knowledgeable and skillful in providing such assistance. The central feature of health care is the personal relationship between a health professional and a person seeking help.[15] Commodities may be

12. *Oxford English* Dictionary, vol. 2 (1961), 687.

13. R. Nozick, *Anarchy, State, and Utopia* (New York: Basic Books, 1974), 160; H.T. Engelhardt, *The Foundations of Bioethics*, 2nd ed. (New York: Oxford University Press, 1996), 381.

14. R. Titmuss, *The Gift Relationship, From Human Blood to Social Policy* (London: London School of Economics, 1997); R. Brecher, "Buying Human Kidneys: Autonomy, Commodity, and Power," *Journal of Medical Ethics* 17 (1991): 99.

15. M.A. Ray, "Health Care and Human Caring in Nursing: Why the Moral Conflict Must Be Resolved," *Family and Community Health* 10 (1987): 35–43; E.D. Pellegrino and D.C. Thomasma,

used in the process of providing care, but the totality of health care itself is not a commodity.

The most common assertion of those who see no objection against classifying health and medical care as commodities is that there is nothing unique about them as human activities. It is an assertion made usually by healthy people, the young, and those who have not thought much about their own vulnerability and finitude. To be fair, there are undoubtedly a few people who steadfastly hold to this view despite everything, even when they themselves, or members of their families, become ill. Still, even they might reexamine that position at the moment when some dear family member is denied access to life-saving or life-enhancing treatments because of the fortuitous operations of the marketplace.

This is happening, today, in communities where hospitals are closing emergency rooms, burn units, premature baby care centers, and neonatal intensive care units to cut losses or to enhance profits.[16] As a result, important care is not available or is so inconvenient or delayed that danger or death might occur. Acute care of this type is a need that may occur anywhere, anytime, and to anyone. If there is a first priority, this surely is one since other needs become insignificant until the emergency is over.

The same can be said for less dramatic medical and healthcare services. Human flourishing can and does occur in the presence of chronic illness, but it is certainly more easily attained when one is healthy. Chronic illness, pain, discomfort, or disability can constrain the most determined and best-adjusted person. For most people, it is difficult or impossible to do the things they want to do or enjoy when they are afflicted by illness. Health is a fundamental requirement for the fulfillment of the human potential and freedom to act and direct one's life. To lack health and to need treatment is to be in a diminished state of human existence—a state quite unlike other deprivations which can be borne if one is healthy.

Serious illness changes our perceptions of ourselves as persons. It forces us to confront the fragility of our own existence. Human finitude is no longer a distinct abstraction, but something illness forces us to confront as a possible present reality. Regardless of whether the illness is serious, if we wish to be treated, we are forced to seek help, to invite and authorize a stranger—the health professional—to probe the secret places of body, mind, and soul. Without this scrutiny, we cannot be helped. To be sure, lawyers are permitted access to some intimate secrets, tax advisors to others, and ministers to still others. But only the physician may need access to the widest range of secrets, since being ill is not confined to the body. It is a disturbance in the whole life-world of the patient.

For the Patient's Good: The Restoration of Beneficence in Health Care (New York: Oxford University Press, 1987).

16. [Peter T. Kilborn, "Fitting Managed Care into Emergency Rooms," *New York Times*, December 28, 1997, 12, https://www.nytimes.com/1997/12/28/us/fitting-managed-care-into-emergency-rooms.htm—Ed.].

What the sick person needs is healing, that is, a restoration of what has been "lost," a reestablishment of the equilibrium that existed before the onset of illness. This "restoration" is not just a return of biological equilibrium.[17] It is rather a restructuring of our whole life world—one with its unique history, set of relationships, and social milieu. In that restoration, there is a meshing of the life and lived worlds of both the doctor and the patient. Healing is achieved by a combination of the physician's biological interventions (drugs, surgery, manipulations) with the healing power of the patient's own body. In the end, where self-healing and medical healing begin and end may be highly problematic, but in either case, the assistance of knowledgeable health professionals is indispensable.

Given the special nature of illness and healing, health care cannot be a commodity. Health care is not a product which the patient consumes and which the doctor produces out of materials of one kind or another. The sick person "consumes" medication and supplies, and expends money for them, but he does not consume health "care" as he would a bag of beans or a six-pack of beer. Health or amelioration of disease may be the end of medicine but health, itself, is not a weighable commodity.

In a commodity transaction, like buying bread, the persons who buy and who sell have no personal interest in each other beyond the transaction. They are focused on the object or product, on the commodity to be traded. Their relationship does not extend beyond the sale or the consumption of that commodity. The medical relationship, in contrast, is intensely personal. Confidence and trust are crucial as is a continuing relationship, at least in general medicine if not in the subspecialties.

There are other intimate professional relationships as well as medicine. But, even in such intimate services as legal or ministerial advice, the sheer totality of engagement with the biological as well as the psycho-social and spiritual, which occurs in medicine, is lacking. This is not to depreciate these other crucial human services. They, too, respond to fundamental human needs: the lawyer to the need for justice; the minister to the need for spiritual reconciliation. Those are human needs of such fundamental significance for human flourishing that they, too, cannot be classified as commodities. However, the universality, unpredictability, inevitability, and intimate nature of the assault of illness on our humanity, the impediments it generates to human flourishing, and the intimate and personal nature of healing give health care a special place even among the helping professions.

Another feature of a commodity is that it is proprietary. Commodities are produced by someone who makes something new out of pre-existing materials. The seller herself, or her agent, owns the goods or commodities she offers for sale. For a price, she transfers ownership to the buyer who consumes the product as he wishes. This is not, and cannot be, the nature of the case with medical or health care. Physicians, nurses, and other health professionals are not the sole owners of their medical knowledge for several good reasons.

17. H.G. Gadamer, *The Enigma of Health, The Art of Healing in a Scientific Age* (Stanford, Calif.: Stanford University Press, 1996).

For one thing, the physician's knowledge comes from many years—and in some cases, centuries—of clinical observations recorded by his or her predecessors. All health professionals have free access to that record across national boundaries. In addition, the most reliable, clinically pertinent medical knowledge comes from autopsies or controlled experiments on other human beings. A research "breakthrough" often results in response to previous work. The investigator uses instruments and methods discovered by others. No research exists in a vacuum. The fruits of research also result from the willingness of our fellow citizens to be research subjects so that others might benefit. Biomedical research is funded by public agencies or by private philanthropies to which all contribute by paying taxes or purchasing products from whose profits the funds for philanthropy derive.

Moreover, the doctor's education depends upon the acquisition of a kind of knowledge and experience which is ethically possible only with society's sanction.[18] In their first years, for example, medical students dissect human cadavers for which they would be criminally prosecuted without social sanction. Later, they are allowed to participate in the care of patients when they are incompletely trained, albeit under supervision of licensed practitioners. Students can only learn procedures and operations by "practice," again, under supervision. They participate in clinical care when, clearly, their skills are rudimentary. They continue to enjoy this privilege in their post-graduate years as residents or fellows.

Society sanctions these practices because the only way future physicians can be trained properly is by "hands-on" experience. Physicians in practice continue to enjoy these privileges in continuing education which is designed to maintain their skills or teach them new procedures. Surgeons develop and maintain their skills by continuing "practice," at first under supervision and, later, independently.

The argument has been made that medical knowledge is proprietary because it has been paid for by tuition and continuing education fees and by the many years of study and demanding work required for a thorough medical education.[19] But even this cannot make medical knowledge proprietary. There is no fee that could buy the privileges or waive the legal penalties that would be imposed if medical students, residents, and fellows did not have social sanction for what they are allowed to do. Moreover, these privileges are accorded not primarily so that doctors or nurses may have a means of livelihood or an "edge" in competing with their fellows. The privileges of medical and nursing education are permitted in exchange for the benefits society accrues from the assurance of a continuing supply of well-trained physicians and other health professionals.

As a result, when they accept the privilege of a medical education, medical and nursing students enter implicitly into a covenant with society to use their knowledge

18. E.D. Pellegrino, "Medical Education," in *Encyclopedia of Bioethics*, vol. 3, rev. ed., ed. W.T. Reich (New York: Simon and Schuster/Macmillan, 1995), 1435–39.

19. R.M. Sade, "Medical Care as a Right: A Refutation," *New England Journal of Medicine* 285 (1971): 1288–92.

for the benefit of the sick. By entering this covenant, they become stewards rather than proprietors of the knowledge they acquire. Their fees and labor entitle them to charge for their personal services—not because they own the knowledge they employ, but because of the effort and time they invest and the danger they may incur in applying that knowledge to particular persons. They are entitled to compensation for their effort. But to paraphrase Plato, they are *true* physicians when they are *healers* first, and *moneymakers* second.[20]

Commodification, Market Ethics, Professional Ethics

Clearly, health and medical care do not fit the conceptual mode of commodities. They center too much on universal human needs which are much more fundamental to human flourishing than any commodity per se. They depend on highly intimate personal inter-relationships to be effective. They are not objects fashioned and owned by health professionals, nor are they consumed by patients like other commodities. Stewardship is a better metaphor than proprietorship for medical knowledge and skill.

All of this might be granted, and still some might hold that even if health and medical care are not commodities in the usual sense, they should be treated as such. In this view, the market is the best mechanism in a free society for the distribution of health care like all human goods. It is the best guarantor of those special aspects of health and medical care we have insisted must be preserved. When consumers are free to choose and providers compete, the interests of all will be better served, especially when costs are high, quality among providers is variable, and resources are scarce.

Let us examine the consequences of such assertions from the point of view of their ethical meaning for persons who are ill and for society in general.

The most immediate and urgent ethical consequences of commodification occur at the bedside at the moment of actual decision making. Here the major issues are divided loyalty, conflicts of interest, conflicts with the traditional ethic of medicine, and challenges to the personal integrity of physicians and nurses. These conflicts have been examined in some detail elsewhere.[21] The focus here will be on commodification and commercialization of the healing relationship and the supplanting of the traditional professional ethic by the ethic of the market and of business.

First of all, the commodification of health and medical care means that the transaction between physicians and patients has become a commercial relationship.

20. Plato, *Republic*, 341c.

21. E.D. Pellegrino, "Managed Care at the Bedside: How Do We Look in the Moral Mirror?" *Kennedy Institute of Ethics Journal* 7 (1997): 321–30; M.A. Rodwin, "Strains in the Fiduciary Metaphor: Divided Physician Loyalties and Obligations in a Changing Health Care System," *American Journal of Law and Medicine* 21 (1995): 241–57; M.A. Rodwin, "Consumer Protection and Managed Care: Issues, Reform Proposals, and Trade-Offs," *Houston Law Review* 32 (1996): 1319–81; and McArthur and Moore, "The Two Cultures and the Health Care Revolution."

That relationship, therefore, will be primarily or solely regulated by the rules of commerce and the laws of torts and contracts rather than the precepts of professional ethics. Profit making and pursuit of self-interest will be legitimated. Inequalities in distribution of services and treatments are not the concerns of free markets. Denial of care for patients who could not pay were not unknown in the past. But they were not legitimated as they are in a free market system where patients are expected to suffer the consequences of a poor choice in healthcare plans, or a decision to go uninsured or to pay only for a plan with lesser levels of coverage. In this view, inequities are unfortunate but not unjust.[22] Some simply are losers in the natural and social lottery.[23] The market ethos does not per se foreclose altruism, but neither does it impose a moral duty to help, especially if helping impinges on the proprietary rights of others without their consent.

In a market-driven economy, commodities are fungible, that is, any one of them can be substituted for any other similar commodity, provided quality and price are the same. In this view of health care, physicians and patients become commodities, too.[24] The identity of physicians, hospitals, clinic locations, and laboratories makes no difference unless a clear quality gap or danger is demonstrated. Patients, too, become fungible. They are "insured lives." They can be traded and bargained for, back and forth, in mergers, network formation, or sales of hospitals and clinics. They are "profit" or "loss" centers, assets when they stay well and pay premiums, and debits if they become ill and need too many services.[25] The "quality" of any group of patients is then measured by their profitability, and this can be a "deal breaker" in mega-mergers, etc., when both doctor and patient become faceless counters in business mega-deals.[26]

When both doctors and patients are fungible, choice of physician has no ultimate weight despite repeated surveys showing it is the most important factor in the therapeutic relationship. Physicians no longer look on patients as "theirs" in the sense that they feel a continuing responsibility for a given patient's welfare. Without this attachment of particular doctors to particular patients, it is easy to justify a strict 9-to-5 workday, signing out to another doctor on short notice, or being "unavailable" for all those personal concerns and worries that beset patients even with non-life-threatening illnesses. Even short-term continuity of care is difficult to come by and, in any case, reckoned as not essential for care to be "delivered."[27]

22. Engelhardt, *The Foundations of Bioethics.*

23. Nozick, *Anarchy, State, and Utopia.*

24. Zoloth-Dorfman and Rubin, "The Patient as Commodity"; W.M. Greenberg et al., "The Hospitalizable Patient as Commodity: Selling in a Bear Market," *Hospital and Community Psychiatry* 40, no. 2 (1989): 184–85; P. Starr, *The Social Transformation of American Medicine* (New York: Basic Books, 1982), 217.

25. Greenberg et al., "The Hospitalizable Patient as Commodity."

26. M.B. Blecher, "Size Does Matter," *Hospitals and Health Networks* (June 20, 1998): 29–36.

27. Flocke et al., "The Impact of Insurance Type."

Similarly, necessary communications between primary care physicians and specialists are hampered, further aggravating the discontinuity.[28]

To remedy this, patients are urged by the corporate ethos and the fungibility of physicians to regard the managed care organization as their "doctor." The corporation will provide. The organization will deliver the commodity and make certain that "someone" is there to deliver it. Many physicians are already socialized into this corporate way of thinking. They place less emphasis on continuity, personal commitment, and personalized relationships with patients than in the past. Physicians who are corporate employees become encultured in an ethic alien to the professional.[29] The ethic of the marketplace and the ethic of the employee begin to displace the more demanding ethic of a profession. The ethic of individual patient welfare based on principles has moved towards greater dependence on the institutions providing care.[30]

There is no room in a free market for the non-player, the person who can't "buy in"—the poor, the uninsured, the uninsurable. The special needs of the chronically ill, the disabled, infirm, aged, and the emotionally distressed are no longer valid claims to special attention. Rather, they are the occasion for higher premiums, more deductibles, or exclusion from enrollment. There is no economic justification for the extra time required to explain, counsel, comfort, and educate these patients and their families since these cost more than they return in revenue.

Despite the boast about prevention under a managed care system, the time required for genuine behavioral change—the essential ingredient in true prevention—is not recompensed. To be sure, immunizations, diet sheets, smoking cessation clinics might be offered. They are not as costly nor as effective as the constant effort required for genuine behavior modification with reference to diet, exercise, stress control, and the like. Effective prevention requires education and counseling, and these are personnel-intensive and rarely profit-making.

The business ethos puts its emphasis on the bottom line, on profit, on an excess of revenue over expenses. The aim of business is to maximize returns to investors.[31] For-profit organizations return those profits to investors, executives, and board members. The care they provide turns out to be more expensive as well.[32] Non-profit organizations use profits to expand services or to "survive" the intense competition that characterizes the healthcare "industry" today. But in both for-profit and non-

28. Z.C. Roulidis and K.A. Schulman, "Physician Communication in Managed Care Organizations: Opinions of Primary Care Physicians," *Journal of Family Practice* 39 (1994): 446–51.

29. McArthur and Moore, "The Two Cultures and the Health Care Revolution."

30. E.J. Emanuel, "Medical Ethics in an Era of Managed Care: The Need for Institutional Structures Instead of Principles in Individual Cases," *Journal of Clinical Ethics* 6, no. 4 (1995): 335–38.

31. M. Friedman, "The Social Responsibility of Business Is to Increase Its Profits," *New York Times Magazine*, September 13, 1970, 33.

32. S. Woolhandler and D. Himmelstein, "Costs of Care at For-Profit and Other Hospitals in the United States," *New England Journal of Medicine* 336 (1997): 769–74.

profit systems, health professionals who contribute to the bottom line are valued; those who do not are devalued or let go.[33]

In a commodity transaction, the ethics of business replaces professional ethics. Business ethics is not to be depreciated. Many businesspeople genuinely seek to be "ethical." A whole literature and a whole new set of experts in business ethics give testimony to a genuine interest in ethical business conduct in general and in health care in particular.[34] Emphasis on worth creation rather than profit maximization and better representation of patient interest are promoted by ethically sensitive corporations.

The question, however, is not the validity of a business ethics, but whether it is appropriate for health and medical care. If there is any weight to the arguments for the special character of health care, then a business ethics is inappropriate and insufficient as a guide for health professionals. Revisions of business ethics are admirable, especially if they ameliorate some of the crasser aspects of for-profit plans. But the problems are of a more fundamental sort not susceptible to cure by changes in management ideology.

The contrasts between business and professional ethics are striking. Business ethics accepts health care as a commodity, its primary principle is non-maleficence, it is investor- or corporate-oriented, its attitude is pragmatic, and it legitimates self-interest, competitive edge, and unequal treatment based on unequal ability to pay. Professional ethics, on the other hand, sees health care not as a commodity but as a necessary human good, its primary principle is beneficence, and it is patient-oriented. It requires a certain degree of altruism and even effacement of self-interest.

When humans are at their most vulnerable and exploitable, they need much more secure protection than a business ethics can afford. Buying an automobile, for example, is a tricky business, to be sure. Much faith must be placed in the manufacturer and the salesperson (a slender reed, indeed). One hopes for an ethical dealer. But the vulnerability of the auto purchaser pales to insignificance when compared to entering an emergency room with a pain in the chest or a fractured skull.

The corrupting power of industrial and business metaphors has been commented upon elsewhere.[35] Suffice it to say that substituting words like "consumers" for patients, "providers" for physicians, "commodities" for healing relationships, or speaking of health care as an "industry," or of "product-lines" or "investment opportunities" inevitably distorts the nature of healing and helping.

33. Kuttner, "Must Good HMOs Go Bad?" 1562–63.

34. N.E. Bowie and R.F. Duska, *Business Ethics*, 2nd ed. (Englewood Cliffs, N.J.: Prentice Hall, 1990); M.M. Blair, *Ownership and Control: Rethinking Corporate Governance for the 21st Century* (Washington. D.C.: Brookings Institute, 1995); R.G. Evans, "We'll Take Care of It for You: Health Care in the Canadian Community," *Daedalus* 117 (1988): 155–89; S.M. Shortell et al., *Remaking Health Care in America: Building Organized Delivery Systems* (San Francisco: Jossey-Bass, 1996).

35. E.D. Pellegrino, "Words *Can* Hurt You: Some Reflections on the Metaphors of Managed Care (First Annual Nicholas J. Pisacano Lecture)," *Journal of the American Board of Family Practice* 6, no. 7 (1994): 505–10.

Euphemisms, if repeated often enough, eventually will shape behavior as though they were true renditions of real world events and states of affairs.

If we treat health care as a commodity, then we are prone to "sell it" like any other commodity—that is, by creating a demand among those who can pay. Getting the competitive edge via advertising is standard commercial practice. The first move in this direction came in 1982 when the Federal Trade Commission decided that the then-standing AMA ethical prohibition against advertising constituted a restraint of trade.[36] The commission treated medicine as a business and not a profession. In effect, the commission ordered the AMA to set its ethical code aside in the interests of commerce. The implications of such an action for the ethical integrity of the profession have not been sufficiently appreciated.

Since the Federal Trade Commission order, the ethos of the advertising world, with its all-too-characteristic seductive promises and often-misleading inducements has dominated the promotion of health care in the media. In the name of competition, everything has become a "P.R." problem and an exercise in spin-control. The claim that advertising provides information upon which consumers (patients) can base their choice of doctors or health plans is no less spurious than it is in the advertisement of cosmetics. Reliable, clear, unambiguous data on coverage and quality are hard to see through the camouflage and persiflage of marketing.

Does Commodification Work?

Some might agree that treating health care as a commodity does, indeed, carry the ethical risks I have detailed above. However, they might insist that, commodity or not, health care is part of the economy, and it is best distributed in a free market. They can point effectively to the dismal failure of government-controlled markets in socialist economies of recent unhappy memory. They might argue that the benefits of a free market might outweigh the dangers. Three of these putative benefits are (1) cost savings which competition would effect, (2) the subscriber or "consumer's" freedom to spend his or her money for health care as he or she wishes, and (3) the satisfaction the patient would enjoy as providers competed for his or her business. Let us examine these presumed benefits in the light of the realities as they are unfolding in managed care, which does, indeed, make health care a commodity and seeks to improve its distribution and price while retaining its quality via the workings of the market.

It is a fact that rationing and limitations on physician decisions have kept costs from escalating. But there are already clear evidences of a reverse trend.[37] Premiums are rising in many states and promise to do so elsewhere. Profits are dropping. Initial savings through mergers, reductions of personnel or acquisitions, and tightened controls on physician decisions are one-time savings. Mergers may be as much signs

36. Federal Trade Commission, "In the Matter of the American Medical Association. Docket 9064, May 19, 1982," *Journal of the American Medical Association* 248 (1982): 981–82.

37. Smith, "Trends in Health Care Coverage"; Anders, *Health Against Wealth.*

of weakness as strength. Treating medical and health care as a commodity subject to market forces must face certain inescapable clinical realities.

For one thing, subscribers cannot all be, and remain, young and healthy forever. As subscribers age, they need and demand more services. As time goes by, initial promises to contain costs or to return a profit become more difficult to keep. Subscribers either must be dropped, selected out at the outset, or corners must be cut and quality endangered. When profits drop, investors in for-profit plans will sell and move to better opportunities; non-profit organizations will face bankruptcy, and for-profit plans will tighten approval requirements for care or for inclusion in the plan. Every plan will scramble to enroll young, healthy people who have little need for care. The poor, the under-insured, and the genuinely sick will be pariahs to be avoided or dis-enrolled in some way.

The promise of freedom of choice is even more illusory.[38] For the largest majority of those insured, their employers (who are the purchasers of the plans) make the first choice—selection of a plan with which to contract for services. This is done primarily on the basis of cost, not of quality. Indeed, neither the employer nor the employee is in a good position to judge outcomes and quality of care except in the grossest terms. The fragility of these choices is manifest in the employer's constant shifting from plan to plan for the "best" buy, which often translates into the cheapest buy. The employee has no choice but to go with the plan or go it alone and buy his or her own insurance.

Once in a plan, freedom is, again, limited—this time in choice of doctor. Yes, there is choice, as long as it is from the panel of approved physicians and a list of approved services. The same limitation on freedom prevails when it comes to choice of hospital, or specialist, or an MRI or lab facility. The advertisements proclaim "choice," but all have fine print that limit those choices.

Even more uncertain is the choice of how best to spend one's money in health care—whether to buy a plan with a high or low deductible, whether to go naked and use one's money for other things, and which of the complex and confusing array of "products" and "packages" best suits one's health needs at one time of life or another. This illusory freedom flies in the face of realities like the difficulty, even for educated people—even health professionals—of comprehending what the convoluted obscurantist language of the contract covers or the impossibility of predicting how much coverage one will, in fact, need, particularly to meet the uncertainty of an unexpected catastrophic illness. Should one buy limited coverage? Should one opt for a high or a low deductible? How are changing needs to be accommodated—like being married, divorced, having children, retiring, choosing a medical savings account, or unexpectedly picking up the care of elderly parents? What do hospitals, physicians, and society do when those who make bad choices appear for care?

38. A.C. Enthoven, *Health Plan: The Only Practical Solution to the Soaring Cost of Medical Care* (Reading, Mass.: Addison-Wesley, 1980); Hertzlinger, *Market-Driven Health Care.*

Libertarians might regale in this richness of choices. Yet, even they would have to admit that bad choices can be made by the most educated people. The libertarian could reply that this is the price of freedom: "Better than someone else making the choice! After all, some people do not value health care above other goods. They may wish to run the risk in order to be able to spend their earnings on more immediate needs or pleasures." This is carrying *caveat emptor* to the extreme without resolving the question of what physicians do when the patient who made a poor choice presents himself or herself with a medical emergency or a serious illness.

In a market-driven ethos, theoretically at least, physicians would be justified in refusing care. They could argue that patients are responsible for their own health, that they must live with their choices, and that to provide care under those circumstances is to distribute other people's wealth without their consent. This is a stark but unavoidable conclusion of a strictly libertarian view of property rights, health care, and social governance. Whether this conclusion is consistent with the most minimal rudiments of professional ethics or with a good or just society is highly debatable.

Many managed care plans measure their success in terms of consumer satisfaction. But the relationship between satisfaction and quality is a tenuous one. Consumer satisfaction is not the whole of health care. The genuine difficulties of measuring quality outcomes are vastly under-rated by the satisfaction criterion. The young, those who have not needed the system, or those who make low demands might very well be satisfied with lower premiums and some of the advertising slogans. This is much less the case with the chronically ill, the aged, and those who make demands on the system. Data are appearing that show these patients have poorer outcomes in managed care systems than in fee-for-service systems.[39] Others go through the "revolving door," leaving their capitated plan for Medicare and Medicaid coverage when they really become ill.

If Not a Commodity, What Is Health Care?

If, as I have argued, health care is not a commodity, if the consequences of treating it as such are morally unpalatable, and if it is a special kind of human activity derived from a universal need of all humans when they become ill, how should it be treated in a just society? Here we enter the complex, much-debated field of distributive justice, in general and in health care, and the very practical issues of healthcare reform and policy. These large subjects are well beyond the scope of this essay. But, by arguing against the commodity notion and the market ethic, I incur a certain obligation to point to the direction in which a morally tenable answer might lie.

Buchanan has summarized the major theories of justice associated with the distribution of health care.[40] He has done so with care and with a fair appraisal of

39. J.E. Ware et al., "Differences in 4-Year Health Outcomes for Elderly and Poor Chronically Ill Patients in HMO and Fee-for-Service Systems: Results from Medical Outcomes Study," *Journal of American Medical Association* 216 (1996): 1039–47.

40. A.E. Buchanan, "Justice and Charity," *Ethics* 97 (1987): 558–75.

the strengths and weaknesses of each. In serial order, Buchanan examines the libertarian theories of Nozick and Engelhardt, the contractarian views of Rawls and Daniels, the egalitarianism of Veatch and Menzel.[41] These theories all depend on whether or not there is a right of the sick to health care, created by the unfortunate circumstance of illness. The limitation of rights and rights language in public life are receiving more attention.[42] Rights language focuses too often on the negative rights of freedom from coercion and not enough on obligation and duties. Given the vulnerable and dependent state of sick persons, it is not intrusion on rights they fear, but abandonment to their fates by their fellow humans. As a result, there is a growing interest in communitarian and common good conceptions of justice.

Buchanan includes a non-rights-based approach which contends, instead, that there is a duty in beneficence to aid the needy and those in distress. In this view, health care is a "collective good." Government enforcement is necessary to ensure that this collective good is provided even to those who may not have a discernible "right" to it. In this approach, beneficence is given at least equal weight with justice, and the collective good at least as much weight as the individual good.[43]

Another challenge to commodification is to regard medical and health care in a societal context as a public work, that is, as an "organized medical response to illness in a social context and to the practice of caring in a community."[44] This notion needs further fleshing out. While it intimates a connection with the idea of a common or civic good, it remains somewhat vague. Importantly, it does return our attention to the fundamental questions about the social ethics of health care.

I believe that the moves to a *prima facie* obligation of beneficence on behalf of the sick[45] or the practice of caring in a community[46] are in the right direction. For a morally defensible policy of healthcare distribution, however, the obligation to provide healthcare needs firmer grounding in a philosophy of medicine and of society. In a theory built on *prima facie* principles, social beneficence could be overridden for good reason by other principles like autonomy or a competing theory of justice like Engelhardt's or Niozick's. A firmer justification can be found in the origins of a moral claim in the phenomena of illness and healing and in the notion

41. Nozick, *Anarchy, State, and Utopia*; Engelhardt, *The Foundations of Bioethics*; J. Rawls, *A Theory of Justice* (Cambridge, Mass.: Harvard University Press, 1971); N. Daniels, *Just Health Care* (Cambridge: Cambridge University Press, 1985); R.M. Veatch, *The Foundations of Justice* (New York: Oxford University Press, 1986); P.T. Menzel, *Medical Costs, Moral Choices* (New Haven, Conn.: Yale University Press, 1983).

42. M.A. Glendon, *Rights Talk: The Impoverishment of Political Discourse* (New York: Free Press, 1991); R. Bellah et al., *The Good Society* (New York: Alfred A. Knopf, 1991).

43. Buchanan, "Justice and Charity," 558.

44. B. Jennings and M.J. Hanson, "Commodity of Public Work? Two Perspectives in Health Care," *Bioethics Forum* (Fall 1995): 8.

45. Buchanan, "Justice and Charity."

46. Jennings and Hanson, "Commodity of Public Work?"

of health care as common good. Together these realities would establish health care as a moral obligation a *good* society owes to all its members.

This is because health, or at least freedom from acute or chronic pain, disability, or disease, is a condition of human flourishing. Human beings cannot attain their fullest potential without some significant measure of health. A good society is one in which each citizen is enabled to flourish, grow, and develop as a human being. A society becomes good if it provides those goods which are most closely linked to being human. Health care is surely one of the first of these goods. It is, to be sure, not the only human good.[47] But other goods, like happiness, wealth, friends, career, etc., are compromised or even impossible without health.

In addition to the claim that arises out of the obligation of a good society to enhance the flourishing of its members, there is the non-proprietary, non-commodity character of medical knowledge alluded to in sections one and two above. The nature of medical knowledge and the way in which it is acquired and transmitted to health professionals make it a collective good on which the members of a society have a substantive claim. Human beings across national boundaries also contribute to the body of knowledge required in health care today. With varying degrees of strength, all humans are linked in a world community which shares in the fruits of medical research.

Health care is both an individual and a social good. A good and well-functioning human society requires healthy members, and healthy members require a good and well-functioning society. This reciprocity of dependence between individual and societal good implies a reciprocity of obligations as well. Aquinas puts the relationship between the individual and common good in this way:

> He who pursues the common good thereby pursues his own good for two reasons. First, because the proper good of the individual cannot exist without the common good of the family or state or realm. . . . Second, since a person is a part of a household or state, he ought to esteem that good for him which provides for the benefit of the community.[48]

Any linkage of a social obligation to provide health care as a human good must engage the question of reciprocal responsibility of citizens to care for their own health. In a common good conception, the self-abuser, the person who refuses to buy insurance he or she can afford, or who refuses to take a genuine part in community life, imposes burdens on his fellow humans. Such persons weaken any claim they might have had to societal resources.[49] Plato warned about

47. Aristotle, *Nicomachean Ethics*, in *The Complete Works of Aristotle*, vol. 2, trans. and ed. J. Barnes (Princeton, N.J.: Princeton University Press, 1984), 1178b30–34.

48. Thomas Aquinas, *Summa Theologiae*, trans. T. Gilby (New York and London: Blackfriars and McGraw-Hill, 1974), 2a2ae, q. 10,2.

49. J. Boyle, "The Concept of Health and the Right to Health Care," *Social Thought* (Summer 1977): 5–17.

hypochondriac or aged members of society whom he believed should not receive care.[50]

This is a harsh judgment that a compassionate society could not impose strictly. It implies refusal to treat the self-abuser in the emergency room whose bleeding from esophageal varices are the result of alcoholic cirrhosis of the liver. Such retributive justice could not be consistent with the primary ethical obligation of physicians as healers. Perhaps a practical compromise is to discriminate against products like alcohol, tobacco, high-risk sports, etc. by taxation or higher premiums but not against persons who do not know how or do not care about their health.[51] After all, where does one draw the line on irresponsibility? How much exercise, weight, or stress-reduction is enough to qualify for participation in the societal good? Who has pursued health so sedulously as to be free of lapses for which he is responsible? Where do we draw the bright line?

The role of government here is not forcefully to redistribute wealth in general or to bring about total equality in all things but to assure that collective goods, or the satisfaction of a common claim on such goods, are justly distributed. This approach does not translate into an unlimited claim on all the health care possible. It is not an invitation to "blank check" medicine. Nor does it mean that health care takes priority at all times and in all places over all other societal goods. Nor does it swallow private property rights in central planning. It does mean that there is a moral obligation of a good society to relieve the sufferings of its citizens, to provide what is needed for the fabric of society to hold together, and to see that the collective interest of society's members in health care is assured. This is, in effect, beneficent justice—justice ordered by the obligation to rescue, sustain, and nourish both society and the individual since each suffers if either is neglected or abused. This is a positive obligation that transcends the more negative or legalistic notion of "rights," but it is not an absolute obligation overriding all other obligations.

Treating health care as a common good implies a notion of the solidarity of humanity, that is, the linkage of humans to each other as social beings. The common good is, however, more than economic good—necessary though this may be in an instrumental sense.[52] It also implies the development of social and governmental institutions designed to promote justice and the well-being of the whole society in essential things like health, safety, environment, and education.

Nowhere does this conception militate against the protection of individual rights. It stands against the absolutization of the Marxian collective as well as the absolutization of Nozickian property rights. Rather, the morally defensible aim is a mixed economy: one in which private property and private enterprise are protected, but there is enough social control to assure justice, especially in those things that

50. D. Heyd, "The Medicalization of Health: Plato's Warning," *Revue Internationale de Philosophie* 193, no. 3 (1995): 375–93.

51. Evans, "We'll Take Care of It for You."

52. Bellah, *The Good Society*; Glendon, *Rights Talk*.

cannot be left to the fortuitous operations of the competitive marketplace.[53] These things must always be few and of such nature that a healthy and well-functioning society could not exist without them.

It is especially important to recognize that rejection of a marketplace ethics for the distribution of health care does not make the market an unethical instrument for the distribution of most other goods and services. As Worland has shown, Adam Smith himself recognized that the market was not the preferred device for providing certain public goods like defense, education, or a transportation infrastructure or for setting rates of interest, etc.[54] Smith had no difficulty reconciling liberty, private property, and government intervention when the rights of a few threatened to endanger the whole. In both *The Wealth of Nations* and *The Theory of Moral Sentiments*,[55] Smith saw a specific role for moral restraint on self-interest and of government's responsibility to prevent monopoly and to administer justice.[56]

In this view, the role of the government is both to protect personal liberty and attend to those common goods that liberty destroys when it becomes license. How this balance is to be achieved is a constant struggle of democratic societies and institutions. What is crucial in health care is that any policy must take cognizance of the common social good, the shared moral claim on medical knowledge, and the special nature of health care as a human activity.

What is equally crucial is that the physician remain truly a physician, concerned with healing and not moneymaking.[57] In any plan in the future, the physician ought not be the gatekeeper, micro-allocator, or rationer. Nor should she or he become provider, insurer, or risk-taker simultaneously.[58] The current move to establish the physician-provider organizations in which employers, hospitals, and corporations "buy" health care from groups of doctors are just as dubious morally as other capitated insurance plans. In addition, they deprive patients of their last advocate since the physician is the healer, the risk-taker, and the profit-maker simultaneously.[59] The assumption that physicians as administrators are more likely to represent patient interests if they own or administer the system is dubious at best. If earlier studies of higher prices and overutilization of doctor-owned radiology and laboratory facilities are accurate, this assumption is likely to be a dangerous myth.

53. J.A. Ryan, *Distributive Justice* (New York: MacMillan, 1916).

54. Worland, "Adam Smith, Economic Justice," 8–9.

55. Smith, *The Wealth of Nations*; Smith, *The Theory of Moral Sentiments*.

56. Worland, "Adam Smith, Economic Justice," 21.

57. Plato, *Republic*, 342c and 342d.

58. B. O'Reilly, "Taking on the HMOs," *Fortune* 137 (1998): 96–100; D. Witten, "Regulation Downstream and Direct Risk Contracting: The Quest for Consumer Protection and a Level Playing Field," *American Journal of Law and Medicine* 23 (1997): 449–86; C. Wang, "Unions and Equity HMOs: Two Sources of Physician Power in a World of Managed Care," *Family Practice Management* (February 1996): 21–27.

59. S. Woolhandler and D. Himmelstein, "Extreme Risk—The New Corporate Proposition for Physicians," *New England Journal of Medicine* 333 (1995): 1706–7.

This is not the place to design a total system of health care, nor to fill in the content of precisely what services constitute a fair share of the common good of health care, nor to speak of the costs, modes of payment, and choices among other societal goods. Obviously, those are the questions most often at issue in policy debates. But, in the end, those are second-order questions. They can be answered properly only in light of the first-order questions: What is health care? What kind of good is it? What moral claim do members of a society have on this good? What are society's obligations, and what are the obligations of the health professional with reference to that good?

Understanding health care to be a commodity takes one down one arm of a bifurcating pathway to the ethic of the marketplace and instrumental resolution of injustices. Taking health care as a human good takes us down a divergent pathway to the resolution of injustice through a moral ordering of societal and individual priorities.

One thing is certain: if health care is a commodity, it is for sale, and the physician is, indeed, a moneymaker; if it is a human good, it cannot be for sale and the physician is a healer. Plato's question admits of only one ethically defensible answer.

> Can we deny, then, said I, that neither does any physician, insofar as he is a physician, seek to enjoin the advantage of the physician but that of the patient?
>
> —*Plato*[60]

[Additional—Ed.] References

Brink, S. (1996). "How Your HMO Can Hurt You." *U.S. News and World Report* 120: 62–64.

Busch, L., and Tanaka, K. (1996). "Rites of Passage: Constructing Quality in a Commodity Subsector." *Science, Technology, and Human Values* 21: 3–27.

Dionne, E.J. (1998). "The Big Idea." *Washington Post*, August 9, 1998, C1–C2.

Elola, J. (1996). "Health Care System Reforms in Western European Countries: The Relevance of Health Care Organization." *International Journal of Health Services* 26: 239–51.

Evan, W.M., and Freeman, R.E. (1988). "A Stakeholder Theory of the Modern Corporation: Kantian Capitalism." In *Ethical Theory and Business*, 3rd ed., ed. T.L. Beauchamp and N.E. Bowie, 97–106. Englewood Cliffs, N.J.: Prentice Hall.

Greenberg, H.M. (1998). "Is American Medicine on the Right Track?" *Journal of the American Medical Association* 279: 426–28.

Iglehart, J.K. (1994). "Health Care Reform and Graduate Medical Education: Part I." *New England Journal of Medicine* 330: 1167–71.

Iglehart, J.K. (1994). "Health Care Reform and Graduate Medical Education: Part II." *New England Journal of Medicine* 332: 407–11.

Outka, G. (1976). "Social Justice and Equal Access to Health Care." In *Social Justice and Equal Access to Health Care*, ed. R. Veatch and R. Branson, 79–126. Cambridge: Ballinger Publishing.

60. Plato, *Republic*, 342c.

Pellegrino, E.D. (2004). "Philosophy of Medicine and Medical Ethics: A Phenomenological Perspective." In *Handbook of Bioethics: Taking Stock of the Field from a Philosophical Perspective*, ed. George Khushf, 183–202. Dordrecht: Kluwer Academic Publishers. [Updated—Ed.]

President's Commission for the Study of Ethical Problems in Medicine and Behavioral Research (1983). *Securing Access to Health Care*, vol. 1. Washington, D.C.: Department of Health. Education and Welfare.

E. The Learning Health Care System

38

Pellegrino and the Learning Health Care System: A Speculative Exploration

F. Daniel Davis, PhD

Introduction

One can comb through Pellegrino's works and *never* read the phrase, "learning health care system" (LHCS). Thus, this essay is an effort to speculate about what he *might* have said about the LHCS model had he been challenged to think and write about it. I begin with a description of the model's history and features, followed by a brief survey of the evolving discussions about ethics in an LHCS. This description and survey provide the groundwork for some speculations about Pellegrino and the LHCS. I suspect that he would be uncomfortable with some features of the LHCS, or at least with their implications, and perhaps endorse, if not embrace others. After all, the first principle for the design of the LHCS is that "care is based on continuous *healing relationships.*"[1]

The healing relationship: in his many essays and books, Pellegrino repeatedly returned to the task of elaborating this end or telos—this normative ideal. Although, on many occasions, he wrote eloquently about the experience of being a patient, his more frequent vantage point was that of a physician who often seemed to be writing almost exclusively to other physicians (and medical ethicists). I do not think that this was a product of chauvinism on his part but rather reflected what he experienced as a moral imperative to speak out to other physicians about his views on their individual and collective obligations, especially in the midst of forces that may frustrate their fulfillment. Such obligations flowed, integrally, from the physician's act of profession—her promise to seek to heal the patient before her.

In the grand scheme of his work, however, he devoted less interest to questions of *how* best to form and sustain healing relationships in the midst of—*perhaps despite*—the frustrating complexities of contemporary health care. For example, he

1. Institute of Medicine Committee on Quality of Health Care in America, *Crossing the Quality Chasm: A New Health System for the 21st Century* (Washington, D.C.: National Academies Press, 2001), 8.

paid scant attention to the increasingly team-oriented nature of healthcare delivery; he wrestled little with the fragmentation of care that patients with chronic disease often experience along the trajectories of their illnesses. These "how" questions are questions about operationalizing ethical norms: about realizing such norms in day-to-day practice. It is fair to say that Pellegrino did not tend, in a systematic way, to these "how" questions. That said, as a good Thomistic-Aristotelian, as the product of a Jesuit education with its emphasis on the moral formation of *contemplatives in action,* Pellegrino would not, I think, dismiss the importance of thinking through these questions about the fate of healing relationships in our current system or in an LHCS. Such thinking is eminently practical and needed. I will not belabor the observation that, at present, our so-called healthcare "system" frustrates the development and sustenance of healing relationships that serve the needs of patients, their loved ones, and the clinicians who care for them. Aspects of the decades-old culture of medicine and health care, coupled with perverse incentives, fragmentation, and the production-line ethos of contemporary care delivery, generate organizational conditions in which relationships between patients and clinicians are all too often marred by preventable harm, ineffective or unnecessary interventions, and dissatisfaction for all. The LHCS is just that: a system with features that, when integrated in practice, are designed to correct or to mitigate these flaws—a system designed to make continuous healing relationships possible.

The LHCS Model

The word "model" here is intended to convey the idea of a system comprised of interrelated features interacting in an exemplary way, a way generating some valuable outcome. The outcome that the LHCS seeks to achieve is safe, effective, and beneficial patient- and family-centered care. Multiple dimensions and implications of the model have been described and analyzed in a series of publications launched in 2007 by the National Academy of Medicine (once known as the Institute of Medicine) (IOM/NAM).

Any thorough account of the LHCS model has to begin earlier, however, with the mid-1990s and the renowned MIT management authority, Peter Senge, whose work has focused on elaborating the concept of a *learning organization.* Writing in his *The Fifth Discipline: The Art and Practice of the Learning Organization,* Senge advances the idea of a learning organization as an antidote to the "the illusion that the world is created of separate, unrelated forces." He writes,

> When we give up this illusion we can then build "learning organizations," organizations where people continually expand their capacity to create the results they truly desire, where new and expansive patterns of thinking are nurtured, where collective aspiration is set free, and where people are continually learning how to learn together. . . . The organizations that will truly excel in the future will be the organizations that discover how to tap people's commitment and capacity to learn at all levels in an organization. . . . What will fundamen-

> tally distinguish learning organizations from traditional authoritarian "controlling organizations" will be the mastery of certain basic disciplines. That is why the "disciplines of the learning organization" are vital.[2]

Senge contends that there are five such disciplines. *Changing our mental models* is the discipline of analyzing and changing "the deeply ingrained assumptions, generalizations, or even pictures or images that influence how we understand the world and how we take action."[3] A second discipline is *building shared vision,* "unearthing shared 'pictures of the future' that foster genuine commitment . . . rather than compliance."[4] *Enabling personal mastery* is the discipline "of continually clarifying and deepening our personal vision, of focusing our energies, of developing patience, and of seeing reality objectively."[5] The fourth discipline is *team learning,* which is built on the insight that groups have intelligence quotients—emotional and otherwise—that are not simply the sum of IQs or EIQs of the individual members; this entails the free and equitable exchange of ideas among members, each of whom is committed not only to his or her individual growth but to the growth of the team.[6] The fifth discipline (alluded to in the title of Senge's book) is *systems thinking:* systems thinking is the integrator of all of the disciplines insofar as it, as Senge says, fuses them into a coherent body of theory and practice. Systems thinking identifies and clarifies the interconnections between and among the organization's component parts in order to change and align them effectively for organizational learning and improvement.[7]

In the mid-2000s, when the IOM/NAM began to "design" the LHCS model, it did so with a debt to Senge: the LHCS focuses on learning as an organizing principle, but in its day-to-day functioning, it also crucially depends upon Senge's five disciplines, especially the fifth discipline of systems thinking. The model itself, as a whole of integrated parts, reflects both the logic and the output of systems thinking, applied to the complex challenges of improving the quality of health care in the United States.

The IOM/NAM sought first, however, to subject those challenges to systematic analysis and in-depth diagnosis. It did so in two reports published on the cusp of the twenty-first century. The first, *To Err Is Human,* was published in 2000.[8] The second, *Crossing the Quality Chasm,* was published in 2001.[9] Lucian Leape, MD, physician-father of the patient safety movement, was the lead author of *To Err Is*

2. Peter Senge, *The Fifth Discipline: The Art and Practice of the Learning Organization* (New York: Doubleday, 1990), 8.

3. Ibid., 11.

4. Ibid.

5. Ibid., 10.

6. Ibid., 12.

7. Ibid., 13.

8. Institute of Medicine Committee on Quality of Health Care in America, *To Err Is Human: Building a Safer Health System* (Washington, D.C.: National Academies Press, 2000).

9. Institute of Medicine, *Crossing the Quality Chasm.*

Human, a groundbreaking study for two reasons. For one, it produced the then-shocking estimate that at least 44,000—and as many as 98,000—people die every year in the US healthcare system as a result of medical errors. (Subsequent studies have pushed the estimate of error-related mortality upward with some arguing that the annual death toll is as much as 400,000 people.) Thus, the effort to quantify the deadly results of medical error is one reason that *To Err Is Human* was—and remains—a landmark.

The second reason is this: Leape and colleagues argued that medical error is only rarely the consequence of an incompetent, distracted, or malignant individual. It is, rather, more often the product of failures in systems and processes and of dysfunctional, rather than effective communication within and between healthcare teams. In recent years, the etiological understanding of medical error has become even more refined: the Joint Commission has concluded that the majority of serious adverse events are due to poor communication, due, in turn, to deficits in psychological safety, mutual accountability and respect, and leadership.

Medical error is but one deficit in the quality of American health care. The others are the focus of the academy's 2001 report, *Crossing the Quality Chasm* (which was explicitly conceived as a sequel to *To Err Is Human). Crossing the Quality Chasm* advances several principal findings. Care is fragmented and uncoordinated. Knowledge about best care is not applied systematically and expeditiously; insights are poorly managed and evidence is poorly used. The potential of information technology remains unrealized—for example in capturing the patient's experience of care. And, incentives are misaligned and the workforce is unprepared. The result is a system riddled with missed opportunities, waste, and harm. Thus, *Crossing the Quality Chasm* paints a sobering, data-driven picture of a dysfunctional, costly healthcare system. Subtitled *A New Health Care System for the 21st Century,* the report, however, looked beyond the dysfunction and envisioned key features of a new, more optimally functional system distinguished by high—and ever improving—quality. Looking forward, the academy argued that such a system should be built and function in accordance with several design principles, the first of which is that "care is based on continuous healing relationships."[10] Other principles reflect the degree to which the new healthcare system for the twenty-first century is patient-centered ("care is customized according to patient needs and values"[11] and "the patient is the source of control"[12]), while others have to do with knowledge and evidence, the primacy of safety and the importance of transparency, and the priority given to cooperation among clinicians.

Finally, the academy argued that such a system should have six overarching aims. It should be safe, effective, patient-centered, timely, efficient, and equitable.[13]

10. Ibid., 8.
11. Ibid.
12. Ibid.
13. Ibid., 6.

These principles and aims for a new healthcare system for the twenty-first century fall short of a complete blueprint for that system, but six years later, in the first of what now totals nearly twenty reports, the academy began to offer just that—a blueprint for what it, with an obvious debt to Senge, dubbed an *LHCS*.

What is an LHCS? In answering this question, it is important to recognize that an LHCS embodies these principles and aims from *Crossing the Quality Chasm*—aims that are realized through disciplines of learning deployed and refined in processes of care. In addition, the LHCS seeks to embed rigorous inquiry—structured, disciplined learning—into routine clinical and operational processes in order to generate insights and knowledge applicable to the continuous improvement of care.[14] In this context, *learning* is a term rich in meaning: it broadly includes initiatives in quality improvement and clinical innovation, in addition to research of all types, especially comparative effectiveness research and pragmatic clinical trials; and it includes, as well, routine observation, retrospective analysis, and the like. All have roles to play in the generation of data that might then be interpreted as evidence for use in healthcare decision making.

Finally, the LHCS "behaves" in distinctive ways. It fully exploits the potential of information technology for digitally capturing the patient's experience of care, for curating the data derivable therefrom, developing that data as evidence, and then delivering, at the point of care, the best available evidence to advance clinical decision making. It seeks to engage patients and their loved ones, not only as active partners in their own care but also as partners in all efforts to improve care. As for clinicians, investigators, and innovators, the LHCS incentivizes them in ways that encourage their integral involvement in waste reduction, continuous improvement, and the delivery of high value care. In addition, in a continuous way, an LHCS assesses multiple dimensions of patient safety, care quality, as well as care costs and outcomes, and does so transparently, making assessment results available to clinicians, payors, and, of course, patients and the public. Finally, the stewards of an LHCS are leaders who are both committed to, and model in their behavior, the mutual respect that, as Leape argued, is at the foundation of an organizational culture that enables safety (both psychological safety for clinicians and, of course, safety for patients), learning, and enhanced quality through collaboration within and across interprofessional teams.[15]

It is an ambitious vision, perhaps relatively simple in design but exceptionally complex in execution. There is, however, emerging evidence of beneficial outcomes when a concerted effort is made to behave, consistently and reliably, in any one of these respects, that is, to embody any one of these features of an LHCS.[16] These features

14. Institute of Medicine Roundtable on Evidence-Based Medicine, *The Learning Healthcare System: Workshop Summary* (Washington, D.C.: The National Academies Press, 2007).

15. Institute of Medicine, *Better Care at Lower Cost: The Path to Continuously Learning Health Care in America* (Washington, D.C.: The National Academies Press, 2013).

16. Sarah M. Greene, Robert J. Reid, and Eric B. Larson, "Implementing the Learning Health System: From Concept to Action," *Annals of Internal Medicine* 157 (2012): 207–10.

target the flaws in traditional systems of care, which can be hierarchical and authoritarian in organization and decision making, which can act like federations of silos rather than integrated parts of wholes, and which can be riddled with often unsafe, toxic cultures of disrespect—ripe conditions for patient harm and less than effective or beneficial care. It would not be an exaggeration to say that such conditions pervade health care in the United States, where hundreds of thousands of people die every year as a result of medical error, where adult patients receive approximately 50 percent of recommended therapies, and where nearly a third of healthcare spending is wasted.[17]

Ethics in a Learning Health Care System

The generative potential of an LHCS is evident in the ensuing scholarly discussions, debates, and practical initiatives spurred by the academy's model. The Association of American Medical Colleges has founded a "research in care" community that seeks to foster the design, implementation, and evaluation of local LHCSs and awards grants to fund such pilots. There are now medical center-based departments of learning sciences and sustained dialogue among individuals working in the related fields of implementation science, healthcare re-engineering, and design thinking. And substantial bodies of literature have developed around key topics, including ethics in an LHCS.

Some of the more prominent voices in discussions about the ethics of LHCSs belong to Faden, Kass, and their colleagues at the Johns Hopkins Berman Institute of Bioethics, who authored and published two landmark, companion articles several years ago in *The Hastings Center Report.* In one, they took on and sought to critique the traditional distinction between research and practice, a distinction laid down as dogma in the Belmont Report but challenged and intentionally blurred by an LHCS, which seeks to embed research—and other forms of learning—*in* practice.[18] In the other, they formulated and described "an ethics framework" for the LHCS, in part motivated by the concern that traditional frameworks are based on the distinction and on flawed or outmoded assumptions, which yield, for example, the practical paradox of a research landscape that is often over-regulated and a clinical landscape that can seem like the wild, wild west.[19]

In fact, subsequent exchanges in the emerging scholarly literature on ethics in an LHCS have unfolded around matters of ethical oversight for the broad range of learning activities—the overarching question being *what type of learning requires*

17. Institute of Medicine, *Crossing the Quality Chasm,* 25–33.

18. Nancy E. Kass, Ruth R. Faden, Steven N. Goodman, Peter Pronovost, Sean Tunis, and Tom L. Beauchamp, "The Research-Treatment Distinction: A Problematic Approach to Determining Which Activities Should Have Ethical Oversight," *The Hastings Center Report* special issue (January–February 2013): S4–S15.

19. Ruth R. Faden, Nancy E. Kass, Steven N. Goodman, Peter Pronovost, Sean Tunis, and Tom L. Beauchamp, "An Ethics Framework for a Learning Health Care System: A Departure from Traditional Research Ethics and Clinical Ethics," *The Hastings Center Report* special issue (January–February 2013): S16–S27.

what type of oversight by an institutional review board or other institutional body?[20] These exchanges have also included some sharp disagreements about the information patients should receive about specific learning activities and about the forms of consent that should or should not be required for a given activity.[21] For example, if a patient with hypertension is receiving treatment that is the focus of a relatively simple, observational study of comparative effectiveness, should she be informed of that study, even though its conduct will be completely invisible to her and she will, in no way, be inconvenienced by it or even notice that it is underway? What if she is to be randomized to one of two treatment arms and thus may end up receiving a medication with which she is unfamiliar? These are obviously very practical questions about what should or must be done in ethically navigating the terrain of an LHCS.

A full engagement with the "state of the discussion" about ethics in an LHCS is beyond my purposes in this essay. Instead, I will focus on four obligations that Kass, Faden, and colleagues include within their framework—and use them as a bridge to the speculative challenge, the ultimate focus of this essay:

(1) Respect the rights and dignity of patients—*an obligation incumbent upon clinicians and healthcare systems.*
(2) Address health inequities—*an obligation incumbent upon clinicians and healthcare systems.*
(3) Conduct continuous learning activities that improve the quality of clinical care and healthcare systems—*an obligation incumbent upon clinicians and healthcare systems.*
(4) Contribute to the common purpose of improving the quality and value of clinical care and healthcare systems—*an obligation incumbent upon patients.*[22]

Speculations on Pellegrino and the LHCS

The natural beginning for this speculative exploration is the first obligation, which is binding on clinicians and healthcare systems. It is the obligation to respect the rights and dignity—rather than the autonomy—of patients. The matter of autonomy is one point of contention between ethics in an LHCS and the ethics of traditional systems of health care. According to Faden and colleagues, in the LHCS, autonomy is not given undue deference or overriding importance: "Respecting autonomy is primarily about allowing persons to shape the basic course of their lives in line with their values and independent of the control of others. Not all healthcare decisions are likely to be attached to a significant autonomy interest of individual patients, and deference of the wrong sort can constitute a moral failure to take ade-

20. Joe V. Selby and Harland M. Krumholz, "Ethical Oversight: Serving the Best Interests of Patients," *The Hastings Center Report* special issue (January–February 2013): S34–S36.

21. Franklin G. Miller and Ezekiel J. Emanuel, "Quality-Improvement Research and Informed Consent," *New England Journal of Medicine* 358, no. 8 (2008): 765–67.

22. Faden et al., "An Ethics Framework."

quate care of patients rather than an instance of showing respect."[23] What would Pellegrino have to say about this first obligation and its supporting argument, which displaces autonomy from the perch of priority that it often enjoys in contemporary medical ethics? Given his (and Thomasma's) arguments in *For the Patient's Good,* I think there can be little doubt that he would concur.[24] In that work, they set out to critique both (physician) paternalism and (patient) autonomy as absolutes. As for the latter, they contend that absolute autonomy simply fails the crucible of reality. The fact of illness diminishes a core element of autonomy, the capacity for self-determination, if only by degrees and if only on a temporary basis. In addition, another core element of autonomy, individualism, tends to ignore the innately social character of being human, of our existence within webs of relationships that structure, inform, and impinge upon our perceptions, hopes, and aspirations as individuals. Thus, there is warrant for the conclusion that Pellegrino would endorse this first obligation.[25]

As for the second obligation, the obligation to address and reduce health inequities, there is solid ground for the assertion that Pellegrino would strongly endorse it. In several essays, he argued, for example, that academic health centers have a duty to improve access, to design and conduct experiments to achieve this aim, and to broadly disseminate lessons learned so that the problem can benefit from solutions forged in real-world conditions. Extending the safety net of care to groups that already suffer other forms of social marginalization was, for Pellegrino, a moral imperative for healthcare systems.[26]

The third obligation, according to Faden and colleagues, is shouldered by clinicians and systems, and it is to "conduct continuous learning activities that improve the quality of clinical care and healthcare systems."[27] This obligation is thus a matter of organizational ethics—of institutional moral agency as Pellegrino would put it. It enjoins clinicians to engage in conscious, outcomes-oriented learning; although not every clinician may wish to become a clinical trialist or an expert in "mixed methods" research, she or he is duty-bound to support this organizational mission. This obligation also enjoins hospitals and health systems to create and sustain the conditions for learning throughout the organization in order to prevent harm and optimize benefit to patients.

Over the course of his life's work, Pellegrino turned repeatedly to the theme of the institutional moral agency of hospitals and academic health centers.[28] His focus was often on the moral quandaries experienced by hospitals and systems at a time when financial and economic pressures prevail and often perversely incent their

23. Ibid.

24. Edmund D. Pellegrino and David C. Thomasma, *For the Patient's Good: The Restoration of Beneficence in Health Care* (New York: Oxford University Press, 1988).

25. Ibid., 18–22.

26. Edmund D. Pellegrino, "Toward an Expanded Medical Ethics: The Hippocratic Ethic Revisited," in *Hippocrates Revisited,* ed. Roger J. Bulger (New York: MEDCOM Press, 1973), 133–47.

27. Faden et al., "An Ethics Framework," S18.

28. Edmund D. Pellegrino, *Humanism and the Physician* (Knoxville: The University of Tennessee Press, 1979).

behavior and, as a result, the behavior of clinicians and staff. The ubiquity and lure of what he called Faustian compacts—with big corporate medicine, with big science—could, he argued, only be countered by institutions with strong moral leadership and investigators and scientists whose character has been formed to place the interests of care and truth ahead of those of self-interest or economic gain. The question of where science, research, and learning fit within the mission of a hospital or health system was not one, however, that Pellegrino treated at great length, but the "evidence"—anecdotal and otherwise—indicate that he would view these functions and activities benignly, if not enthusiastically. He was proud of his own past work as a "bench scientist" and led more than a few research institutions over the years. He supported and championed the progress of biomedical science and research.

We come finally to the fourth obligation, which, in an LHCS, *rests with patients* for its fulfillment. In such a system, Faden and colleagues argue, patients have an affirmative duty to participate in system efforts to improve the quality of care. This argument has gained traction in recent years apart from discussions about the ethics of the LHCS: several prominent bioethicists have advanced the idea of a "new" research ethic that is *participatory* rather than *protectionist* and that couples the mandate to provide rational, risk-based protections to participants with the injunction to facilitate well-designed studies that hold the promise of gains in new knowledge for improved care—in part, because they have engaged patients in the research itself.[29] With such arguments, these authors have sought to correct the excesses of a protectionist regime of research oversight, along with misconceptions about the risks, benefits, and burdens associated with different types of research, innovation, and quality improvement.

Research ethics was not a primary focus of interest for Pellegrino, whose preoccupation was the healing relationship between a patient in need of healing and a physician, an individual who professes, who promises to heal. On the relatively few occasions when he did turn to the ethics of the investigator-research participant relationship, he tackled the moral dilemmas generated by the physician who mingles the roles of clinician-caregiver and investigator.[30] For him, the physician's fundamental duty is to benefit the patient—to realize or achieve the patient's good through clinically wise, compassionate acts of healing. He was thus especially keen to identify the worldly threats to the integrity of the physician's profession if she or he is also interested in the pursuit of knowledge that may or may not be of benefit to the patient. Such conflicts of obligation were, for him, defining tests of personal and professional integrity. It also should come as no surprise that he explored and delineated the virtues of the virtuous investigator as the ultimate guarantor of the ethical validity of the conduct of a given research study or clinical trial. The virtue ethic that he (and David Thomasma) had

29. G. Owen Shaefer, Ezekiel J. Emanuel, and Alan Wertheimer, "The Obligation to Participate in Biomedical Research," *Journal of the American Medical Association* 302, no. 1 (2009): 67–72; Rosemond Rhodes, "In Defense of the Duty to Participate in Biomedical Research," *American Journal of Bioethics* 8, no. 10 (2008): 37–38.

30. Pellegrino, *Humanism and the Physician,* 130–40.

developed for the clinical practice of medicine had its parallel in a virtue ethic for the investigator, one emphasizing the cultivation of habits of integrity, truthfulness, methodological discipline, openness and skepticism, and mutual respect.

For Pellegrino, however, even the character of the most ethical investigator cannot mitigate the harsh reality: clinical investigation is an *invasion.* "Clinical research," he argues, "is an invasion of the integrity and privacy of the person who is the research subject. This invasion necessarily places the person at some risk for a result that, by the very nature of research, must be problematic. Nothing in the nature of science or its internal need for freedom and autonomy constitutes an autonomous moral right to engage in human experimentation. Rather, clinical research is a privilege accorded the investigator by social consensus. Society permits experimentation with humans because of the benefits that the whole society, as well as the experimental subject, may gain from new medical knowledge. The clinical investigator thus enters a covenant with society when he or she accepts the privilege of experimentation involving fellow human beings" (see chap. 10 above).[31]

There is much in this passage with which to agree, just as there is much to question. Given what we know about the thin evidence base that supports most clinical judgments and given what we know about the risks of iatrogenic harm, is there not in the urgency of these problems an ethical rationale that has the force of a moral imperative—an imperative for both physicians and patient to engage in learning activities that seek, always and ever, to improve care? To put it another way, from the vantage point of the LHCS, notions of privilege and of autonomous moral rights seem inapt for the investigator, as does the notion that clinical research is always an invasion of the integrity and the privacy of the patient. From that vantage point, we can envision a different possibility: that physician and patient encounter each other as partners and collaborators in care and its ongoing improvement. Ultimately, though, I find it difficult to imagine that Pellegrino would, after reasoned dialogue, completely reject the idea that the common good embraces research and its fruits—and thus looks to individual physicians and individual patients for its achievement.

We can acknowledge all of this and still wonder whether Pellegrino would fully embrace the model in all of its characteristic features. It does, after all, pose a direct challenge to the traditional, hierarchical cultures of medicine and health care, in which he was, himself, formed and toward whose flaws he was either largely silent or only mildly critical. And thus, the question of whether he would or would not welcome a system that depends upon fundamental change in its supporting culture will have to remain open and unanswered. *However,* if he were here with us today, I would urge him to do just that: embrace the model of an LHCS as the *needed context* for what the academy called "continuous healing relationships" in *Crossing the Quality Chasm,* especially given the deep resonance between this design principle and his own philosophy and ethics of medicine.

31. Edmund D. Pellegrino, "Beneficence, Scientific Autonomy, and Self-Interest: Ethical Dilemmas in Clinical Research," *Georgetown Medicine* 1, no. 1 (1991): 21–22.

Contributors

F. Daniel Davis, PhD
Chief Bioethics Officer
Geisinger Health System
Danville, Penn.

G. Kevin Donovan, MD, MA
Professor Emeritus and immediate past Director
Pellegrino Center for Clinical Bioethics
Georgetown University
Washington, D.C.

Kevin T. FitzGerald, SJ, PhD, PhD
John A. Creighton University Chair
Chair of the Department of Medical Humanities
Associate Professor
Creighton University School of Medicine
Creighton University
Omaha, Neb.

John Keown, DPhil, PhD, DCL
Rose F. Kennedy Professor of Christian Ethics
Kennedy Institute of Ethics
Georgetown University
Washington, D.C.

Aaron Kheriaty, MD
Fellow and Director, Bioethics and American Democracy Program
Ethics and Public Policy Center
Washington, D.C.

David G. Miller, PhD
Associate Director and Director of Academic Programs
Pellegrino Center for Clinical Bioethics
Adjunct Assistant Professor of Family Medicine
Georgetown University
Washington, D.C.

Allen H. Roberts, II, MD, MDiv, MA (Bioethics)
Associate Medical Director
Professor of Medicine
MedStar Georgetown University
Faculty member
Pellegrino Center for Clinical Bioethics
Georgetown University
Washington, D.C.

Virginia Ashby Sharpe, PhD
Retired Acting Executive Director
National Center for Ethics in Health Care
Veterans Health Administration (VHA)
Washington, D.C.

Myles N. Sheehan, SJ, MD
Director
Pellegrino Center for Clinical Bioethics
David Lauler Chair in Catholic Health Care Ethics
Professor of Medicine
Georgetown University
Washington, D.C.

Claudia Ruiz Sotomayor, MD, DBe
Chief of the Ethics Consultation Service
Pellegrino Center for Clinical Bioethics
Assistant Professor of Internal Medicine
Georgetown University
Washington, D.C.

Daniel P. Sulmasy, MD, PhD
Director
Kennedy Institute of Ethics
André Hellegers Professor of Biomedical Ethics
Departments of Medicine and Philosophy
Georgetown University
Washington, D.C.
Faculty
Pellegrino Center for Clinical Bioethics
Georgetown University
Washington, D.C.

Joseph Tham, LC, MD, PhD
Faculty of Bioethics
Pontifical Athenaeum Regina Apostolorum
Rome, Italy

Sarah Vittone, DBe, MA, MSN, RN
Associate Professor
Director Undergraduate Studies and Master Entry Programs
School of Nursing
Georgetown University
Washington, D.C.
Clinical Ethicist and Faculty
Pellegrino Center for Clinical Bioethics
Georgetown University
Washington, D.C.

Index